Carbonaceous Quantum Dots: Synthesis And Applications

Edited by

Kulvinder Singh
Department of Chemistry
DAV College
Chandigarh
India

&

Vineet Kumar
School of Bioengineering and Biosciences
Lovely Professional University
Jalandhar, Punjab
India

Carbonaceous Quantum Dots: Synthesis and Applications

Editors: Kulvinder Singh and Vineet Kumar

ISBN (Online): 978-981-5136-26-5

ISBN (Print): 978-981-5136-27-2

ISBN (Paperback): 978-981-5136-28-9

First published in 2023.

need for a court order if at any point you breach any terms of this License Agreement. In no event will any delay or failure by Bentham Science Publishers in enforcing your compliance with this License Agreement constitute a waiver of any of its rights.

3. You acknowledge that you have read this License Agreement, and agree to be bound by its terms and conditions. To the extent that any other terms and conditions presented on any website of Bentham Science Publishers conflict with, or are inconsistent with, the terms and conditions set out in this License Agreement, you acknowledge that the terms and conditions set out in this License Agreement shall prevail.

Bentham Science Publishers Pte. Ltd.
80 Robinson Road #02-00
Singapore 068898
Singapore
Email: subscriptions@benthamscience.net

BENTHAM
SCIENCE

CONTENTS

FOREWORD i

PREFACE ii

LIST OF CONTRIBUTORS iv

CHAPTER 1 INTRODUCTION TO CARBONACEOUS QUANTUM DOTS 1
Abhinay Thakur, Ashish Kumar and *Sumayah Bashir*
- **INTRODUCTION** 1
  - Quantum Dots (QD) 1
- **CARBONACEOUS QUANTUM DOTS (CQD)** 2
  - Discovery of CQDs 2
  - Chemical Structure of CQDs 3
- **SYNTHETIC TECHNIQUES** 4
- **PHYSICAL AND CHEMICAL PROPERTIES** 5
  - Absorbance 5
  - Photoluminescence (PL) 5
  - Electroluminescence (ECL) 6
  - Up-conversion Photoluminescence (UCPL) 6
- **APPLICATIONS** 6
  - Electrocatalysis 6
  - Biomedical Applications 7
    - *Bioimaging* 7
    - *Targeted Drug Delivery* 7
    - *Nanomedicine* 7
    - *Biosensing* 8
  - Photocatalysis 8
  - Cytotoxicity 9
  - Sensor 9
  - Optronics 10
    - *Dye-sensitized Solar Cells (DSCs)* 10
    - *Light-emitting Devices (LED)* 10
- **CHARACTERIZATION OF CARBON QUANTUM DOTS** 10
- **TEM** 11
- **XRD** 11
- **FTIR** 11
- **NMR** 11
  - UV Spectroscopy 11
- **PL** 12
- **ADVANTAGES OF CARBON QUANTUM DOTS** 12
- **CHALLENGES AND PROSPECTIVE** 12
- **CONCLUSION** 13
- **REFERENCES** 14

CHAPTER 2 SYNTHESIS OF CARBONACEOUS QUANTUM DOTS 20
Munish Kumar
- **INTRODUCTION** 20
  - Chemical States of CQDs 21
  - Structure of CQDs 21
  - Synthesis Method 22
  - Top-down Strategy 22

Bottom-up Strategy ... 23
SEVERAL METHODS UTILIZED UNDER THE TOP-DOWN AND BOTTOM-UP STRATEGY FOR THE PREPARATION OF CQDS ... 24
Arc Discharge Method ... 24
Laser Ablation Method ... 25
Acidic Oxidation Method ... 26
Combustion/Thermal Method ... 27
Microwave Pyrolysis Method ... 28
Hydrothermal/Solvothermal Method ... 30
Electrochemistry Method ... 32
CONCLUSION ... 33
REFERENCES ... 33

CHAPTER 3 PROPERTIES OF CARBONACEOUS QUANTUM DOTS ... 38
Devika Vashisht, Priyanka, Aseem Vashisht, Shweta Sharma and *Surinder Kumar Mehta*
INTRODUCTION ... 38
Optical Features ... 40
Photoluminescence (PL) ... 40
Up Conversion Photoluminescence ... 41
Anti-bleaching and Photo-bleaching Properties ... 41
pH-dependent Features ... 42
Phosphorescence ... 42
Chemiluminescence ... 42
Adsorption ... 44
Electrical Properties ... 45
Electrochemical Luminescence (ECL) ... 45
Photo-induced Electron Transfer ... 46
Biological Properties ... 47
Toxicity ... 48
CONCLUSION ... 49
REFERENCES ... 49

CHAPTER 4 CHARACTERIZATION OF CARBONACEOUS QUANTUM DOTS ... 53
Abhinay Thakur, Harpreet Kaur, Ashish Kumar and *Sumayah Bashir*
INTRODUCTION ... 53
Carbonaceous Quantum Dots (CQDs) ... 53
CHARACTERIZATION OF CQDS ... 56
Fluorescence Spectroscopy ... 57
Fourier Transform Infrared Spectroscopy (FTIR) ... 59
Transmission Electron Microscopy (TEM) ... 61
Atomic Force Microscopy (AFM) ... 62
Optical Analysis using Photoluminescence (PL) Spectrophotometer ... 64
Light Absorption by UV-Vis Spectrophotometer ... 67
Powder X-Ray Diffraction Analysis (XRD) ... 69
Raman Spectroscopy ... 72
CHALLENGES AND FUTURE OUTLOOKS ... 72
CONCLUSION ... 73
REFERENCES ... 73

CHAPTER 5 APPLICATION OF CARBONACEOUS QUANTUM DOTS IN BIOMEDICAL 78

Yarima S. García, Adelmo M. Estrada, Eneida R. López, Ivan H. Herrera and *Jorge L. Neves*
INTRODUCTION 78
Carbon Quantum Dots for Theranostic Development 81
Carbon Quantum Dots and their Application in Drug Delivery 82
Carbon Quantum Dots as Contrast Agents 84
Carbon Quantum Dots in Gene Therapy 86
CONCLUSION 87
REFERENCES 87

CHAPTER 6 APPLICATION OF CARBONACEOUS QUANTUM DOTS IN SOLAR CELLS 94
Ekta Sharma and *Vaishali Thakur*
INTRODUCTION 94
Carbonaceous Quantum Dots in Photocatalysis 96
Electron Mediator 96
Photosensitizer 96
Spectral Converter 97
Sole Photocatalyst 98
Carbonaceous Quantum Dots in Photoelectrochemical Water Splitting 98
Carbonaceous Quantum Dots in Photovoltaic Solar Cells 101
CONCLUSION 103
REFERENCES 103

CHAPTER 7 CARBONACEOUS QUANTUM DOTS AS EFFICIENT ZERO-DIMENSIONAL NANOMATERIALS FOR SENSING APPLICATIONS 110
Alma Mejri, Abdelmoneim Mars and *Hamza Elfil*
INTRODUCTION 110
Photoluminescence (Bio) Sensors 112
Heavy Metal Ions Sensing 112
Organic Compound Sensing 117
Biological Molecules Sensing 119
CHEMILUMINESCENCE (BIO) SENSORS 122
ELECTROCHEMICAL (BIO) SENSORS 127
Heavy Metal Ions Sensing 127
Organic Toxic Compounds Sensing 128
Biological Compounds Sensing 129
ELECTROCHEMILUMINESCENCE (BIO) SENSORS 132
CONCLUSION 135
REFERENCES 136

CHAPTER 8 APPLICATION OF QUANTUM DOTS IN WASTEWATERS TREATMENT 144
Garima Kumari, Eder Lima, Mika Sillanpaa, Nitesh Kumar, Anupam Guleria, Dinesh Kumar, Atma Rai and *Ashish Guleria*
INTRODUCTION 144
APPLICATIONS OF CQDS FOR WASTEWATERS REMEDIATION 146
Carbon Quantum Dots for Removal of Heavy Metals from Wastewater 148
CQDs for Sensing of Heavy Metals in Water 153
CARBON QUANTUM DOTS FOR REMOVAL AND PHOTOCATALYTIC DEGRADATION OF ORGANIC POLLUTANTS 159
CONCLUSION 163
REFERENCES 164

CHAPTER 9 APPLICATION OF CARBONACEOUS QUANTUM DOTS IN ENERGY STORAGE 178
Vaishali Thakur and *Ekta Sharma*
INTRODUCTION 178
CQDs in the Energy Storage and Conversion Applications 180
CQDs in Supercapacitors 180
CQDs in Photovoltaics 181
CQDs in Dye-sensitized-solar Cells (DSSCs) and Quantum-dot-sensitized Solar Cells (QDSSCs) 182
CQDs in Perovskite Solar Cells (PSCs) 183
CQDs in Lithium-ion Batteries 183
CQDs in Hydrogen Evolution Reaction 184
CQDs in Oxygen Evolution Reaction (OER) 185
CONCLUSION 185
REFERENCES 186

CHAPTER 10 FUTURE PROSPECT OF CARBONACEOUS QUANTUM DOTS 192
Devika Vashisht, Aseem Vashisht, Shweta Sharma and *Surinder Kumar Mehta*
INTRODUCTION 192
Photodetectors 193
UV Photodetectors 193
Broadband Photodetectors 194
Solar Cells 195
Heterostructured Solar Cells 195
Dye-sensitized Solar Cells 195
Light-emitting Diodes 195
Wearable and Flexible Electronic Devices 196
Hydrogen Fuels 196
Wastewater Treatment and Purification 198
CONCLUSION AND FUTURE PERSPECTIVE 199
REFERENCES 199

SUBJECT INDEX 204

FOREWORD

Carbon-based quantum dots (CQDs) have got a lot of attention because of their bright luminescence and strong solubility. CQDs are the potential replacement for the toxic metal-based quantum dots. CQDs are characteristically quasispherical carbon nanoparticles composed of amorphous to crystalline carbon bases. CQDs can be prepared simply and inexpensively by multiple techniques such as the arc-discharge method, microwave pyrolysis, hydrothermal method and electrochemical synthesis. Additionally, CQDs exhibit excellent physical and chemical characteristics like significant crystallization, excellent dispersibility, and photoluminescence attributes. In this book, in-depth deliberation on several potent applications, such as biomedical, water treatment, energy storage, solar cells *etc*., has been provided. These applications would affect human life quality significantly and have the potential to draw substantial commercial interest.

This book provides a comprehensive guide for researchers and students to understand the overview, properties, synthesis route and applications of carbonaceous quantum dots. This class of quantum dots has recently gained a lot of interest in various fields of basic science and technology. This could be a handbook for various professionals, researchers and students working in the field of biomedical application, green energy, sensing, energy storage, water treatment, *etc*. Most importantly, this book has proven helpful to postgraduate students in their academic studies.

Dai-Viet N. Vo
Institute of Applied Technology and Sustainable Development
Nguyen Tat Thanh University
Ho Chi Minh City 755414, Vietnam

PREFACE

Carbonaceous quantum dots (CQDs) are an intriguing category of carbon nanoparticles with diameters of about 10 nm. Several sustainable and green characteristics of CQDs, such as low toxicity, chemical inertness, superior biocompatibility, photo-induced electron transfer, and extremely controllable photoluminescence behavior, render them to be an appealing candidate in numerous potent fields. The production of CQDs offers the benefits of basic, inexpensive synthesis processes as well as a large selection of synthetic raw materials. CQD synthesis techniques can be broadly divided into "top-down" and "bottom-up" approaches. In the former, carbonaceous materials are reduced or subdivided using chemical, electrochemical, or physical methods. The latter is accomplished either by step-by-step chemical fusion of small aromatic molecules or by pyrolysis or carbonization of small organic molecules. The worldwide significance and attention of CQDs are attributed to their unique advantages in terms of properties, such as low toxicity, chemical stability, biocompatibility, redox properties, excellent luminescence, tunable and comparable photoluminescence, fluorescent sensing, excellent solubility, and effortless environmental friendly synthesis process. In this book, the selected chapters intend to illustrate the significant potential of CQDs in numerous sectors, such as biomedical, solar cell, sensing, water treatment, and energy storage/generation, with an emphasis on prospects. This book provides a comprehensive guide for researchers and students to understand the overview, properties, synthesis route and applications of carbonaceous quantum dots. This could be a handbook for various professionals, researchers and students working in the field of biomedical application, green energy, sensing, energy storage, water treatment, *etc*. Most importantly, this book has proven helpful to postgraduate students in their academic studies. The authors of this book's chapters are renowned professionals who are enthusiastic about the rapidly emerging, cross-disciplinary fields of research, particularly material science.

We express our gratitude to all the contributors. We would like to thank Prof. (Dr) Andreas Constantinou for writing the foreword. Finally, it is a profound pleasure to thank Bentham Science for taking up the publication of this book. We hope that this book will provide recent scientific knowledge on the chemotherapeutic potential of natural products and the techniques employed for the detection of cancer and other diseases and will lead to new discussions about the global scope of nutraceuticals.

Abhinay Thakur *et al.*, in Chapter 1, discusses the basic introduction of CQDs with a round note on its potent utilization in several fields; DevikaVashisht *et al.*, in Chapter 2, elaborate on the properties of CQDs, including phosphorescence, chemiluminescence, adsorption, electrical properties, electrochemical luminescence, photo-induced electron transfer, *etc.*; Munish Kumar in Chapter 3 discusses arc discharge method, laser ablation method, acidic oxidation method and combustion/thermal method; Abhinay Thakur *et al.* in Chapter 4 mentioned about the characterization techniques of CQDs, including fluorescence spectroscopy, Fourier transform infrared spectroscopy, transmission electron microscopy, atomic force microscopy, optical analysis using photoluminescence spectrophotometer, light absorption by UV-vis spectrophotometer, *etc.*; Yarima S. García *et al.*, in Chapter 5, discusses the potential of CQDs in the biomedical field; Ekta Sharma and Vaishali Thakur, in Chapter 6, elaborated on the significance of CQDs in the solar cell; Alma Mejri *et al.*, in Chapter 7, enlighten the potential of CQDs in sensing, Garima Kumari *et al.*, in Chapter 8, mentioned about the capability of CQDs in wastewater treatment, Vaishali Thakur and Ekta Sharma, in Chapter 9, analyzed the potential of CQDs in energy storage; in last, DevikaVashisht *et al.*, in Chapter 10, deeply enlighten the future aspects of CQDs in various fields.

I anticipate that the readers will find these insights and examination crucial and will inspire them to explore more about further research on Carbonaceous quantum dots in several other potent domains. I am grateful for the prompt initiatives and support provided by the editorial personnel, especially Ms. Humaira Hashmi and Ms. Rabia Maqsood at Bentham Science Publishers.

Kulvinder Singh
Department of Chemistry
DAV College
Chandigarh
India

&

Vineet Kumar
School of Bioengineering and Biosciences
Lovely Professional University
Jalandhar, Punjab
India

List of Contributors

Abhinay Thakur	Department of Chemistry, School of Chemical Engineering and Physical Sciences, Lovely Professional University, Phagwara, Punjab, India
Abdelmoneim Mars	Laboratory of Desalination and Natural Water Valorisation (LADVEN, Water Researches and Technologies Center (CERTE), Techno-park BorjCedria, BP 273, 8020, Soliman, Tunisia
Adelmo M. Estrada	Center for Advanced Studies, HAV, Cuba and Federal University of Pernambuco, PE, Brazil
Alma Mejri	Laboratory of Desalination and Natural Water Valorisation (LADVEN, Water Researches and Technologies Center (CERTE), Techno-park BorjCedria, BP 273, 8020, Soliman, Tunisia
Anupam Guleria	Centre of Biomedical Research, SGPGIMS Campus, Lucknow, Uttar Pradesh, India
Ashish Kumar	NCE, Department of Science and Technology, Government of Bihar, India
Ashish Guleria	Department of Applied Sciences, WIT Dehradun, India
Aseem Vashisht	Punjab State Council for Science and technology Sector - 26, Chandigarh, India
Atma Rai	Department of Applied Sciences, WIT Dehradun, Uttarakhand, India
Devika Vashisht	Department of Chemistry and Centre of Advanced Studies in Chemistry, Panjab University, Chandigarh, India
Dinesh Kumar	Centre of Biomedical Research, SGPGIMS Campus, Lucknow, Uttar Pradesh, India
Eder Lima	Institute of Chemistry, Federal University of Rio Grande, Rio Grande, Brazil
Ekta Sharma	Department of Chemistry, School of Basic and Applied Sciences, Maharaja Agrasen University, Baddi, India
Eneida R. López	Center for Advanced Studies, HAV, Cuba and Federal University of Pernambuco, PE, Brazil
Garima Kumari	Department of Biotechnology, Eternal University, Baru Sahib, Sirmaur, Himachal Pradesh, India
Hamza Elfil	Laboratory of Desalination and Natural Water Valorisation (LADVEN, Water Researches and Technologies Center (CERTE), Techno-park BorjCedria, BP 273, 8020, Soliman, Tunisia
Harpreet Kaur	Department of Chemistry, School of Chemical Engineering and Physical Sciences, Lovely Professional University, Phagwara, Punjab, India
Ivan H. Herrera	Center for Advanced Studies, HAV, Cuba and Federal University of Pernambuco, PE, Brazil
Jorge L. Neves	Center for Advanced Studies, HAV, Cuba and Federal University of Pernambuco, PE, Brazil
Mika Sillanpaa	Department of Green Chemistry, Aarhus University, Nordre Ringgade, Denmark

Munish Kumar	Department of Chemistry & Centre for Advanced Studies in Chemistry, Panjab University, Sector-14, Chandigarh, India
Nitesh Kumar	Department of Chemistry, Govt. College Jukhala, Pradesh, India
Priyanka	Department of Chemistry, Centre of Advanced Studies in Chemistry Punjab University, Chandigarh, India
Shweta Sharma	Institute of Forensic Science and Criminology, Panjab University, Chandigarh, India
Sumayah Bashir	Department of Chemistry, Central University of Kashmir, Kashmir, India
Surinder Kumar Mehta	Department of Chemistry and Centre of Advanced Studies in Chemistry, Panjab University, Chandigarh, India University of Laddakh, Leh, India
Sumayah Bashir	Department of Chemistry, Central University of Kashmir, Kashmir, India
Vaishali Thakur	Department of Chemistry, School of Basic and Applied Sciences, Maharaja Agrasen University, Baddi, India
Yarima S. García	Center for Advanced Studies, HAV, Cuba and Federal University of Pernambuco, PE, Brazil

CHAPTER 1

Introduction to Carbonaceous Quantum Dots

Abhinay Thakur[1], **Ashish Kumar**[2,*] and **Sumayah Bashir**[3]

[1] *Department of Chemistry, School of Chemical Engineering and Physical Sciences, Lovely Professional University, Phagwara, Punjab, India*

[2] *NCE, Department of Science and Technology, Government of Bihar, India*

[3] *Department of Chemistry, Central University of Kashmir, Kashmir, India*

Abstract: Carbonaceous quantum dots (CQDs), relatively small carbon nanoparticles (<10 nm in size), have sparked the attention over the last few decades for their potential as a promising resource in various fields, such as biomedical, solar cells, sensors, water treatment, energy generation storage because of their benign, abundant, low preparation costs, small size, non-hazardous nature, high biocompatibility, high water solubility and effective alteration nature. Numerous applications in optronics, catalysis, and sensing are made possible by the excellent electronic characteristics of CQDs as electron acceptors and donors that cause photocatalytic activity and electrochemical luminosity. This feature series aims to assess the current status of CQDs by discussing the literature in this field and deliberate the basics, applicability and advancements in the field of CQDs in both scientific and technology circles.

Keywords: Biocompatibility, Carbonaceous quantum dots, Carbon dots, Carbon nanoparticles.

INTRODUCTION

Quantum Dots (QD)

Currently, possibilities to assess the features linked to the unity of quantum-constrained features have been made possible by advancements in the manufacture of high-quality quantum dots (QDs) [1 - 3]. Acclaimed as a milestone in nanotechnology, QDs are semiconductors inorganic crystalline with adjustable quantities of electrons that can exist in distinct quantum systems. The atom arrangement in QDs is identical to that in bulk materials, but due to the 3-dimensional truncation, there are more atoms on their surfaces. Since QDs frequently have small sizes and a diversity of varying element proportions, they might exhibit luminous features [4]. QDs in particular, feature unique luminous

* **Corresponding author Ashish Kumar:** NCE, Department of Science and Technology, Government of Bihar, India; Tel: +91- 94784-78088, E-mail: drashishchemlpu@gmail.com

Kulvinder Singh & Vineet Kumar (Eds.)

qualities, a size-dependent emission wavelength, broad excitation spectra, and electronic characteristics like broad and persistent absorption spectrum, confined dispersion, and good light persistence [5]. After a few nanoseconds, they absorbed white light and reemitted distinct shades based on the band gaps of the materials. Furthermore, by following the quantum confinement principle, the configuration of QDs could be readily controlled. The emission and absorption spectra corresponding to the energy band gap of QDs are governed by the quantum entrapment principle, which is the energy required to stimulate electrons from the electronic range to greater energy states [6]. This excitation results in the spontaneous formation of an electron-hole pair, which enables it to discharge energy in the character of fluorescent photons. QDs could be thought of as synthetic atoms that can produce different energy states, and by adjusting their diameters, their band gap could be accurately adjusted. Smaller nanocrystals have wider band gaps, and larger nanocrystals have smaller band gaps. The usage of QDs in devices such as telecom optics, light-emitting diodes, and medicinal applications is another possibility. QDs have stable and tuneable wavelengths [7].

CARBONACEOUS QUANTUM DOTS (CQD)

Discovery of CQDs

Since carbon is normally considered a dark substance, it used to be difficult to imagine that it might be miscible in aqueous and even gloriously illuminated [8]. CDs (probably known as CQD) were found unintendedly in 2004 at the time of filtration of single-wall carbon nanotubes (SWCNTs). They were initially referred to as "carbon nanoparticles," but the name "carbon dots" was later adopted because it evoked similar properties as same that of inorganic QD. Two years after constructing stable photoluminescent carbon nanoparticles of various diameters in 2006, Sun *et al.* gave them the name "carbon quantum dots." Within a year, Sun *et al.* reported water-dispersible CDs that had been passivized using poly-propionylethylenimine-co-ethylenimine [9 - 11]. The CDs were used to detect human breast malignant MCF-7 cells and showed dual photon-induced luminescence spectra. The potential for carbon dots in biological applications has attracted significant interest. Another benefit of carbon dots in the context of nanoparticle applications is their biocompatibility. As they are predominantly composed of the abundant and non-toxic substance carbon, carbon dots stand out from other nanoparticle families due to their unique geometric and electrical characteristics [12].

CQDs are spherically symmetrical, have a size of less than 10 nm, and can have amorphous or crystalline structures. Photoluminescence and wavelength-dependent emission are interesting properties of both amorphous and crystalline materials, as are high solubility, minimal hazardous, ease of synthesis, and biocompatibility [13]. Due to their broad range of technical applications in various domains, including photocatalysis, solar cells, LED devices, sensors, bioimaging, and drug delivery systems, CQDs gradually became a focus of discussion among researchers. Traditional QD formulations frequently contained cadmium, however, toxicity spurred on by cadmium ions that spilled signaled the development of the more compatible QD. The goal was to create cadmium-free QDs (CFQDs) with excellent chemical resilience, minimal cytotoxicity, and simplicity in pharmaceutical activities as the demand for more biocompatible QDs expanded. As a result, different QDs were created, including graphene QDs, silicon QDs, and carbon QDs [14, 15]. CQDs compensate for the toxicological, ecological risk, and biological deficiencies of traditional semiconductor quantum dots whilst inheriting their outstanding photonic features. CQDs are also easily interface functionalized and prepared on a massive volume, and they also exhibit strong solubility in water, chemical resistance, and light absorption resilience.

In this chapter, we will critically elaborate on the advantages, limitations, and potential results of the physical and chemical features that are the main focus of our investigation. We will also discuss several syntheses and characterization methodologies. We believe that by providing guidance on the fundamentals of CQDs from a perspective viewpoint, this chapter can be helpful.

Chemical Structure of CQDs

Due to their intense brightness and robust dispersion, carbon-based quantum dots have drawn considerable scrutiny. The architectures of carbon-based quantum dots control their many characteristics [16]. The CQD surface's numerous carboxyl moieties have excellent biocompatibility and water solubility. CQDs can also be used for surface modification and chemical treatment with various elastomeric, microbiological, regenerative, or inorganic substances. Surface passivation can enhance the CQDs' physical and fluorescent characteristics. CQDs are highly conductive, have a stable chemical structure, and exhibit strong photochemical resilience. CQDs are spherical carbon nanoparticles that can be made with or without a crystal structure [17]. Approximately 0.34 nm separates the sections of CQDs, which is compatible with the crystalline graphite spacing of (002). A system of interconnected or altered chemical functional moieties including oxygen- and amino-based clusters, could be found at the interface of CQDs. The substituent of CQDs could be identified using the matrix-assisted laser desorption ionization time-of-flight (MALDI-TOF) method relying on the chemical and physical configuration of the CQDs. CQDs are created *via* bottom-up techniques by dehydrating polycyclic aromatic molecules and carbonizing them [18]. The anticipated model architecture for the CQD DFT experiment is shown in Fig. (**1**). To investigate the feasibility of band edge location tweaking, the CQD configuration was synthesized with the distal carbon atoms linked by.

hydrogen, hydroxy and carboxylate [19]. Fig. (**2**) depicts the one-step hydrothermal-extraction method's role in the synthesis of CQD.

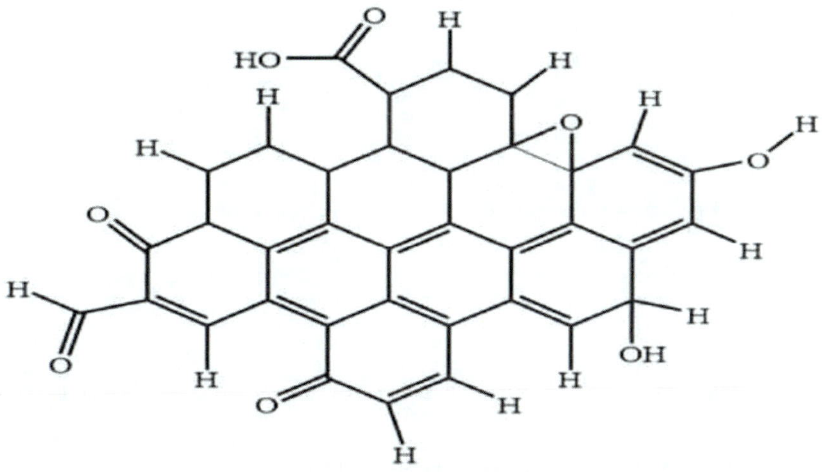

Fig. (1). The chemical structure of CQDs. (Source adapted from the ref [19]).

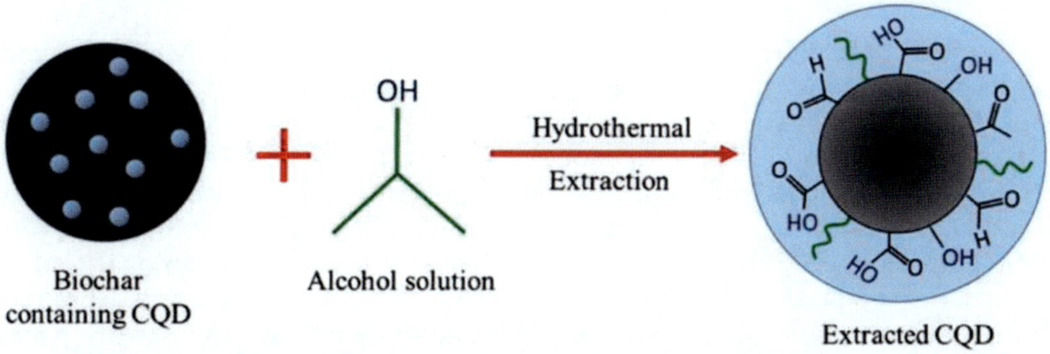

Fig. (2). A potential method for extracting CQD (Source adapted from the ref [19]).

SYNTHETIC TECHNIQUES

Despite the creation of CQDs, authors have developed numerous methods for obtaining them. Among top-bottom techniques, arc discharge, laser ablation, and acidic oxidation might be noted. The microwave pyrolytic procedure, the templates technique, the hydrothermal/solvothermal procedures, and the plasma ablation procedures are examples of bottom-up techniques. Chapter 3 contains a thorough discussion of each of the mentioned synthetic approaches. Fig. (**3**) displays different synthetic methods that use a Top-Down and Bottom-Up procedure [20].

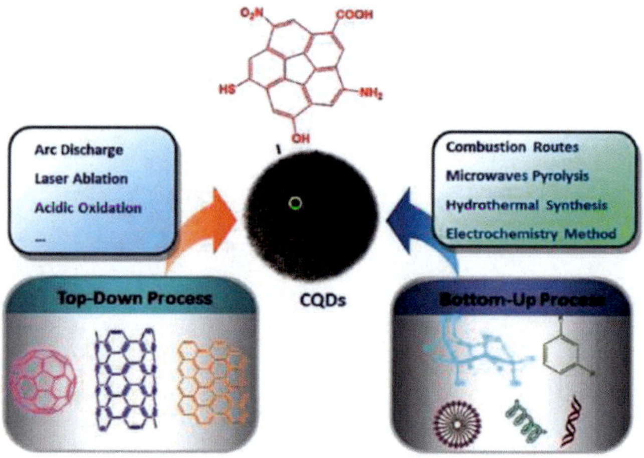

Fig. (3). Various synthetic techniques utilizing the Top-Down and Bottom-Up methods. (Source adapted from the ref [20]).

PHYSICAL AND CHEMICAL PROPERTIES

Absorbance

To determine the photonic absorption maxima of CQDs in the UV visible field, the p-p transitions of sp^2 conjugated carbon and the n-p transitions of hybridization involving heteroatoms such as S, N and P are frequently used. Surface passivation or alteration processes may change the absorption property [21, 22]. Researchers developed a straightforward hydrothermal process for producing green, red, and blue luminous CQDs using three isomers of phenylenediamines. Similar patterns were apparent in the UV-visible absorption spectrum of the produced CQDs. These three CQDs' absorption shifts were red-shifted, demonstrating that their electronic energy band gap was narrower than those of their predecessors.

Photoluminescence (PL)

Among the fascinating characteristics of CQDs, both in terms of fundamental physics and functional applications, is photoluminescence [23, 24]. One typical characteristic of the PL for CQDs is the clear dependence between the emission intensity and wavelength. This peculiar occurrence might be caused by an optical matrix of distinct nanoparticles or CQDs with various luminescent entrapment on the substrate. The broad and excitation-dependent PL emitted spectra may show different PL emission particle sizes. Researchers investigated the emission characteristics of CQDs when exposed to 470 nm light at various concentrations [25]. The PL frequency of the liquid first elevated and subsequently reduced as the CQD concentration was raised.

Electroluminescence (ECL)

It is not surprising that CQDs have inspired attention in ECL studies because semiconductor nanocrystals are widely recognized for displaying electroluminescence (ECL), which has applications in electrolytic domains. The controlling power of CQD-based light-emitting diodes (LEDs) could be used to control the emission spectrum [26 - 28]. The identical CQDs' shade bistable ECL, which ranged between blue to white, can be seen at different operating voltages. To investigate the luminescent process of CQDs, the investigators proposed novel hypotheses relying on the concatenated p domain's band gap dispersion and an additional interface defect's edge influence. The PL features of CQD fluorescence intensity from the linked p region are controlled by the quantum confinement effect (QCE) of p-conjugated electrons in the sp^2 atomic structure and could be altered by their magnitude, border arrangement, and structure. Fluorescence emission is caused by surface flaws in CQDs like sp^2 and sp^3 hybridized carbon, and these flaws can even affect fluorescence intensity and peak region [29, 30].

Up-conversion Photoluminescence (UCPL)

The multi-photon activation phase, in which two or more photons are absorbed simultaneously, causes light to be emitted at a minimal wavelength in contrast to the excitation wavelength, which is responsible for CQDs' UCPL properties [31]. The UCPL of CQDs offers unique avenues for two-photon luminescence microscopy cell imaging and also exceedingly efficient catalyst construction for biotechnology and power technological applications. On NIR stimulation, the PL spectrum revealed a heterogeneous emission spike at 540 nm that does not vary with variation in the excitation wavelength. According to the heterogeneous emission site, the emission happens through the least single state regardless of the stimulation source.

APPLICATIONS

The wide range of features of carbon-based QDs makes them suitable for a wide range of applications. These applications would affect human life quality and have the potential to draw substantial commercial interest. In-depth deliberation on these listed applications will be given in the forthcoming chapters.

Electrocatalysis

Due to the emergence of challenging environmental issues, CQDs have gained a curiosity in the fields of energy storage and transmission. The abundant functional moieties on the substrate of CQDs (-COOH, -OH, $-NH_2$, *etc.*) could be used as active coordination sites for transition metal ions [32]. By encouraging electron transfer through internal interactions, heteroatom-loaded CQDs containing different components may significantly enhance their electrocatalytic activity. Some inorganic substances, including metal sulfides, layered double-hydroxides (LDHs) and metal phosphides, respectively, hybridized into CQDs. Due to the following factors, CQDs could be used as effective electrocatalysts for OER, ORR, HER, and CO2RR, among others: (1) the low cost and ease of access of CQDs compared to the precious metals; (2) CQDs offer more potent catalytic reaction centers; (3) hybrids made from CQDs have improved electronic conductivity; and (4) favorable charge transfer during electrocatalytic reactions [2].

Biomedical Applications

Bioimaging

CQDs are crucial in biomedical applications, with bioimaging being one of the most relevant. The process of capturing images of living things using methods including X-rays, magnetic resonance imaging (MRI) and ultrasound is known as bioimaging. Additionally, three-dimensional structural data could be determined using it. CQDs offer various benefits beyond semiconductor QDs, including biomedical applications, minimal neurotoxicity, and excellent PL. These qualities make CQDs, particularly effective for the *in vitro* and *in vivo* imaging of biological processes. It's crucial to remember that CQDs are not hazardous; rather, the passivating reagent on their surface causes toxicity [33]. Surface passivating compounds with minimal cytotoxicity could be employed effectively for *in vivo* imaging at larger doses.

Targeted Drug Delivery

CQDs are potential candidates for secure, efficient, and targeted delivery due to their low cytotoxicity [34]. Theranostic compounds, which are medicines that could be utilized for both diagnostic and therapeutic objectives, are excellent options for treating CQDs. A multipurpose theranostic agent (CD-Oxa) was created when the interfaces of C-dots containing amine groups were coupled with an antitumor drug [oxidized oxaliplatin, oxa (IV)-COOH].

Nanomedicine

CQDs are a safer alternative to other fluorescent nanomaterials because they are tiny fluorescent NPs. Since CQDs do not cause any toxicity in animals, they have a lot of potential in nanomedicine [35 - 50]. For example, CQDs were injected into mice in an experiment, and the organs and internal functions of the mice were evaluated after four weeks, with the conclusion that there was no significant impact. Because of their small impact and low cytotoxicity, they can be used in *in vivo* research. Highly biocompatible CQDs supported by prothrombin time assays revealed that CQDs had no impact on the thrombin activity in plasma specimens [51]. The future of CQDs in photodynamic therapy is bright. CQDs successfully suppress MCF-7 and MDA-MB-231 tumor cells. Due to their ability to induce ROS and their ability to localize themselves specifically in malignancies, CQDs also are potential photocatalysts. CQDs' ability to act as nanocarriers makes them useful for monitoring and transporting genes and medications, as per branched polyethylenimine. CQDs have a lot of potential for gene transmission. In HeLa cells, CQDs can also be employed to achieve regulated drug administration. In order to induce regulated drug administration in HeLa cells, doxorubicin can be put into CQDs. However, it is unclear if CQDs will target a disease state directly, which limits their therapeutic efficacy [52].

Biosensing

CQDs could be utilized in a number of biosensing applications. The excellent water dispersion, flexibility in surface functionalization, improved cell penetration, minimal cytotoxicity, and excellent biocompatibility of CQDs offer them suitable biosensors. Visual cellulose, copper, sugar, nucleic, copper, sodium, and phosphate tracking are possible with CQD-based biosensors [20]. CQDs are a useful fluorescence sensing reagent for finding a single match in nucleic acid. The definition provides for the pi-pi attachment of fluorescently tagged single-stranded DNA (ssDNA) to CQDs, assisted by substantial fluorescence quench, and subsequent hybridization with the goal to generate double-stranded DNA. The consequence was that the ssDNA was desorbed from the CQDs' interface, which was accompanied by a further resurrection of fluorescence that allowed for DNA-prying. CQD-based FRET was used to confirm the detection and imaging of mitochondrial H_2O_2 [5]. CQDs contribute to energy transfer and serve as a transporter for the sensing system. This immunosensor can be used as a model for the creation of immunoassays to distinguish analytes with preferred antigens and antibodies. CQDs are additionally utilized as a fluorescence sensors to identify minute bioanalytics [53].

Photocatalysis

Photocatalysis has recently exploded in popularity as a green alternative to natural synthesis. The realization that sunlight is a necessary and infinite source of energy has roused consciousness in the photocatalytic process to some extent [54]. Natural substances may suffer from the enhanced viability of UV and short-wavelength visible light. They have a great chance of being used as photocatalysts in biological synthesis operations, given the established capacity to employ a

CQD solution to enable long-wavelength light and energy transformation. According to a recent survey, CQDs in the scope of 1-4 nm are efficient NIR light-determined photocatalysts for particularly oxidizing alcohols into benzaldehydes with elevated transformation aptitude (92 percent) and selectivity. This is because of their astounding catalytic intervention for H_2O_2 deterioration and NIR light-determined electron transfer activity (100 percent) [55].

Cytotoxicity

Numerous studies have recently focused on luminous CQD-based bio-probes with good stability. However, when utilizing functionalized CQDs in live structures, tissues, and animals, biocompatibility is a significant challenge. Recent years have seen the completion of comprehensive cytotoxicity studies on both functionalized and pure CQDs. By arc-discharging graphite rods and then refluxing them in HNO_3 for 12 hours, a team of scientists created CQDs for cytotoxicity testing [56]. Up to 0.4 mg ml 1, unaltered CQDs did not seem to be harmful to cells. A human kidney cell line was used in a different cytotoxicity investigation to evaluate luminous CQDs that had been electrochemically created. The results showed that the CQDs had no appreciable impact on cell survivability. The PEG-coated CQDs in all accessible diameters were benign and biocompatible even at far elevated levels than required for cell imaging and associated purposes. According to all of the research, CQDs possess a great deal of promise for *in vitro* and *in vivo* imaging techniques, and it is anticipated that in the nearish term, ubiquitous QDs and FDA-approved dyes will transition out CQDs as optical imaging reagents [57].

Sensor

Most CQD sensors work by quenching or inhibiting the fluorescence emission of the CQD in the vicinity of an analyte, which could be a metal ion (like Hg^{2+}) or another molecule. The quenching of the fluorescence emission could be brought on by the charge transfer mechanism [58]. The quenching processes in CQDs include photoinduced electron transmission, internal filter impact, fluorescence resonance energy transfer (FRET), statically quenched, and dynamic quenching. Whenever a quencher engages using a CQD, a nonfluorescent ground-state complex is produced, causing a dynamic quench in the CQD. In the event of a dynamic quench, the development of the ground-state complex may alter the CQDs' absorption spectra. The integrity of the groundstate compound may weaken during the dynamic quench, which will cause the impact to reduce. A charge transmission or energy transmission interaction between the CQDs and the quencher causes the exciting phase of CQDs to revert to its initial state during a procedure known as dynamic quenching.

Optronics

Dye-sensitized Solar Cells (DSCs)

DSCs have attracted a lot of interest due to their adaptability, affordability, and processing simplicity. Although DSCs gain from the range of organic dyes and perform admirably, their broad application may be constrained by factors such as the photobleaching of organic dyes, the expensive price and cytotoxicity of dyes incorporating ruthenium, as well as the flammable electrolyte. CQDs constructed from a range of inexpensive sources that have consistent light absorption seem promising for DSCs [59, 60]. The addition of CQDs to the dye/semiconductor complex increased the complex's photoelectric conversion efficiency by seven times.

Light-emitting Devices (LED)

CQDs are a popular material for LEDs because of their consistent light emission, relatively inexpensive, and environmental protection. Nitrogen-rich CQDs release a broad and intense visible light when exposed to UV light, which may be beneficial in phosphor systems. It is possible to make large-scale (20 x 20 cm) free-standing luminous coatings using the PMMA matrix embedded with CQDs. To prevent solid-state quenching, the polymer matrix can spread the CQDs and offer mechanical assistance. The resulting films have a lot of potential in large-scale flexible solid-state lighting projects since they are inexpensive, completely flexible, expandable, thermally consistent, environmentally sustainable, and mechanically durable [61]. InGaN blue LEDs could be employed as illuminators in white LEDs, and the resulting coatings could be utilized as color-converting phosphors. There are CQD-based LEDs that can transition color when the driving current changes. In these machines, which are produced by a solution-based technique, a CQD emissive barrier is positioned over an organic void transmission barrier and an organic or inorganic electron transmission barrier. By modifying the system's architecture and adding current density, it is feasible to produce multicolor emissions from the identical CQDs in the colors blue, magenta, cyan, and white [62].

CHARACTERIZATION OF CARBON QUANTUM DOTS

Several techniques, such as X-ray diffraction (XRD), nuclear magnetic resonance (NMR), transmission electron microscope (TEM), ultraviolet (UV) spectroscopy, Fourier-transform infrared spectroscopy (FTIR), and PL, could be utilized to classify C-dots with the aim of obtaining information about the synthetic properties of C-dots. Chapter 4 includes a detailed description of the techniques mentioned below.

TEM

Due to the 0.1–0.2 nm divine precision of TEM, the samples' ultrastructure can be categorized. TEM is highly sought after in research and development departments in the fields of medicine, pharmacology, bioengineering, and other fields. This technique can be used to investigate the morphology of CQDs in order to learn more about their shape, scale, and dispersion. C-dots are frequently characterized using the TEM. The fine structure of C-dots could also be examined using high-resolution TEM.

XRD

To characterize C-dots and determine their particle diameter, aspect integrity, and crystalline structure, employ XRD. XRD is another method used to identify the crystalline phases of CQDs. It is possible to examine C-dots with various architectures and optical characteristics.

FTIR

FTIR has also been utilized to determine the presence of functional moieties on the surface of CQDs. CQDs are mostly made up of oxygen, carbon, and hydrogen. Since the partial oxidation of a carbon precursor results in the abundance of carboxyl or carboxylic acid molecules, epoxy/ether and hydroxyl groups on the interface of CQDs, FTIR is a helpful tool for analyzing these oxygen-containing groups [63]. Improvements must be done with C-dots prior to application in order to balance out prospective reservoirs on the energy interface, reduce cytotoxicity, and increase fluorescence QY. Infrared spectroscopy could be used to assess modified CQDs and determine whether or not they have been sufficiently passivated [36, 50].

NMR

The structural knowledge of CQDs is frequently obtained using an NMR strategy. NMR determines the hybrid forms of C-atoms in the crystalline network and the binding mode across several carbon atoms. The best example to use in this field is 13C NMR spectroscopic estimations [64, 65].

UV Spectroscopy

CQDs prepared using various techniques typically display strong (UV) absorption, but the locations of UV absorption spikes are entirely different for different techniques used to prepare CQDs. NIR, UV (350 nm), visible (400-700 nm), and C-dots transmit at 3.8, 1.5-3, and 1.2 nm, respectively.

PL

PL is the most intriguing characterization of CQDs from the standpoint of property and application. CQDs can refract inputs from particles of various dimensions because of their optical characteristics. Each CQD also includes a number of emissive locations. Studies on the optical characteristics of small-sized CQDs are dubious because it is unclear what exactly causes PL. A distinctive quality of the PL of CQDs is the considerable ex-dependence of the emission intensity and wavelength. These CQDs produce PL radiation that is -dependent and has a broad emission spectrum spanning from 430 to 580 nm. The dazzling and vivid PL of CQDs can be used to explain the existence of a surface energy entrapment set up by surface functionalization [66 - 68].

ADVANTAGES OF CARBON QUANTUM DOTS

a. CQDs are low-cost and accessible, rendering them an enhancing option among nanocarbon members.
b. As opposed to organic dyes and conventional QDs, CQDs have a higher photostability due to their stability and composition.
c. When compared to other cadmium-based QDs and organic dyes, CQDs exhibit a wider excitation spectrum and a sharper emission spike.
d. C-dots are an ideal choice for use in biosensors, drug administration, and bioimaging due to their exceptional biologic features, including hydrophilic nature, low cytotoxicity, chemical inertness, and strong biocompatibility.
e. As contrasted to several QD, CQD has a higher luminescence.
f. As contrasted to several other organic dyes and cadmium-based QDs, CQDs have high aqueous stability.
g. Because of their superior electronic characteristics as electron generators and acceptors, carbon-based QDs have a wide range of applications in optoelectronic devices, catalysts, and sensing.
h. C-dots have very high chemical stability as compared to other QDs (traditional or metallic).

CHALLENGES AND PROSPECTIVE

The properties and advantages of emerging CQDs have been discussed in this chapter with respect to various types of applicability. Chemical resilience and minimal cytotoxicity are two exciting advantages of CQDs, which are QDs with a carbon center and oxygen-containing functional moieties on their substrate. Among the many uses for CQDs are bioimaging, drug/gene administration, detectors, optoelectronic devices, LEDs, photovoltaics, and electrochemical water splitting. The optical characteristics of CQDs are influenced by surface treatment and heteroatom doping, and additional study is required in several areas [69].

CQDs are also a viable low-cost substitute for semiconductor QDs in a number of applications, such as ultra-sensitive detectors for metallic ions, biological molecules, chemicals, and inorganic compounds. CQDs, on the other hand, can be produced cost-effectively and simply with minimal toxicity. To make CQDs smaller than 10 nm, several top-down and bottom-up methods have been utilized. Before the development of CQD in the aforementioned districts, there are still some challenges that needed to be overcome by additional study. The investigation must focus on how these elements impact CQD characteristics because CQD interface composition and impurity doping significantly influence CQD features. Therefore, more investigation is required to address the low QY, moderate light, and non-uniform electro-optic features of the CQDs [70]. If these difficulties can be overcome in the upcoming years, CQDs may replace semiconductor QDs in a variety of applications due to their low cost and minimal cytotoxicity, and they are often on the edge of becoming economically feasible in an array of applicants.

CONCLUSION

Due to their superior optical qualities and biocompatibility, CQDs have shown great promise as replacements for hazardous metal-based quantum dots in biological contexts and living things. In addition, the employment of CQDs as a catalyst, supercapacitors, and other applications shows their ubiquity of application potential and adaptability as flexible materials in the creation of sophisticated feasible intelligent sensors for the future. Furthermore, the application of CQDs as photocatalytic cations benefits from the generally easy and cost-effective synthesis of CQDs from a diversity of biomass carbon sources and has a lot of potential in terms of large-scale production, especially for technological purposes.

In this chapter, we have elaborated the basic introduction to the CQDs, emphasizing their physical and chemical properties, and advantages contrasted by their applications.

a. The appreciable attributes of CQDs are advantageous for biomedical and scientific applications.
b. CQDs offer a wide range of adequacy for bioimaging and drug delivery studies *in vitro* and *in vivo*.
c. Due to their bioactivity, nontoxicity, solid PL, synthesized nature, and stability, C-dots have emerged as an intriguing material for bioimaging and the identification of diverse analytes.

While the use of CQDs in electrocatalysis is still in its infancy. CQDs can also be created electrochemically, which results in uniformly sized CQDs and, more crucially, enables CQDs to collaborate with other traditional electrocatalysts in a one-pot green chemistry method. CQDs offer significant electrocatalytic capability due to their large surface area, high electrical conductivity, and rapid charge transmission. To further exploit the promise of these progressively significant carbon compounds, scientists predict the development of more affordable, straightforward, and innovative synthesis procedures ahead.

To summarise, carbon quantum dots are only in their infancy, but their potential is enormous. Researchers believe that the field of research should now move on from designing synthetic methods and demonstrating sensing principles alone. Due to their exceptional light absorption capacity and distinct photo-induced electron transmission capabilities, CQDs are also regarded as a top choice. Due to CQDs' excellent QY, photo- and chemical durability, and non-blinking nature, investigators may build extremely precise biosensors that function in a range of settings. Additionally, new methods and techniques for recognizing and classifying the characteristics of emergent CQDs in the physiological matrix must be created in order to minimize community distrust and tackle the hazards and prospective dangers of such disorders.

REFERENCES

[1] Wu, P.; Xu, Y.; Zhan, J.; Li, Y.; Xue, H.; Pang, H. The research development of quantum dots in electrochemical energy storage, small. In: *Nano small micro*; wiley, **2018**; 14, pp. 1-24. [http://dx.doi.org/10.1002/smll.201801479]

[2] Wang, Y.; Hu, A. Carbon quantum dots: Synthesis, properties and applications. *J. Mater. Chem. C Mater. Opt. Electron. Devices,* **2014**, *2*(34), 6921-6939. [http://dx.doi.org/10.1039/C4TC00988F]

[3] Li, C.; Wang, Y.; Jiang, H.; Wang, X. Review—Intracellular Sensors Based on Carbonaceous Nanomaterials: A Review. *J. Electrochem. Soc.,* **2020**, *167*(3), 037540. [http://dx.doi.org/10.1149/1945-7111/ab67a3]

[4] Molaei, M.J. A review on nanostructured carbon quantum dots and their applications in biotechnology, sensors, and chemiluminescence. *Talanta,* **2019**, *196*, 456-478. [http://dx.doi.org/10.1016/j.talanta.2018.12.042] [PMID: 30683392]

[5] Li, M.; Chen, T.; Gooding, J.J.; Liu, J. Review of carbon and graphene quantum dots for sensing. *ACS Sens.,* **2019**, *4*(7), 1732-1748. [http://dx.doi.org/10.1021/acssensors.9b00514] [PMID: 31267734]

[6] Shen, T.; Yang, L.; Pam, M.E.; Shi, Y.; Yang, H.Y. Quantum dot-carbonaceous nanohybrid composites: preparation and application in electrochemical energy storage. *J. Mater. Chem. A Mater. Energy Sustain.,* **2020**, *8*(43), 22488-22506. [http://dx.doi.org/10.1039/D0TA07674K]

[7] Ashouri, R.; Ghasemipoor, P.; Rasekh, B.; Yazdian, F.; Mofradnia, S.R.; fattahi, M. The effect of ZnO-based carbonaceous materials for degradation of benzoic pollutants: A review. *Int. J. Environ. Sci. Technol.,* **2019**, *16*(3), 1729-1740. [http://dx.doi.org/10.1007/s13762-018-2056-5]

[8] Iravani, S.; Varma, R.S. Green synthesis, biomedical and biotechnological applications of carbon and graphene quantum dots. A review. *Environ. Chem. Lett.,* **2020**, *18*(3), 703-727. [http://dx.doi.org/10.1007/s10311-020-00984-0] [PMID: 32206050]

[9] Shah, A.; Zahid, A.; Subhan, H.; Munir, A.; Iftikhar, F.J.; Akbar, M. Heteroatom-doped carbonaceous electrode materials for high performance energy storage devices. *Sustain. Energy Fuels,* **2018**, *2*(7), 1398-1429. [http://dx.doi.org/10.1039/C7SE00548B]

[10] Dou, D.; Duan, J.; Zhao, Y.; He, B.; Tang, Q. Cubic carbon quantum dots for light-harvesters in mesoscopic solar cells. *Electrochim. Acta,* **2018**, *275*, 275-280. [http://dx.doi.org/10.1016/j.electacta.2018.04.124]

[11] Cui, L.; Li, C.; Tang, B.; Zhang, C. Advances in the integration of quantum dots with various nanomaterials for biomedical and environmental applications. *Analyst,* **2018**, *143*(11), 2469-2478. [http://dx.doi.org/10.1039/C8AN00222C] [PMID: 29736519]

[12] Huang, R.; Wang, L.; Zhang, Q.; Chen, Z.; Li, Z.; Pan, D.; Zhao, B.; Wu, M.; Wu, C.M.L.; Shek, C.H. Irradiated graphene loaded with SnO_2 quantum dots for energy storage. *ACS Nano,* **2015**, *9*(11), 11351-11361. [http://dx.doi.org/10.1021/acsnano.5b05146] [PMID: 26434377]

[13] Molaei, M.J. Carbon quantum dots and their biomedical and therapeutic applications: A review. *RSC Advances,* **2019**, *9*(12), 6460-6481. [http://dx.doi.org/10.1039/C8RA08088G] [PMID: 35518468]

[14] Saikia, M.; Das, T.; Dihingia, N.; Fan, X.; Silva, L.F.O.; Saikia, B.K. Formation of carbon quantum dots and graphene nanosheets from different abundant carbonaceous materials. *Diamond Related Materials,* **2020**, *106*, 107813. [http://dx.doi.org/10.1016/j.diamond.2020.107813]

[15] Namdari, P.; Negahdari, B.; Eatemadi, A. Synthesis, properties and biomedical applications of carbon-based quantum dots: An updated review. *Biomed. Pharmacother.,* **2017**, *87*, 209-222. [http://dx.doi.org/10.1016/j.biopha.2016.12.108] [PMID: 28061404]

[16] Sinha, A.; Dhanjai, R.; Jain, R.; Zhao, H.; Karolia, P.; Jadon, N. Voltammetric sensing based on the use of advanced carbonaceous nanomaterials: A review. *Mikrochim. Acta,* **2018**, *185*(2), 89. [http://dx.doi.org/10.1007/s00604-017-2626-0] [PMID: 29594390]

[17] Saud, P.S.; Pant, B.; Alam, A.M.; Ghouri, Z.K.; Park, M.; Kim, H.Y. Carbon quantum dots anchored TiO_2 nanofibers: Effective photocatalyst for waste water treatment. *Ceram. Int.,* **2015**, *41*(9), 11953-11959. [http://dx.doi.org/10.1016/j.ceramint.2015.06.007]

[18] Cruz, A.A.C.; Freire, R.M.; Carneiro, S.V.; Ferreira, O.P.; Fechine, P.B.A. Carbon quantum dots: A safe tool to learn about quantum phenomenon in nanomaterials. *J. Lab. Chem. Educ.,* **2017**, *5*(3), 48-54. [http://dx.doi.org/10.5923/j.jlce.20170503.02]

[19] Jamaludin, N.; Tan, T.L.; Zaman, A.S.K.; Sadrolhosseini, A.R.; Rashid, S.A. Acid-free hydrothermal-extraction and molecular structure of carbon quantum dots derived from empty fruit bunch biochar. *Materials,* **2020**, *13*(15), 3356. [http://dx.doi.org/10.3390/ma13153356] [PMID: 32751070]

[20] Wang, X.; Feng, Y.; Dong, P.; Huang, J. A mini review on carbon quantum dots: Preparation, properties, and electrocatalytic application. *Front Chem.,* **2019**, *7*, 671. [http://dx.doi.org/10.3389/fchem.2019.00671] [PMID: 31637234]

[21] Tian, X.; Zeng, A.; Liu, Z.; Zheng, C.; Wei, Y.; Yang, P.; Zhang, M.; Yang, F.; Xie, F. Carbon quantum dots: *In vitro* and *in vivo* studies on biocompatibility and biointeractions for optical imaging. *Int. J. Nanomedicine,* **2020**, *15*, 6519-6529.

[http://dx.doi.org/10.2147/IJN.S257645] [PMID: 32943866]

[22] Molaei, M.J. Principles, mechanisms, and application of carbon quantum dots in sensors: A review. *Anal. Methods,* **2020**, *12*(10), 1266-1287. [http://dx.doi.org/10.1039/C9AY02696G]

[23] Tajik, S.; Dourandish, Z.; Zhang, K.; Beitollahi, H.; Le, Q.V.; Jang, H.W.; Shokouhimehr, M. Carbon and graphene quantum dots: a review on syntheses, characterization, biological and sensing applications for neurotransmitter determination. *RSC Advances,* **2020**, *10*(26), 15406-15429. [http://dx.doi.org/10.1039/D0RA00799D] [PMID: 35495425]

[24] Flores-Pacheco, A.; Álvarez-Ramos, M.E.; Ayón, A. *Down-shifting by quantum dots for silicon solar cell applications,* **2019**. [http://dx.doi.org/10.1016/B978-0-08-102762-2.00013-6]

[25] de Oliveira, B.P.; da Silva Abreu, F.O.M. Carbon quantum dots synthesis from waste and by-products: Perspectives and challenges. *Mater. Lett.,* **2021**, *282*, 128764. [http://dx.doi.org/10.1016/j.matlet.2020.128764]

[26] Chandrasekar, S.; Kasi Viswanathan, P.; Uthirakumar, P.; Montanari, G.C. Investigations on novel carbon quantum dots covered nanofluid insulation for medium voltage applications. *J. Electr. Eng. Technol.,* **2020**, *15*(1), 269-278. [http://dx.doi.org/10.1007/s42835-019-00316-5]

[27] Farshbaf, M.; Davaran, S.; Rahimi, F.; Annabi, N.; Salehi, R.; Akbarzadeh, A. Carbon quantum dots: Recent progresses on synthesis, surface modification and applications. *Artif. Cells Nanomed. Biotechnol.,* **2018**, *46*(7), 1331-1348. [http://dx.doi.org/10.1080/21691401.2017.1377725] [PMID: 28933188]

[28] Pillar-Little, T.J. Carbon quantum dots: Bridging the gap between chemical structure and material properties. *theses diss. -chemistry,* **2018**.

[29] Ray, S.C.; Jana, N.R. Application of carbon-based nanomaterials as drug and gene delivery carrier. In: *Carbon Nanomat. for Biol. and Medical Applic*; Elsevier, **2017**; pp. 163-203. [http://dx.doi.org/10.1016/B978-0-323-47906-6.00005-9]

[30] Wang, Y.; Hu, A. Carbon quantum dots: Synthesis, properties and applications. *J. Mater. Chem. C Mater. Opt. Electron. Devices,* **2014**, *2*(34), 6921-6939. [http://dx.doi.org/10.1039/C4TC00988F]

[31] Yoo, D.; Park, Y.; Cheon, B.; Park, M.H. Carbon dots as an effective fluorescent sensing platform for metal ion detection. *Nanoscale Res. Lett.,* **2019**, *14*(1), 272. [http://dx.doi.org/10.1186/s11671-019-3088-6] [PMID: 31410663]

[32] Rani, U.A.; Ng, L.Y.; Ng, C.Y.; Mahmoudi, E. A review of carbon quantum dots and their applications in wastewater treatment. *Adv. Colloid Interface Sci.,* **2020**, *278*, 102124. [http://dx.doi.org/10.1016/j.cis.2020.102124] [PMID: 32142942]

[33] Pan, M.; Xie, X.; Liu, K.; Yang, J.; Hong, L.; Wang, S. Fluorescent carbon quantum dots-synthesis, functionalization and sensing application in food analysis. *Nanomaterials,* **2020**, *10*(5), 930. [http://dx.doi.org/10.3390/nano10050930] [PMID: 32403325]

[34] Wan, C.F.; Cui, Y.; Gai, W.X.; Cheng, Z.L.; Chung, T.S. Nanostructured membranes for enhanced forward osmosis and pressure-retarded osmosis. In: *Sustainable Nanoscale Engineering*; Elsevier Inc., **2019**; pp. 373-394. [http://dx.doi.org/10.1016/B978-0-12-814681-1.00014-X]

[35] Thakur, A.; Kumar, A. Corrosion inhibition activity of Pyrazole derivatives on metals and alloys in acidic environment : A review. *J. Emerg. Technol. Innov. Res.,* **2019**, *6*, 339-343.

[36] Bashir, S.; Thakur, A.; Lgaz, H.; Chung, I.M.; Kumar, A. Corrosion inhibition efficiency of bronopol on aluminium in 0.5 M HCl solution: Insights from experimental and quantum chemical studies. *Surf. Interfaces,* **2020**, *20*, 100542.

[http://dx.doi.org/10.1016/j.surfin.2020.100542]

[37] Kaur, M.; Thakur, A.; Sharma, V.; Kumar, A. Tricine as an effective corrosion inhibitor for aluminium in acidic medium. *Think India J.,* **2019**, *22*, 3200-3219.

[38] Sandilya, S.; Thakur, A.; Suresh singh, S.; Kumar, A. Recent advances, synthesis and characterization of bio-based based polymers from natural sources. *Plant Cell Biotechnol. Mol. Biol.,* **2021**, *22*, 70-93.

[39] Thakur, A.; Kumar, A.; Kaya, S.; Vo, D.V.N.; Sharma, A. Suppressing inhibitory compounds by nanomaterials for highly efficient biofuel production: A review. *Fuel,* **2022**, *312*, 122934. [http://dx.doi.org/10.1016/j.fuel.2021.122934]

[40] Thakur, A.; Kumar, A. Sustainable inhibitors for corrosion mitigation in aggressive corrosive media: A comprehensive study. *J. Bio- Tribo-Corrosion.,* **2021**, *7*, 1-48. [http://dx.doi.org/10.1007/s40735-021-00501-y]

[41] Thakur, A.; Kumar, A. Potential of weeds extract as a green corrosion inhibitor on mild steel: A review. *Think India J.,* **2019**, *22*, 3226-3240.

[42] Thakur, A.; Kumar, A. Nanotechnology and its significance in food industries : An Overview. *J. Emerg. Technol. Innov. Res.,* **2019**, *6*, 344-348.

[43] Thakur, A.; Kumar, A. Potential of natural product extract as green corrosion inhibitors for steel : A review. *J. Gujarat Res. Soc.,* **2019**, *21*, 200-212.

[44] Prakash, S.; Srivastava, S. Impact of climate change on biodiversity: An overview. *International Journal of Biological Innovations,* **2019**, *1*(2), 60-65. [http://dx.doi.org/10.46505/IJBI.2019.1205]

[45] Bashir, S.; Thakur, A.; Lgaz, H.; Chung, I.M.; Kumar, A. Corrosion inhibition performance of acarbose on mild steel corrosion in acidic medium: An experimental and computational study. *Arab. J. Sci. Eng.,* **2020**, *45*(6), 4773-4783. [http://dx.doi.org/10.1007/s13369-020-04514-6]

[46] Kumar, A.; Thakur, A. Encapsulated nanoparticles in organic polymers for corrosion inhibition. In: *Corrosion Protection at the Nanoscale*; Elsevier Inc., **2020**; pp. 345-362. [http://dx.doi.org/10.1016/B978-0-12-819359-4.00018-0]

[47] Parveen, G.; Bashir, S.; Thakur, A.; Saha, S.K.; Banerjee, P.; Kumar, A. Experimental and computational studies of imidazolium based ionic liquid 1-methyl- 3-propylimidazolium iodide on mild steel corrosion in acidic solution. *Mater. Res. Express,* **2020**, *7*(1), 016510. [http://dx.doi.org/10.1088/2053-1591/ab5c6a]

[48] Thakur, A.; Kaya, S.; Abousalem, A.S.; Sharma, S.; Ganjoo, R.; Assad, H.; Kumar, A. Computational and experimental studies on the corrosion inhibition performance of an aerial extract of Cnicus Benedictus weed on the acidic corrosion of mild steel. *Process Saf. Environ. Prot.,* **2022**, *161*, 801-818. [http://dx.doi.org/10.1016/j.psep.2022.03.082]

[49] Verma, C.; Quraishi, M.A.; Ebenso, E.E. A review on ammonia derivatives as corrosion inhibitors for metals and alloys. *Green Energy and Technology,* **2020**, *07*, 49-67. [http://dx.doi.org/10.1007/978-3-030-35106-9_3]

[50] Bashir, S.; Thakur, A.; Lgaz, H.; Chung, I.M.; Kumar, A. Computational and experimental studies on Phenylephrine as anti-corrosion substance of mild steel in acidic medium. *J. Mol. Liq.,* **2019**, *293*, 111539. [http://dx.doi.org/10.1016/j.molliq.2019.111539]

[51] Gayen, B.; Palchoudhury, S.; Chowdhury, J. Carbon dots: A mystic star in the world of nanoscience. *J. Nanomater.,* **2019**, *2019*, 1-19. [http://dx.doi.org/10.1155/2019/3451307]

[52] Qu, D.; Wang, X.; Bao, Y.; Sun, Z. Recent advance of carbon dots in bio-related applications. *Journal*

of Physics: Materials, **2020**, *3*(2), 022003.
[http://dx.doi.org/10.1088/2515-7639/ab7cb9]

[53] Tuerhong, M.; Xu, Y.; Yin, X-B. Review on carbon dots and their applications. *Chin. J. Anal. Chem.,* **2017**, *45*(1), 139-150.
[http://dx.doi.org/10.1016/S1872-2040(16)60990-8]

[54] Thangaraj, B.; Solomon, P.R.; Ranganathan, S. Synthesis of carbon quantum dots with special reference to biomass as a source : A review. *Curr. Pharm. Des.,* **2019**, *25*(13), 1455-1476.
[http://dx.doi.org/10.2174/1381612825666190618154518] [PMID: 31258064]

[55] Singh, I.; Arora, R.; Dhiman, H.; Pahwa, R. Carbon quantum dots: Synthesis, characterization and biomedical applications. *Turkish Journal of Pharmaceutical Sciences,* **2018**, *15*(2), 219-230.
[http://dx.doi.org/10.4274/tjps.63497] [PMID: 32454664]

[56] Jelinek, R. Carbon quantum dots. In: *Synthesis, Properties and Applicatons*; Springer, **2017**.
[http://dx.doi.org/10.1007/978-3-319-43911-2]

[57] Sagbas, S.; Sahiner, N. Carbon dots: Preparation, properties, and application. In: *Nanocarbon and its Composites*; Elsevier Ltd., **2018**; pp. 651-676.
[http://dx.doi.org/10.1016/B978-0-08-102509-3.00022-5]

[58] Scaria, J.; Karim, A.V.; Divyapriya, G.; Nidheesh, P.V.; Suresh Kumar, M. Carbon-supported semiconductor nanoparticles as effective photocatalysts for water and wastewater treatment. In: *Nano-Materials as Photocatalysts for Degradation of Environmental Pollutants*; Elsevier Inc., **2019**; pp. 245-278.
[http://dx.doi.org/10.1016/B978-0-12-818598-8.00013-4]

[59] Jiang, R.; Wu, D.; Lu, G.; Yan, Z.; Liu, J. Modified 2D-2D $ZnIn_2S_4$/BiOCl van der Waals heterojunctions with CQDs: Accelerated charge transfer and enhanced photocatalytic activity under vis- and NIR-light. *Chemosphere,* **2019**, *227*, 82-92.
[http://dx.doi.org/10.1016/j.chemosphere.2019.04.038] [PMID: 30986605]

[60] Feng, S.; Chen, T.; Liu, Z.; Shi, J.; Yue, X.; Li, Y. Z-scheme CdS/CQDs/g-C_3N_4 composites with visible-near-infrared light response for efficient photocatalytic organic pollutant degradation. *Sci. Total Environ.,* **2020**, *704*, 135404.
[http://dx.doi.org/10.1016/j.scitotenv.2019.135404] [PMID: 31806342]

[61] Wang, T.; Liu, X.; Ma, C.; Zhu, Z.; Liu, Y.; Liu, Z.; Wei, M.; Zhao, X.; Dong, H.; Huo, P.; Li, C.; Yan, Y. Bamboo prepared carbon quantum dots (CQDs) for enhancing Bi3Ti4O12 nanosheets photocatalytic activity. *J. Alloys Compd.,* **2018**, *752*, 106-114.
[http://dx.doi.org/10.1016/j.jallcom.2018.04.085]

[62] Behnood, R.; Sodeifian, G. Synthesis of N doped-CQDs/Ni doped-ZnO nanocomposites for visible light photodegradation of organic pollutants. *J. Environ. Chem. Eng.,* **2020**, *8*(4), 103821.
[http://dx.doi.org/10.1016/j.jece.2020.103821]

[63] Parveen, G.; Bashir, S.; Thakur, A.; Saha, S.K.; Banerjee, P.; Kumar, A. Experimental and computational studies of imidazolium based ionic liquid 1-methyl- 3-propylimidazolium iodide on mild steel corrosion in acidic solution. *Mater. Res. Express,* **2020**, *7*(1), 016510.
[http://dx.doi.org/10.1088/2053-1591/ab5c6a]

[64] Pandey, F.P.; Rastogi, A.; Singh, S. Optical properties and zeta potential of carbon quantum dots (CQDs) dispersed nematic liquid crystal 4′- heptyl-4-biphenylcarbonitrile (7CB). *Opt. Mater.,* **2020**, *105*, 109849.
[http://dx.doi.org/10.1016/j.optmat.2020.109849]

[65] Kulandaivalu, T.; Abdul Rashid, S.; Sabli, N.; Tan, T.L. Visible light assisted photocatalytic reduction of CO2 to ethane using CQDs/Cu2O nanocomposite photocatalyst. *Diam. Rela. Mate.,* **2019**, *91*, 64-73.
[http://dx.doi.org/10.1016/j.diamond.2018.11.002]

[66] Xingyuan, Du.; Liu, Lu.; Dong, Zhe.; Zheng, Cui.; Xiaochen, Sun.; Wu, Di.; Zhe, Ma.; Zhang, Fang.; Yang, Liu.; Yonglei, An. Accelerated redox cycles of Fe(III)/Fe(II) and Cu(III)/Cu(II) by Photo-Induced Electron from N-CQDs for Enhanced Photo-Fenton Capability of CuFe-LDH. *Catalysts,* **2020**, *10*(9), 960.

[67] Zhao, X.; Yang, H.; Cui, Z.; Wang, X.; Yi, Z. Growth process and CQDs-modified Bi 4 Ti 3 O 12 square plates with enhanced photocatalytic performance. *Micromachines,* **2019**, *10*(1), 66. [http://dx.doi.org/10.3390/mi10010066] [PMID: 30669287]

[68] Janus, Ł.; Radwan-Pragłowska, J.; Piątkowski, M.; Bogdał, D. Smart, tunable CQDs with antioxidant properties for biomedical applications—ecofriendly synthesis and characterization. *Molecules,* **2020**, *25*(3), 736.
[http://dx.doi.org/10.3390/molecules25030736] [PMID: 32046279]

[69] Song, J.; Zhao, L.; Wang, Y.; Xue, Y.; Deng, Y.; Zhao, X.; Li, Q. Carbon quantum dots prepared with chitosan for synthesis of CQDS/auNPs for iodine ions detection. *Nanomaterials,* **2018**, *8*(12), 1043. [http://dx.doi.org/10.3390/nano8121043] [PMID: 30551611]

[70] Wei, X.; Wang, Y.; Huang, Y.; Fan, C. Composite ZIF-8 with CQDs for boosting visible-light-driven photocatalytic removal of NO. *J. Alloys Compd.,* **2019**, *802*, 467-476. [http://dx.doi.org/10.1016/j.jallcom.2019.06.086]

CHAPTER 2

Synthesis of Carbonaceous Quantum Dots

Munish Kumar[1,*]

[1] *Department of Chemistry & Centre for Advanced Studies in Chemistry, Panjab University, Sector-14, Chandigarh , India*

Abstract: New class of nanocarbon materials, such as luminescent carbon quantum dots (CQDs) has gained a great deal of interest in the area of electrocatalysis, solar cells, bioimaging nanomedicine, a chemical sensor and a light-emitting diode. CQDs exhibit good physio-chemical properties, such as photoluminescence, high crystallization and good dispersibility. The rapid electron transfer, small size and superconductivity of CQDs provide the CQDs-based composite offering enhanced catalytic activity and electric conductivity. However, additional active moieties are present on the surface, which might aid in the formation of multi-component electrically activated catalysts. Additionally, the multi-component catalysts' internal interactions promote charge transfer and catalytic efficiency, both of which are essential for electrochemistry. Therefore, keeping in mind the importance of CQDs, they are synthesized on the basis of two approaches: Top-down and Bottom-up. The bulk material is reduced in size by utilizing chemical and physical processes in the top-down approach. On the contrary, in the bottom-up method, the atoms are assembled and converted into CQDs using polymerization and carbonization through a chemical reaction. Hence, in this chapter, we will discuss the synthesis techniques for CQDs, such as hydrothermal/solvothermal method, laser ablation, arc-discharge method, acidic oxidation, thermal/combustion routes, electrochemical method and microwave pyrolysis method.

Keywords: CQDs, Catalyst, Polymerization, Synthesis, Solvothermal.

INTRODUCTION

At present, carbon-based nanomaterials (NMs), including fullerenes, graphene, nanodiamonds and carbon nanotubes, have attracted scientists' attention worldwide [1 - 4]. However, there are some difficulties observed in the separation and preparation of nanodiamonds, due to less water solubility of carbon nanotubes, graphene and fullerenes, which make these nanocrystals less fluorescence in the visible region and limit their application. However, CQDs are carbon-based nanomaterials having noble zero-dimensional small size and strong

* **Corresponding author Munish Kumar:** Department of Chemistry & Centre for Advanced Studies in Chemistry, Panjab University, Sector-14, Chandigarh, India; Email: munishthukral7@gmail.com

Kulvinder Singh & Vineet Kumar (Eds.)

fluorescence nature [5]. Also, carbon quantum dots are known as carbon dots in some cases. The photoelectrochemical characteristics and dimensions of all quantum dots (including those made of graphene, carbon nano, and polymer) are comparable, but their internal structures and external chemical moieties vary. Mainly, CQDs are spherical and monodisperse in nature, and the surface is covered with a carbon-based chain, having more oxygen groups present in the skeleton [6]. In order to make these materials fluorescent and to tune the electronic structure, the surface moiety and size must be carefully familiar. CQDs' superior optical qualities make up for the material's shortcomings in regard to contamination, cytotoxic and ecological impact. Also, carbon quantum dots have shown excellent chemical stability, good water solubility, easily functionalized, scale up the reaction and photobleaching resistance [7]. Due to the superior characteristics of CQDs, they have been broadly utilized in nanomedicine, biology, photoelectrocatalysis and chemical sensing [8 - 11].

Hence, in this chapter, we will discuss the synthesis techniques for CQDs, such as hydrothermal/solvothermal method, laser ablation, acidic oxidation, arc-discharge method, thermal/ combustion routes, electrochemical method and microwave pyrolysis method. But before it, we will be discussing the chemical states and structures of the CQDs.

Chemical States of CQDs

CQDs can exist in various chemical states, depending on their synthesis method, functionalization, and environmental conditions. The most common chemical states of CQDs include sp2-hybridized carbon atoms, oxygen-containing functional groups, and surface defects. The sp2-hybridized carbon atoms are the core building blocks of CQDs and are responsible for their electronic properties, such as conductivity and fluorescence. The sp2 carbon atoms in CQDs form a conjugated network that allows for the delocalization of π electrons, leading to their unique optical properties. The size and shape of the sp2 domains in CQDs can also affect their fluorescence properties. Oxygen-containing functional groups, such as hydroxyl, carboxyl, and carbonyl groups, are commonly present on the surface of CQDs. These functional groups can be introduced intentionally during synthesis or can result from surface oxidation. The presence of oxygen-containing functional groups can affect the surface charge and chemical reactivity of CQDs, making them more suitable for various applications, such as sensing and catalysis. Surface defects, such as vacancies, dangling bonds, and edge sites, can also influence the electronic and catalytic properties of CQDs. Surface defects can act as active sites for chemical reactions and can enhance the charge transfer between CQDs and their surroundings.

Structure of CQDs

The structure of CQDs is another important factor that determines their physiochemical properties. The structure of CQDs can be classified based on their size, shape, crystallinity, and morphology.

Size: The size of CQDs typically ranges from a few nanometers to tens of nanometers. The size of CQDs can affect their fluorescence properties, with smaller CQDs exhibiting higher quantum yields due to the quantum confinement effect.

Shape: CQDs can have various shapes, such as spherical, elliptical, triangular, and hexagonal. The shape of CQDs can influence their surface area and packing density, affecting their chemical reactivity and stability.

Crystallinity: CQDs can exhibit different degrees of crystallinity, ranging from amorphous to highly crystalline. The degree of crystallinity can affect their electronic and optical properties, with highly crystalline CQDs exhibiting better conductivity and fluorescence properties.

Morphology: CQDs can have various morphologies, such as core-shell, hollow, and porous structures. The morphology of CQDs can affect their surface area, porosity, and surface charge, affecting their catalytic and sensing properties.

The structure of CQDs can also be influenced by their synthesis method. Various synthesis methods, such as hydrothermal/solvothermal, arc-discharge, thermal/combustion routes, electrochemical, and microwave pyrolysis, can result in CQDs with different structures and properties, which will be discussed beneath.

Synthesis Method

Therefore, keeping in mind the importance of CQDs, they are synthesized on the basis of two approaches: Top-down and Bottom-up [12, 13]. The bulk material is transformed into small-sized CQDs utilizing chemical and physical processes in the top-down technique. On the contrary, in the bottom-up method, the atoms are assembled and converted into CQDs using polymerization and carbonization through a chemical reaction.

Top-down Strategy

The top-down strategy is one of the most promising approaches for the preparation of CQDs. This strategy involves the fragmentation of bulk materials into small particles. In the case of CQDs, the starting material is a carbonaceous material that can be derived from natural sources such as wood, lignin, and cellulose, or from synthetic sources such as graphite and carbon black. There are two main top-down strategies for the preparation of CQDs: mechanical and chemical fragmentation. Mechanical fragmentation involves the grinding or ball milling of the carbonaceous material into small CQDs. The ball milling process involves the use of high-energy ball milling equipment to generate shear forces that break the carbonaceous material into small particles. The advantage of the mechanical fragmentation approach is that it is relatively simple and cost-effective, but the size and shape of the CQDs produced by this method are not well controlled, and the process may generate defects in the CQDs, leading to reduced quantum yield. Chemical fragmentation involves the oxidative or hydrothermal treatment of the carbonaceous material. The oxidative treatment involves the treatment of the carbonaceous material with oxidizing agents such as nitric acid or hydrogen peroxide to break the material into small particles. The hydrothermal treatment involves the treatment of the carbonaceous material with water or other solvents at high temperatures and pressures to generate small CQDs. The advantage of the chemical fragmentation approach is that it allows for better control of the size and shape of the CQDs produced. The process also allows for the functionalization of the CQDs with different groups to improve their properties. However, the process involves the use of toxic chemicals, and the preparation of the CQDs may require multiple steps. In addition to these two main strategies, other techniques have also been reported for the preparation of CQDs, including laser ablation, electrochemical fragmentation, and plasma-induced fragmentation. These methods have their advantages and disadvantages, and the choice of the preparation method depends on the specific application of the CQDs.

Bottom-up Strategy

To prepare CQDs, bottom-up strategies have been developed, which involve the synthesis of carbon-based materials using organic precursors, followed by controlled manipulation of their size and surface properties to create quantum dots. Hydrothermal and solvothermal methods are two common bottom-up approaches. In the hydrothermal method, a precursor solution containing carbon-rich compounds, such as glucose or citric acid is heated at high temperature and pressure in a sealed container, leading to the formation of small carbon clusters that grow into larger particles under hydrothermal conditions. Similarly, in the solvothermal method, a solvent-based precursor is used to synthesize CQDs at high temperatures and pressure. The choice of solvent can influence the size and surface properties of the resulting CQDs. Other bottom-up approaches include the electrochemical method, microwave-assisted method, and laser ablation method. The electrochemical method involves the electrochemical reduction of a carbon-based precursor in solution, providing precise control over the size and shape of the resulting CQDs. The microwave-assisted method utilizes microwave radiation to heat and reacts with the precursor solution, resulting in the rapid formation of CQDs. The laser ablation method involves the use of a laser to ablate a carbon-based target material, producing highly uniform CQDs with precise control over their size and surface properties. Overall, bottom-up strategies provide a powerful approach for the preparation of CQDs, allowing for precise control over their size, shape, and surface properties. This level of control is essential for applications in various fields, including biomedicine, optoelectronics and environmental monitoring. The continued development of bottom-up strategies will likely lead to new and exciting applications for CQDs in the future.

SEVERAL METHODS UTILIZED UNDER THE TOP-DOWN AND BOTTOM-UP STRATEGY FOR THE PREPARATION OF CQDS

Arc Discharge Method

The arc discharge method is another commonly used technique for the preparation of carbonaceous quantum dots. This method involves the use of an electric arc to vaporize carbon-containing materials, such as graphite or carbon nanotubes, in an inert gas atmosphere. The vaporized carbon atoms then condense and form nanoparticles, which are collected on a substrate. The mechanism of the arc discharge method involves several steps, including the vaporization of carbon-containing materials, nucleation and growth of nanoparticles, and their subsequent collection on a substrate. The arc discharge method begins with the vaporization of a carbon-containing material, such as graphite or carbon nanotubes. A high electric current is passed through the material in an inert gas atmosphere, causing the carbon atoms to be vaporized and ionized. The ionized carbon atoms then condense and form nanoparticles in the gas phase. The vaporized carbon atoms then undergo nucleation to form small clusters of carbon atoms. Nucleation occurs when the concentration of carbon atoms in the gas phase exceeds their solubility limit. The small clusters of carbon atoms serve as the starting point for nanoparticle growth. Once the small clusters of carbon atoms have formed, they begin to grow with the addition of more carbon atoms from the gas phase. The growth of the nanoparticles is determined by the availability of carbon atoms and the reaction conditions, such as temperature and pressure. The nanoparticles can grow in size until they reach their maximum size, which is determined by the reaction conditions and the size of the precursor clusters. After the nanoparticles have grown to their desired size, they are collected on a substrate. The substrate can be a variety of materials, such as silicon, glass, or metal, and is chosen based on the intended application of the carbonaceous quantum dots. The nanoparticles can be collected on the substrate by various methods, such as filtration, centrifugation, or deposition. The resulting carbonaceous quantum dots produced by the arc discharge method exhibit unique properties, such as high purity, uniform size distribution, and excellent photoluminescence. The optical properties of the carbonaceous quantum dots are dependent on their size and composition. The size of the nanoparticles can be controlled by adjusting the reaction conditions, such as temperature and pressure. The composition of the nanoparticles can be tailored by choosing specific carbon-containing materials, such as graphite or carbon nanotubes, as the precursor. Xu *et al.* synthesized CQDs in 2004 using the arc discharge technique [8]. Unfortunately, three distinct forms of quantum dots with varying sizes and luminous properties were initially produced during the production of single-walled carbon nanotubes using the arc discharge technique. Therefore, the obtained nanoparticles can emit orange, blue and yellow fluorescence at 365 nm. Additionally, a hydrophilic moiety was functionalized onto the interface of CQDs. By using this arc discharge process, fine nanoparticles were created that had excellent water dispersion [8]. Therefore, Sharma *et al.* follow this method by the use of gas plasma liberated through a sealed reactor, the bulk material (carbon precursor) decomposed and reorganized at the anodic electrode [14]. The reactor's temperature may increase to as much as 4000K in order to create gas plasma under an electric current. The carbon vapor assembles to create CQDs at the cathode [15].

Laser Ablation Method

The laser ablation method is another approach to prepare carbonaceous quantum dots. In this method, a high-powered laser beam is focused on a carbon-containing target, causing vaporization and fragmentation of the target material into nanoparticles. These nanoparticles are then collected and purified to obtain carbonaceous quantum dots. The mechanism of the laser ablation method involves several steps, including laser-induced vaporization, nanoparticle formation, and purification. The laser ablation method begins with the use of a high-powered laser beam to vaporize a carbon-containing target material. The laser beam is typically a pulsed or continuous-wave laser, and the target material can be any carbon-containing material, such as graphite, carbon nanotubes, or carbon black. The laser beam is focused on the target material, which causes rapid heating and vaporization of the material. The vaporized carbon atoms and clusters then condense and form nanoparticles in the gas phase. The size and shape of the nanoparticles are determined by the laser energy, pulse duration, and the physical properties of the target material. The nanoparticles can range from a few nanometers to hundreds of nanometers in size and can have a variety of shapes, including spherical, rod-like, or irregular. The nanoparticles are then collected and purified to obtain carbonaceous quantum dots. The purification process involves the removal of impurities, such as residual target material, metal catalysts, or other contaminants. The purification can be achieved through various methods, such as centrifugation, dialysis, or chromatography.

The resulting carbonaceous quantum dots produced by the laser ablation method exhibit unique properties, such as narrow size distribution, high purity, and excellent photoluminescence. The optical properties of the carbonaceous quantum dots are dependent on their size and composition, which can be controlled by adjusting the laser parameters and choosing specific target materials. The laser ablation method has several advantages, such as its ability to produce high-quality carbonaceous quantum dots with excellent photoluminescence, narrow size distribution, and high purity. Additionally, this method is versatile and can be used to prepare carbonaceous quantum dots from a wide range of target materials [16 - 18]. However, the laser ablation method also has some limitations, such as the high cost of the equipment and the limited scalability of the method [19, 20]. These synthesized quantum dots exhibited tunable and good visible photoluminescence. To change the photoluminescence properties of the CQDs, Hu *et al.* synthesized the quantum dots by modifying the surface state in the proper organic solvent during the laser irradiation process [21]. The synthesized carbon quantum dots exhibited good water solubility, fluorescence properties and narrow size distribution. Therefore, the laser ablation method is an effective process for the manufacture of carbon quantum dots. The major advantage of this method is that it does not use hazardous chemicals [22]. Also, continuous stirring is required in this method to avoid any precipitation [23]. The size and morphology of the CQDs could be modulated by changing the parameter such as pulse repetition rate, pulse duration, laser irradiation time, laser wavelength and laser power.

Acidic Oxidation Method

The acidic oxidation method is another approach for the preparation of carbonaceous quantum dots. In this method, a carbon-containing material is subjected to an oxidative treatment using an acidic solution. The oxidative treatment leads to the formation of carbonaceous quantum dots, which can be further purified and characterized for various applications. The mechanism of the acidic oxidation method involves several steps, including oxidation, fragmentation, and purification. In the first step, the carbon-containing material is subjected to an oxidative treatment using an acidic solution. The acidic solution typically contains strong oxidizing agents, such as nitric acid, sulfuric acid, or a mixture of both. The oxidizing agents break down the carbon-containing material, leading to the formation of carbonaceous fragments. The oxidative treatment causes the fragmentation of the carbonaceous fragments into smaller pieces, leading to the formation of carbonaceous quantum dots. The size of the carbonaceous quantum dots can be controlled by varying the oxidative conditions, such as the concentration of the oxidizing agent, reaction temperature, and reaction time. After the formation of the carbonaceous quantum dots, the solution is subjected to a purification step to remove any impurities, such as unreacted starting material, or any by-products of the oxidation process. The purification can be achieved using various methods, such as centrifugation, dialysis, or chromatography.

The resulting carbonaceous quantum dots produced by the acidic oxidation method exhibit unique properties, such as excellent photoluminescence, narrow size distribution, and good biocompatibility. The optical properties of the carbonaceous quantum dots can be controlled by adjusting the oxidation conditions and the carbon-containing material used as a starting material. The acidic oxidation method has several advantages, such as its low cost, simplicity, and scalability. Additionally, this method is versatile and can be used to prepare carbonaceous quantum dots from a wide range of carbon-containing materials. However, the acidic oxidation method also has some limitations, such as the potential formation of toxic by-products during the oxidation process [24]. Moreover, the fluorescence nature of these quantum dots also increases [25]. Yang *et al.* synthesized the heteroatom-doped quantum dots using the acidic oxidation method and subsequently, the hydrothermal reduction approach. They firstly oxidized the Chinese ink with an acidic solution (H_2SO_4, HNO_3 and $NaClO_3$) to form the carbon quantum dots, and this oxidized nanomaterial subsequently reacted with different element sources (Sodium hydrosulfide, dimethylformamide and sodium selenide) through hydrothermal reaction to obtain the sulfur, nitrogen and selenium doped CQDs [26]. The synthesized doped CQDs showed higher quantum yield, tunable photoluminescence performance and longer fluorescence lifetime than the un-doped carbon quantum dots. However, the heavily doped heteroatoms can affect the tunable photoluminescence owing to the distinction in the electronegativity of nitrogen, sulfur and selenium. Also, the electronegativity of the element which is available on the interface of the CQDs plays an important role when used as electrocatalysts during the electrocatalytic activity. On the other hand, the heavy-doped CQDs have the capability to interact with Ni^{2+}, Fe^{3+} and Co^{2+} to produce single-atom catalysts [26].

Combustion/Thermal Method

The combustion/thermal method is a simple and efficient technique for the preparation of carbonaceous quantum dots. In this method, a carbon-containing precursor is heated to high temperatures in the presence of an oxidizing agent, leading to the formation of carbonaceous quantum dots. The mechanism of the combustion/thermal method involves several steps, including precursor combustion, nanoparticle formation, and purification. In the first step, a carbon-containing precursor is heated to high temperatures in the presence of an oxidizing agent. The precursor is typically a hydrocarbon or carbohydrate, such as sucrose, glucose, or citric acid. The oxidizing agent can be air, oxygen, or a mixture of both. The high-temperature combustion of the precursor leads to the formation of carbonaceous fragments. The carbonaceous fragments then undergo nucleation and growth to form carbonaceous nanoparticles. The size and shape of the nanoparticles can be controlled by varying the combustion conditions, such as the temperature, heating rate, and duration. After the formation of the carbonaceous nanoparticles, the solution is subjected to a purification step to remove any impurities, such as unreacted starting material or by-products of the combustion process. The purification can be achieved using various methods, such as centrifugation, dialysis, or chromatography.

The resulting carbonaceous quantum dots produced by the combustion/thermal method exhibit unique properties, such as excellent photoluminescence, narrow size distribution, and good stability. The optical properties of the carbonaceous quantum dots can be controlled by adjusting the combustion conditions and the carbon-containing precursor used. The combustion/thermal method has several advantages, such as its simplicity, efficiency, and scalability. Additionally, this method is versatile and can be used to prepare carbonaceous quantum dots from a wide range of carbon-containing precursors [27 - 29]. Xu *et al.* first proposed the combustion/thermal routes for the fabrication of CQDs, and after that, many researchers followed the same route for the synthesis of quantum dots. For instance, Li *et al.* synthesized fluorescent gadolinium quantum dots by the use of citric acid and the surface of the quantum dots was modulated with a carboxyl group at high temperature. The dimension of the quantum dots was found to be 8.5 nm, and the surface was rich in carboxyl moiety. Therefore, these quantum dots were found to be good for the electrocatalytic process in an aqueous solution because they have the ability *i.e*, the adsorption of water molecules through oxygen-containing moieties [30].

Microwave Pyrolysis Method

The microwave pyrolysis method is a relatively new technique for the preparation of carbonaceous quantum dots. In this method, a carbon-containing precursor is heated using microwave radiation in the presence of a suitable solvent or surfactant, leading to the formation of carbonaceous quantum dots. The mechanism of the microwave pyrolysis method involves several steps, including precursor decomposition, nucleation, and growth. In the first step, the carbon-containing precursor is heated using microwave radiation in the presence of a suitable solvent or surfactant. The precursor undergoes decomposition to form carbonaceous fragments. The carbonaceous fragments then undergo nucleation to form carbonaceous nanoparticles. The size and shape of the nanoparticles can be controlled by varying the reaction conditions, such as the precursor concentration, reaction temperature, and duration. The carbonaceous nanoparticles undergo further growth to form carbonaceous quantum dots. The growth can be controlled by adjusting the reaction conditions and the precursor concentration.

The resulting carbonaceous quantum dots produced by the microwave pyrolysis method exhibit unique properties, such as good photoluminescence, narrow size distribution, and high stability. The optical properties of the carbonaceous quantum dots can be controlled by adjusting the reaction conditions and the precursor used. The microwave pyrolysis method has several advantages, such as its simplicity, efficiency, and scalability. Additionally, this method is versatile and can be used to prepare carbonaceous quantum dots from a wide range of carbon-containing precursors [31]. For rapid synthesis and commercialization, the microwave pyrolysis strategy has been widely used [32 - 35]. In 2009, Zhu *et al.* synthesized the carbon quantum dots by dispersing the poly (ethylene glycol) and saccharide in water using a microwave pyrolysis strategy. The synthesized quantum dots exhibited an excitation-dependent photoluminescence property [36]. The microwave pyrolysis strategy is a fast, simple and, most importantly, environment-friendly approach for CQD fabrication. Moreover, the surface of the quantum dots was rich in oxygen-containing moiety, which can coordinate with a metal ion to form the carbon-based electrocatalyst. Therefore this approach can design many carbon-based electrocatalysts. In the microwave method, polymerization and carbonization occur [37]. Also, the high reaction rate and short duration of this method make it effective because carbon material absorbs electromagnetic energy and transforms it into heat energy [38]. As a result, the molecular spinning of the polar solvent molecules generates thermal energy through the interaction of microwave radiation with the electric dipole moment. As a result, during microwave-assisted heating, the carbon material receives thermal energy through radiant heat [39]. In one report, the ascorbic acid was mixed with polyethylene glycol solution under microwave reaction (200 °C, 700 W) for 1 minute, culminating in the fabrication of CQDs through dehydration [40]. Further, the quantum dots were purified by dialysis for 24 h [41]. In other reports, the CQDs were produced from a lemon by dissolving into the water, and blended substances were heated in a microwave for a few minutes. The obtained quantum dots were purified through dialysis [42]. The fundamental benefit of microwave heating is that it provides consistent energy, resulting in the carbon quantum dots' homogeneous dimension. By altering the heating duration in the presence of a catalyst, the size of the quantum dots may also be adjusted [23, 43].

Hydrothermal/Solvothermal Method

Hydrothermal synthesis is a popular and effective method for the preparation of carbonaceous quantum dots. The mechanism of hydrothermal synthesis involves several steps, including the nucleation of precursor molecules, growth of nuclei, aggregation, and stabilization of the nanoparticles [44]. The precursor solution is the starting material for hydrothermal synthesis. In the case of carbonaceous quantum dots, the precursors are typically carbon-containing molecules, such as carbohydrates, amino acids, or polymers. The precursor molecules are dissolved in a solvent, typically water, and placed in a reactor vessel. The solution is then heated to a high temperature and pressurized to induce the hydrothermal reaction [45]. During the heating and pressurization of the precursor solution, the solubility of the precursor molecules decreases, causing the molecules to come together and form nuclei. The nuclei are small clusters of precursor molecules that serve as the starting point for nanoparticle growth. The nucleation process is highly dependent on the reaction conditions, such as temperature, pressure, and concentration of precursor molecules. After nucleation, the nuclei begin to grow with the addition of precursor molecules. The growth rate is determined by the availability of precursor molecules and the reaction conditions. As the nanoparticles grow, their surface area increases, and they become more reactive. The growth process is also affected by the presence of other species in the solution, such as surfactants, which can affect the size and shape of the nanoparticles. As the nanoparticles continue to grow, they can begin to aggregate or clump together. The aggregation process can be controlled by adjusting the reaction conditions or adding stabilizing agents, such as polymers or surfactants. Stabilizing agents help to prevent aggregation by coating the surface of the nanoparticles and repelling each other. Once the nanoparticles have reached their desired size, they must be stabilized to prevent further growth or aggregation. Stabilization can be achieved by removing the solvent or adding stabilizing agents, such as polymers or surfactants. The stabilizing agents form a protective layer around the nanoparticles, preventing them from interacting with each other and maintaining their stability in solution. The resulting carbonaceous quantum dots produced by hydrothermal synthesis exhibit unique properties, including high water solubility, low toxicity, and excellent photoluminescence. The optical properties of carbonaceous quantum dots are dependent on their size, shape, and composition. The size of the nanoparticles can be controlled by adjusting the reaction conditions, such as temperature, pressure, and concentration of precursor molecules. The shape of the nanoparticles can be influenced by the presence of surfactants or other species in the solution. The composition of the nanoparticles can be tailored by choosing specific precursor molecules or adding dopants to the solution. Overall, hydrothermal synthesis is an effective method for preparing carbonaceous quantum dots with excellent optical and physicochemical properties. The mechanism of hydrothermal synthesis involves several steps, including nucleation, growth, aggregation, and stabilization of the nanoparticles. By controlling the reaction conditions, the size, shape, and composition of the carbonaceous quantum dots can be tailored for specific applications, such as bioimaging, sensing, and catalysis [46]. At various wavelengths, Hola *et al.* synthesized full-color carbon quantum dots under hydrothermal conditions with controllable fluorescence by changing the quantity of graphitic nitrogen [47]. Moreover, Lu *et al.* changed the inner composition of the carbon quantum dots with carbon and nitrogen-rich biomolecules under hydrothermal conditions [48]. Therefore, the hydrothermal method can be used to design and fabricate novel electrocatalysts by changing the electronic structures and doping composition. In one report, carbon quantum dots were synthesized by warming ammonium citrate ($C_6H_{11}NO_7$) at 180 °C for 3 h [49]. In another report, green fluorescent quantum dots were synthesized by dissolving the sodium hydroxide, L-cysteine and hydrogen peroxide in Milli Q water at 90 °C for 12 h [50]. At high pressure and temperature, the interaction of water molecules with precursors occurs, and when heat is given to the system, the pressure raise, which ultimately deoxidizes the water molecules [51]. Similarly, citrate-capped CQDs were synthesized and deposition of citric acid ($C_6H_8O_7$) occurred at elevated temperatures to produce a high energy density of quantum dots [52, 53].

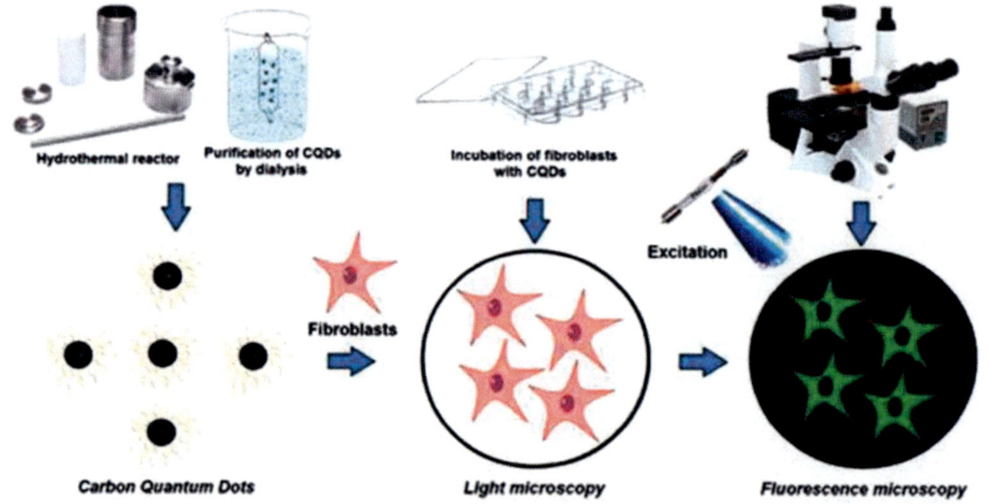

Fig. (1). A hydrothermal method for the synthesis of carbon quantum dots. Adopted from the MDPI material journal [54].

Electrochemistry Method

Electrochemistry is a method used to prepare carbonaceous quantum dots (CQDs). This method involves the use of an electrochemical cell containing an electrode (anode and cathode) and an electrolyte solution containing the carbon source. The mechanism of CQD synthesis by electrochemistry involves the electrochemical oxidation of the carbon source at the anode. During this process, carbon atoms are oxidized to form carbon ions, which then undergo further oxidation to form carbon radicals. These carbon radicals then react with each other to form carbon clusters, which are the precursors of CQDs.

The carbon clusters are then reduced at the cathode, where they form CQDs by a mechanism that is still under investigation. One possible mechanism is the electrochemical reduction of the carbon clusters to form CQDs, which involves the transfer of electrons from the cathode to the carbon clusters. The reduction process can be controlled by adjusting the electrochemical parameters, such as the applied voltage, current density, and electrolyte composition. The CQDs synthesized by electrochemistry have unique optical and electronic properties that make them suitable for a wide range of applications, including bioimaging, drug delivery, sensing, and catalysis. The electrochemistry method provides a simple, low-cost, and scalable approach for the preparation of CQDs, which has attracted significant interest in recent years [55]. Electrochemical methods for synthesizing CQDs have several advantages over other methods. Firstly, they offer precise control over the size, morphology, and surface chemistry of CQDs by tuning the electrochemical parameters, such as applied potential, current density, and electrolyte composition. Secondly, electrochemistry can be performed in aqueous solutions at room temperature and atmospheric pressure, making it a simple, low-cost and environmentally friendly method. Finally, the electrochemical method can also be easily scaled up for mass production. The electrochemical method can also be modified to incorporate other elements or functional groups into CQDs. For example, doping with nitrogen, sulfur, or phosphorus can enhance the fluorescence properties of CQDs. Additionally, functionalization with biomolecules, such as proteins, peptides, or DNA, can confer specific targeting and sensing capabilities to CQDs. There are several challenges that need to be addressed in the electrochemical synthesis of CQDs. One of the challenges is the formation of undesired by-products during the electrochemical reaction, which can affect the quality and yield of CQDs. Another challenge is the lack of a standardized protocol for the electrochemical synthesis of CQDs, which can lead to variability in the properties of CQDs synthesized by different groups. Also, we can tune the PL (photo-luminescent) performance and particle size of the carbon quantum dots which have been synthesized by the electrochemical method [56]. Hou *et al.* synthesized the blue emission carbon quantum dots (size= 2.4 nm) by electrochemical carbonization of sodium citrate ($Na_3C_6H_5O_7$) and urea (CH_4N_2O) in MilliQ and synthesized quantum dots used as detection of Hg^{2+} ion in wastewater [57].

CONCLUSION

In this chapter, we have discussed the different synthetic techniques for CQDs, such as hydrothermal/solvothermal method, laser ablation, acidic oxidation, arc-discharge method, thermal/ combustion routes, electrochemical method and microwave pyrolysis method. Also, this material (CQDs) can be utilized in various applications, including drug delivery systems, bio-imaging, electrocatalysis and sensor. In conclusion, the field of nanocarbon materials has expanded to include a new class of materials known as carbon quantum dots (CQDs). These materials have gained considerable attention in various fields, such as electrocatalysis, solar cells, bioimaging, nanomedicine, chemical sensors, and light-emitting diodes, due to their unique physio-chemical properties, such as photoluminescence, high crystallization, and good dispersibility. CQDs have shown promising results in electrochemistry, where their rapid electron transfer, small size, and superconductivity provide the basis for enhanced catalytic activity and electric conductivity. The active moieties present on the surface of CQDs promote the formation of multi-component electrically activated catalysts, leading to improved charge transfer and catalytic efficiency.

REFERENCES

[1] Rao, R.; Pint, C.L.; Islam, A.E.; Weatherup, R.S.; Hofmann, S.; Meshot, E.R.; Wu, F.; Zhou, C.; Dee, N.; Amama, P.B.; Carpena-Nuñez, J.; Shi, W.; Plata, D.L.; Penev, E.S.; Yakobson, B.I.; Balbuena, P.B.; Bichara, C.; Futaba, D.N.; Noda, S.; Shin, H.; Kim, K.S.; Simard, B.; Mirri, F.; Pasquali, M.; Fornasiero, F.; Kauppinen, E.I.; Arnold, M.; Cola, B.A.; Nikolaev, P.; Arepalli, S.; Cheng, H.M.; Zakharov, D.N.; Stach, E.A.; Zhang, J.; Wei, F.; Terrones, M.; Geohegan, D.B.; Maruyama, B.; Maruyama, S.; Li, Y.; Adams, W.W.; Hart, A.J. Carbon nanotubes and related nanomaterials: Critical advances and challenges for synthesis toward mainstream commercial applications. *ACS Nano,* **2018**, *12*(12), 11756-11784.
[http://dx.doi.org/10.1021/acsnano.8b06511] [PMID: 30516055]

[2] Lin, H.S.; Jeon, I.; Xiang, R.; Seo, S.; Lee, J.W.; Li, C.; Pal, A.; Manzhos, S.; Goorsky, M.S.; Yang, Y.; Maruyama, S.; Matsuo, Y. Achieving high efficiency in solution-processed perovskite solar cells using C_{60}/C_{70} mixed fullerenes. *ACS Appl. Mater. Interfaces,* **2018**, *10*(46), 39590-39598.
[http://dx.doi.org/10.1021/acsami.8b11049] [PMID: 30259726]

[3] Clancy, A.J.; Bayazit, M.K.; Hodge, S.A.; Skipper, N.T.; Howard, C.A.; Shaffer, M.S.P. Charged carbon nanomaterials: Redox chemistries of fullerenes, carbon nanotubes, and graphenes. *Chem. Rev.,* **2018**, *118*(16), 7363-7408.
[http://dx.doi.org/10.1021/acs.chemrev.8b00128] [PMID: 30109931]

[4] Georgakilas, V.; Perman, J.A.; Tucek, J.; Zboril, R. Broad family of carbon nanoallotropes: classification, chemistry, and applications of fullerenes, carbon dots, nanotubes, graphene, nanodiamonds, and combined superstructures. *Chem. Rev.,* **2015**, *115*(11), 4744-4822.
[http://dx.doi.org/10.1021/cr500304f] [PMID: 26012488]

[5] Semeniuk, M.; Yi, Z.; Poursorkhabi, V.; Tjong, J.; Jaffer, S.; Lu, Z.H.; Sain, M. Future perspectives

and review on organic carbon dots in electronic applications. *ACS Nano,* **2019**, *13*(6), 6224-6255. [http://dx.doi.org/10.1021/acsnano.9b00688] [PMID: 31145587]

[6] Lim, S.Y.; Shen, W.; Gao, Z. Carbon quantum dots and their applications. *Chem. Soc. Rev.,* **2015**, *44*(1), 362-381.
[http://dx.doi.org/10.1039/C4CS00269E] [PMID: 25316556]

[7] Yang, S.; Sun, J.; Li, X.; Zhou, W.; Wang, Z.; He, P.; Ding, G.; Xie, X.; Kang, Z.; Jiang, M. Large-scale fabrication of heavy doped carbon quantum dots with tunable-photoluminescence and sensitive fluorescence detection. *J. Mater. Chem. A Mater. Energy Sustain.,* **2014**, *2*(23), 8660-8667.
[http://dx.doi.org/10.1039/c4ta00860j]

[8] Xu, X.; Ray, R.; Gu, Y.; Ploehn, H.J.; Gearheart, L.; Raker, K.; Scrivens, W.A. Electrophoretic analysis and purification of fluorescent single-walled carbon nanotube fragments. *J. Am. Chem. Soc.,* **2004**, *126*(40), 12736-12737.
[http://dx.doi.org/10.1021/ja040082h] [PMID: 15469243]

[9] Yuan, T.; Meng, T.; He, P.; Shi, Y.; Li, Y.; Li, X.; Fan, L.; Yang, S. Carbon quantum dots: An emerging material for optoelectronic applications. *J. Mater. Chem. C Mater. Opt. Electron. Devices,* **2019**, *7*(23), 6820-6835.
[http://dx.doi.org/10.1039/C9TC01730E]

[10] Yuan, F.; Li, S.; Fan, Z.; Meng, X.; Fan, L.; Yang, S. Shining carbon dots: Synthesis and biomedical and optoelectronic applications. *Nano Today,* **2016**, *11*(5), 565-586.
[http://dx.doi.org/10.1016/j.nantod.2016.08.006]

[11] Sun, X.; Lei, Y. Fluorescent carbon dots and their sensing applications. TrAC Trend. *Anal. Chem.,* **2017**, *89*, 163-180.

[12] Wang, X.; Feng, Y.; Dong, P.; Huang, J. A mini review on carbon quantum dots: Preparation, properties, and electrocatalytic application. *Front Chem.,* **2019**, *7*, 671.
[http://dx.doi.org/10.3389/fchem.2019.00671] [PMID: 31637234]

[13] Rani, U.A.; Ng, L.Y.; Ng, C.Y.; Mahmoudi, E. A review of carbon quantum dots and their applications in wastewater treatment. *Adv. Colloid Interface Sci.,* **2020**, *278*, 102124.
[http://dx.doi.org/10.1016/j.cis.2020.102124] [PMID: 32142942]

[14] Arora, N.; Sharma, N.N. Arc discharge synthesis of carbon nanotubes: Comprehensive review. *Diamond Related Materials,* **2014**, *50*, 135-150.
[http://dx.doi.org/10.1016/j.diamond.2014.10.001]

[15] Yatom, S.; Bak, J.; Khrabryi, A.; Raitses, Y. Detection of nanoparticles in carbon arc discharge with laser-induced incandescence. *Carbon,* **2017**, *117*, 154-162.
[http://dx.doi.org/10.1016/j.carbon.2017.02.055]

[16] Kuzmin, P.G.; Shafeev, G.A.; Bukin, V.V.; Garnov, S.V.; Farcau, C.; Carles, R.; Warot-Fontrose, B.; Guieu, V.; Viau, G. Silicon nanoparticles produced by femtosecond laser ablation in ethanol: Size control, structural characterization, and optical properties. *J. Phys. Chem. C,* **2010**, *114*(36), 15266-15273.
[http://dx.doi.org/10.1021/jp102174y]

[17] Xiao, J.; Liu, P.; Wang, C.X.; Yang, G.W. External field-assisted laser ablation in liquid: An efficient strategy for nanocrystal synthesis and nanostructure assembly. *Prog. Mater. Sci.,* **2017**, *87*, 140-220.
[http://dx.doi.org/10.1016/j.pmatsci.2017.02.004]

[18] Doñate-Buendia, C.; Torres-Mendieta, R.; Pyatenko, A.; Falomir, E.; Fernández-Alonso, M.; Mínguez-Vega, G. Fabrication by laser irradiation in a continuous flow jet of carbon quantum dots for fluorescence imaging. *ACS Omega,* **2018**, *3*(3), 2735-2742.
[http://dx.doi.org/10.1021/acsomega.7b02082] [PMID: 30023850]

[19] Sun, Y.P.; Zhou, B.; Lin, Y.; Wang, W.; Fernando, K.A.S.; Pathak, P.; Meziani, M.J.; Harruff, B.A.; Wang, X.; Wang, H.; Luo, P.G.; Yang, H.; Kose, M.E.; Chen, B.; Veca, L.M.; Xie, S.Y. Quantum-

sized carbon dots for bright and colorful photoluminescence. *J. Am. Chem. Soc.,* **2006**, *128*(24), 7756-7757.
[http://dx.doi.org/10.1021/ja062677d] [PMID: 16771487]

[20] Li, X.; Wang, H.; Shimizu, Y.; Pyatenko, A.; Kawaguchi, K.; Koshizaki, N. Preparation of carbon quantum dots with tunable photoluminescence by rapid laser passivation in ordinary organic solvents. *Chem. Commun.,* **2011**, *47*(3), 932-934.
[http://dx.doi.org/10.1039/C0CC03552A] [PMID: 21079826]

[21] Hu, S.L.; Niu, K.Y.; Sun, J.; Yang, J.; Zhao, N.Q.; Du, X.W. One-step synthesis of fluorescent carbon nanoparticles by laser irradiation. *J. Mater. Chem.,* **2009**, *19*(4), 484-488.
[http://dx.doi.org/10.1039/B812943F]

[22] Singh, A.; Mohapatra, P.K.; Kalyanasundaram, D.; Kumar, S. Self-functionalized ultrastable water suspension of luminescent carbon quantum dots. *Mater. Chem. Phys.,* **2019**, *225*, 23-27.
[http://dx.doi.org/10.1016/j.matchemphys.2018.12.031]

[23] Das, R.; Bandyopadhyay, R.; Pramanik, P. Carbon quantum dots from natural resource: A review. *Mater. Today Chem.,* **2018**, *8*, 96-109.
[http://dx.doi.org/10.1016/j.mtchem.2018.03.003]

[24] Shen, P.; Xia, Y. Synthesis-modification integration: One-step fabrication of boronic acid functionalized carbon dots for fluorescent blood sugar sensing. *Anal. Chem.,* **2014**, *86*(11), 5323-5329.
[http://dx.doi.org/10.1021/ac5001338] [PMID: 24694081]

[25] Zhang, Q.; Sun, X.; Ruan, H.; Yin, K.; Li, H. Production of yellow-emitting carbon quantum dots from fullerene carbon soot. *Sci. China Mater.,* **2017**, *60*(2), 141-150. b
[http://dx.doi.org/10.1007/s40843-016-5160-9]

[26] Yang, S.; Sun, J.; Li, X.; Zhou, W.; Wang, Z.; He, P.; Ding, G.; Xie, X.; Kang, Z.; Jiang, M. Large-scale fabrication of heavy doped carbon quantum dots with tunable-photoluminescence and sensitive fluorescence detection. *J. Mater. Chem. A Mater. Energy Sustain.,* **2014**, *2*(23), 8660-8667.
[http://dx.doi.org/10.1039/c4ta00860j]

[27] Li, H.; Kang, Z.; Liu, Y.; Lee, S.T. Carbon nanodots: Synthesis, properties and applications. *J. Mater. Chem.,* **2012**, *22*(46), 24230-24253. a
[http://dx.doi.org/10.1039/c2jm34690g]

[28] Guo, Y.; Zhang, L.; Cao, F.; Leng, Y. Thermal treatment of hair for the synthesis of sustainable carbon quantum dots and the applications for sensing Hg2+. *Sci. Rep.,* **2016**, *6*(1), 35795.
[http://dx.doi.org/10.1038/srep35795] [PMID: 27762342]

[29] Thoda, O.; Xanthopoulou, G.; Vekinis, G.; Chroneos, A. Review of recent studies on solution combustion synthesis of nanostructured catalysts. *Adv. Eng. Mater.,* **2018**, *20*(8), 1800047.
[http://dx.doi.org/10.1002/adem.201800047]

[30] Li, S.; Zhou, S.; Li, Y.; Li, X.; Zhu, J.; Fan, L.; Yang, S. Exceptionally high payload of the IR780 iodide on folic acid-functionalized graphene quantum dots for targeted photothermal therapy. *ACS Appl. Mater. Interfaces,* **2017**, *9*(27), 22332-22341.
[http://dx.doi.org/10.1021/acsami.7b07267] [PMID: 28643511]

[31] Schwenke, A.M.; Hoeppener, S.; Schubert, U.S. Synthesis and modification of carbon nanomaterials utilizing microwave heating. *Adv. Mater.,* **2015**, *27*(28), 4113-4141.
[http://dx.doi.org/10.1002/adma.201500472] [PMID: 26087742]

[32] In, I.; Park, S.Y.; Lim, D.; Park, J.E.; Kim, Y.K.; Chae, A. Simple microwave-assisted synthesis of amphiphilic carbon quantum dots from A3/B2 polyamidation monomer set. *ACS Appl. Mater. Interfaces,* **2017**, *9*, 327883-327893.
[PMID: 29310436]

[33] Rai, S.; Singh, B.K.; Bhartiya, P.; Singh, A.; Kumar, H.; Dutta, P.K.; Mehrotra, G.K. Lignin derived reduced fluorescence carbon dots with theranostic approaches: Nano-drug-carrier and bioimaging. *J.*

Lumin., **2017**, *190*, 492-503.
[http://dx.doi.org/10.1016/j.jlumin.2017.06.008]

[34] Jiang, K.; Wang, Y.; Gao, X.; Cai, C.; Lin, H. Facile, quick, and gram-scale synthesis of ultralong-lifetime room-temperature-phosphorescent carbon dots by microwave irradiation. *Angew. Chem. Int. Ed.,* **2018**, *57*(21), 6216-6220.
[http://dx.doi.org/10.1002/anie.201802441] [PMID: 29637675]

[35] Shen, Z.; Zhang, C.; Yu, X.; Li, J.; Wang, Z.; Zhang, Z.; Liu, B. Microwave-assisted synthesis of cyclen functional carbon dots to construct a ratiometric fluorescent probe for tetracycline detection. *J. Mater. Chem. C Mater. Opt. Electron. Devices,* **2018**, *6*(36), 9636-9641.
[http://dx.doi.org/10.1039/C8TC02982B]

[36] Zhu, H.; Wang, X.; Li, Y.; Wang, Z.; Yang, F.; Yang, X. Microwave synthesis of fluorescent carbon nanoparticles with electrochemiluminescence properties. *Chem. Commun.,* **2009**, *4*(34), 5118-5120.
[http://dx.doi.org/10.1039/b907612c] [PMID: 20448965]

[37] Alarfaj, N.; El-Tohamy, M.; Oraby, H. CA 19-9 pancreatic tumor marker fluorescence immunosensing detection *via* immobilized carbon quantum dots conjugated gold nanocomposite. *Int. J. Mol. Sci.,* **2018**, *19*(4), 1162.
[http://dx.doi.org/10.3390/ijms19041162] [PMID: 29641488]

[38] Yang, P.; Zhu, Z.; Chen, M.; Chen, W.; Zhou, X. Microwave-assisted synthesis of xylan-derived carbon quantum dots for tetracycline sensing. *Opt. Mater.,* **2018**, *85*, 329-336.
[http://dx.doi.org/10.1016/j.optmat.2018.06.034]

[39] Singh, R.K.; Kumar, R.; Singh, D.P.; Savu, R.; Moshkalev, S.A. Progress in microwave-assisted synthesis of quantum dots (graphene/carbon/semiconducting) for bioapplications: A review. *Mater. Today Chem.,* **2019**, *12*, 282-314.
[http://dx.doi.org/10.1016/j.mtchem.2019.03.001]

[40] Kaur, M.; Kaur, M.; Sharma, V.K. Nitrogen-doped graphene and graphene quantum dots: A review onsynthesis and applications in energy, sensors and environment. *Adv. Colloid Interface Sci.,* **2018**, *259*, 44-64.
[http://dx.doi.org/10.1016/j.cis.2018.07.001] [PMID: 30032930]

[41] Wang, Q.; Wang, G.; Liang, X.; Dong, X.; Zhang, X. Supporting carbon quantum dots on NH2-MI--125 for enhanced photocatalytic degradation of organic pollutants under a broad spectrum irradiation. *Appl. Surf. Sci.,* **2019**, *467-468*, 320-327.
[http://dx.doi.org/10.1016/j.apsusc.2018.10.165]

[42] Monte-Filho, S.S.; Andrade, S.I.E.; Lima, M.B.; Araujo, M.C.U. Synthesis of highly fluorescent carbon dots from lemon and onion juices for determination of riboflavin in multivitamin/mineral supplements. *J. Pharm. Anal.,* **2019**, *9*(3), 209-216.
[http://dx.doi.org/10.1016/j.jpha.2019.02.003] [PMID: 31297299]

[43] Khan, Z.M.S.H.; Rahman, R.S.; Shumaila, I.S.; Islam, S.; Zulfequar, M. Hydrothermal treatment of red lentils for the synthesis of fluorescent carbon quantum dots and its application for sensing Fe3+. *Opt. Mater.,* **2019**, *91*, 386-395.
[http://dx.doi.org/10.1016/j.optmat.2019.03.054]

[44] Wang, G.; Guo, Q.; Chen, D.; Liu, Z.; Zheng, X.; Xu, A.; Yang, S.; Ding, G. Facile and highly effective synthesis of controllable lattice sulfur-doped graphene quantum dots *via* hydrothermal treatment of durian. *ACS Appl. Mater. Interfaces,* **2018**, *10*(6), 5750-5759.
[http://dx.doi.org/10.1021/acsami.7b16002] [PMID: 29350521]

[45] Anwar, S.; Ding, H.; Xu, M.; Hu, X.; Li, Z.; Wang, J.; Liu, L.; Jiang, L.; Wang, D.; Dong, C.; Yan, M.; Wang, Q.; Bi, H. Recent advances in synthesis, optical properties, and biomedical applications of carbon dots. *ACS Appl. Bio Mater.,* **2019**, *2*(6), 2317-2338.
[http://dx.doi.org/10.1021/acsabm.9b00112] [PMID: 35030725]

[46] Zhu, S.; Meng, Q.; Wang, L.; Zhang, J.; Song, Y.; Jin, H.; Zhang, K.; Sun, H.; Wang, H.; Yang, B.

Highly photoluminescent carbon dots for multicolor patterning, sensors, and bioimaging. *Angew. Chem. Int. Ed.,* **2013**, *52*(14), 3953-3957.
[http://dx.doi.org/10.1002/anie.201300519] [PMID: 23450679]

[47] Holá, K.; Sudolská, M.; Kalytchuk, S.; Nachtigallová, D.; Rogach, A.L.; Otyepka, M.; Zbořil, R. Graphitic nitrogen triggers red fluorescence in carbon dots. *ACS Nano,* **2017**, *11*(12), 12402-12410.
[http://dx.doi.org/10.1021/acsnano.7b06399] [PMID: 29136460]

[48] Lu, S.; Sui, L.; Wu, M.; Zhu, S.; Yong, X.; Yang, B. Graphitic nitrogen and high-crystalline triggered strong photoluminescence and roomtemperature ferromagnetism in carbonized polymer dots. *Adv. Sci.,* **2019**, *6*(2), 1801192.
[http://dx.doi.org/10.1002/advs.201801192] [PMID: 30693180]

[49] Ratnayake, S.P.; Mantilaka, M.M.M.G.P.G.; Sandaruwan, C.; Dahanayake, D.; Murugan, E.; Kumar, S.; Amaratunga, G.A.J.; de Silva, K.M.N. Carbon quantum dots-decorated nano-zirconia: A highly efficient photocatalyst. *Appl. Catal. A Gen.,* **2019**, *570*, 23-30.
[http://dx.doi.org/10.1016/j.apcata.2018.10.022]

[50] Liang, J.Y.; Han, L.; Liu, S.G.; Ju, Y.J.; Gao, X.; Li, N.B.; Luo, H.Q. Green fluorescent carbon quantum dots as a label-free probe for rapid and sensitive detection of hematin. *Spectrochim. Acta A Mol. Biomol. Spectrosc.,* **2019**, *212*, 167-172.
[http://dx.doi.org/10.1016/j.saa.2019.01.001] [PMID: 30639601]

[51] Pan, D.; Zhang, J.; Li, Z.; Wu, M. Hydrothermal route for cutting graphene sheets into blue-luminescent graphene quantum dots. *Adv. Mater.,* **2010**, *22*(6), 734-738.
[http://dx.doi.org/10.1002/adma.200902825] [PMID: 20217780]

[52] Gai, W.; Zhao, D.L.; Chung, T.S. Thin film nanocomposite hollow fiber membranes comprising Na^+-functionalized carbon quantum dots for brackish water desalination. *Water Res.,* **2019**, *154*, 54-61.
[http://dx.doi.org/10.1016/j.watres.2019.01.043] [PMID: 30771707]

[53] Czajczyńska, D.; Nannou, T.; Anguilano, L.; Krzyżyńska, R.; Ghazal, H.; Spencer, N.; Jouhara, H. Potentials of pyrolysis processes in the waste management sector. *Energy Procedia,* **2017**, *123*, 387-394.
[http://dx.doi.org/10.1016/j.egypro.2017.07.275]

[54] Janus, Ł.; Radwan-Pragłowska, J.; Piątkowski, M.; Bogdał, D. Facile synthesis of surface-modified carbon quantum dots (cqds) for biosensing and bioimaging. *Materials,* **2020**, *13*(15), 3313.
[http://dx.doi.org/10.3390/ma13153313] [PMID: 32722356]

[55] Deng, J.; Lu, Q.; Mi, N.; Li, H.; Liu, M.; Xu, M.; Tan, L.; Xie, Q.; Zhang, Y.; Yao, S. Electrochemical synthesis of carbon nanodots directly from alcohols. *Chemistry,* **2014**, *20*(17), 4993-4999.
[http://dx.doi.org/10.1002/chem.201304869] [PMID: 24623706]

[56] Ahirwar, S.; Mallick, S.; Bahadur, D. Electrochemical method to prepare graphene quantum dots and graphene oxide quantum dots. *ACS Omega,* **2017**, *2*(11), 8343-8353.
[http://dx.doi.org/10.1021/acsomega.7b01539] [PMID: 31457373]

[57] Hou, Y.; Lu, Q.; Deng, J.; Li, H.; Zhang, Y. One-pot electrochemical synthesis of functionalized fluorescent carbon dots and their selective sensing for mercury ion. *Anal. Chim. Acta,* **2015**, *866*, 69-74.
[http://dx.doi.org/10.1016/j.aca.2015.01.039] [PMID: 25732694]

CHAPTER 3

Properties of Carbonaceous Quantum Dots

Devika Vashisht[1], **Priyanka**[1], **Aseem Vashisht**[2], **Shweta Sharma**[3] and **Surinder Kumar Mehta**[1,*]

[1] *Department of Chemistry and Centre of Advanced Studies in Chemistry, Panjab University, Chandigarh, India*

[2] *Punjab State Council for Science and technology Sector - 26 , Chandigarh, India*

[3] *Institute of Forensic Science and Criminology, Panjab University, Chandigarh, India*

Abstract: Carbonaceous quantum dots (CQDs) is defined as a subclass of carbon nanoparticles, which offer a size of around 10 nm, and have unique characteristics and a wide range of applications in diverse fields. CQDs have attained widespread attention due to their excellent abilities in several domains, including sensing, nanomedicine and environmental remediation. The mode of synthesis for CQDs is quite simple and inexpensive *via* methods such as microwave pyrolysis, arc-discharge, *etc*. CQDs are entitled to diverse physical, chemical and biological properties. Besides this, CQDs have various functional groups present on their surface that improve the properties, specifically the catalytic performance by a phenomenon called charge transfer. The physical, optical, electrical, and biological features of CQDs are explored in this chapter.

Keywords: Biological properties, Carbon quantum dots, Electrical properties, Optical properties, Photoluminescence.

INTRODUCTION

Carbonaceous quantum dots (CQDs) are nanoparticles of carbon with a very small size from 2 to 10 nm and have exceptional properties such as excellent conductivity, chemical stability, eco-friendly nature and strong photoluminescence (PL) emission. Xu *et al*. [1] first discovered the CQDs in 2004 while purifying carbon nanotubes. CQDs are well known for their tunable features, enabling applications in biomedicines, photocatalytic, sensing and optoelectronics applications. They improve the characteristics of semiconductor quantum dots as well as fill the gaps due to the inadequacy of conventional

* **Corresponding author Surinder Kumar Mehta:** Department of Chemistry and Centre of Advanced Studies in Chemistry, Panjab University, Chandigarh, India; E-mail: skmehta@pu.ac.in, ss

Kulvinder Singh & Vineet Kumar (Eds.)

materials in terms of cytotoxicity, biohazard and eco-friendliness. CQDs have fascinating applications in various fields due to their good aqueous solubility and tunable fluorescence properties. During the past few years, various researchers have made substantial progress in the synthesis of the CQDs. In contrast to the traditional semiconductor QDs, the fluorescent CQDs are superior owing to their excellent solubility in an aqueous medium, resistance to photo-bleaching and facile modification.

The 'top-down' and 'bottom-up' are the two well-known approaches by which CQDs are synthesized, as shown in Fig. (**1**) [1]. In the top-down approach, using chemical or physical methods, the large-sized macromolecules are disintegrated into small-sized CQDs. On the other side, in a bottom-up approach, the tiny molecules carbonize *via* polymerization resulting in CQDs through a chemical reaction. Some of the methods are chemical ablation, microwave irradiation, laser ablation, hydrothermal treatment, *etc.*

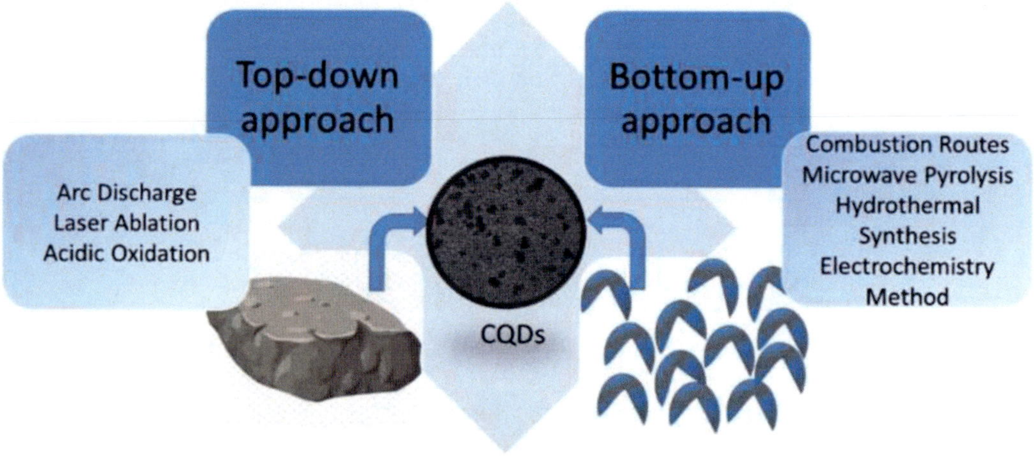

Fig. (1). Classification of methods of synthesis of CQDs [1].

The structure of the CQDs may vary depending on the mode of synthesis followed. Generally, CQDs are spherically shaped with a distance of 0.34 nm between two layers corresponding to (002) plane. Materials used to fabricate CQDs include different sources of carbon, such as citric acid, ammonium citrate, graphite, sodium hydroxide and some natural sources, such as food wastes, grass, denatured milk, broccoli *etc*. As of now, exceptional progress has been made in exploring the properties of luminescent CQDs. Thus, this chapter summarizes various properties and features of CQDs.

Optical Features

The CQDs have various applications due to their remarkable optical properties. The important optical properties have been discussed further.

Photoluminescence (PL)

The excellent feature with respect to fundamental and application-oriented perspectives is PL. In comparison to organic fluorophores, the CQDs exhibit excellent photostability and non-blinking fluorescence. According to a literature report [2], CQDs synthesized by laser ablation display PL intensity which is reduced by just 4.5% even after 4 h of irradiation. Another special property of CQDs is that λ_{ex} depends on the emission wavelength owing to quantum impact and emissive pits on its interface. Additionally, multicolor imaging technology can be achieved by using excitation-dependent PL. The fluorescent features of CQDs could be altered by the modification in experimental synthesis procedures, as shown in Fig. (**2**) [2]. The photoluminescence behavior of CQDs is strongly determined by size, crystallinity and functional groups. Generally, naked CQDs possess low fluorescence however; the fluorescence quantum yield is increased by chemical modification using organic moieties [2].

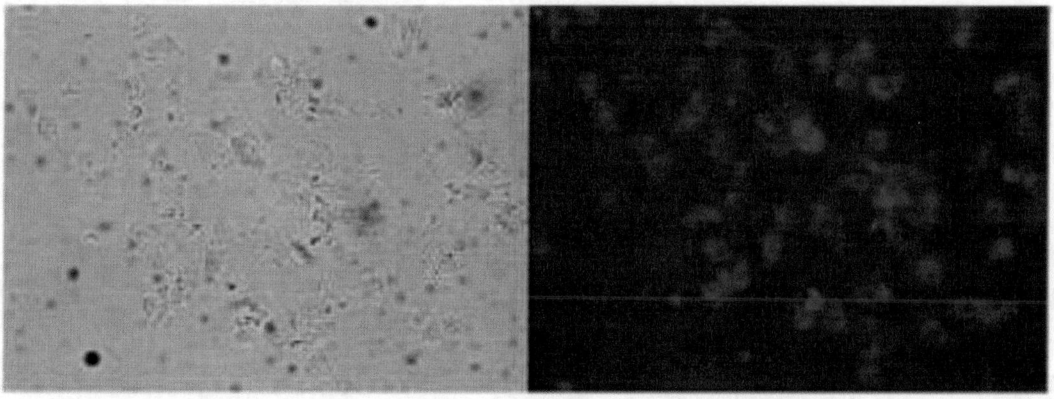

Fig. (2). Cellular scans of HUVEC experimented with carbon dots following 12 h in a spark and fluorescent field [2] .

Numerous mechanisms have been reported to be responsible for the fluorescence of CQDs, including surface defects, quantum size defects and molecular states. Surface defects refer to the defects formed due to oxidation. The quantum size effect indicates the effect of radiative electron-hole pair recombination [3]. CQDs are known to possess certain specific mechanisms based on photoluminescence which have been described below.

Up Conversion Photoluminescence

Up-conversion fluorescence is the term for the phenomenon where the emission wavelength is shorter than that of the excitation wavelength. Due to the concurrent absorbance of two or even multiple photons, the emission wavelength is shorter than the excitation wavelength. This unique property opens prospects for application in bioscience technology. Generally, CQDs synthesized by ultrasonic treatment show up-conversion fluorescence [4 - 6]. Su *et al.* [4] reported that CQDs made across ammonium hydroxide and glucose showed up-conversion emission in the wavelength band nearby 300-600 nm when stimulated from 650-1000 nm. The up-conversion fluorescence is an exciting feature; however, recently, a report by Wen and co-workers [1] pointed out that 'the fluorescence monochromator's second diffraction element leakage, the up-conversion, which results from regular fluorescence, is excited by the spectrophotometer'. They identified up-conversion photoluminescence by measuring the intensity of excitation dependence on the fluorescence. They also found that by adding a long pass filter to the excitation circuit, any form of incorrect excitation may be removed. Hu *et al.* demonstrated the Upconversion Luminescence behavior of β-Ba_2ScAlO_5: Yb^{+3}/Er^{+3} Phosphor by the Intermediate Band through Ca^{+2} Doping, as shown in Fig. (**3**) [2].

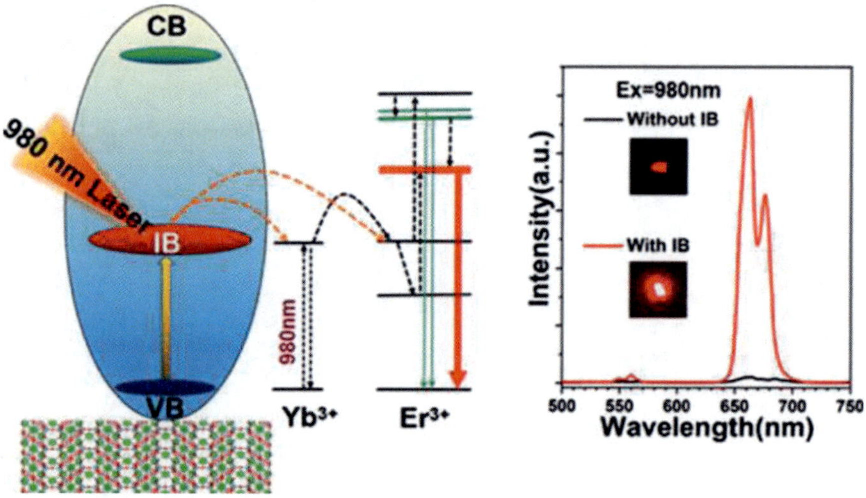

Fig. (3). Upconversion Luminescence of β-Ba_2ScAlO_5: Yb^{+3}/Er^{+3} Phosphor by the Intermediate Band through Ca^{+2} Doping [2] .

Anti-bleaching and Photo-bleaching Properties

The application of CQDs dependent on fluorescence mechanism requires them to be of long emissive lifetimes and stable photoluminescence signals. Literature reports have shown the CQDs to have strong photoluminescence varying from 6 h to 1 year [3, 4] in an aqueous solution. The changes in pH over a wide range can be resisted by CQDs, which designates exceptional impedance to photo-bleaching. To exemplify, Wang *et al.* [5] and Zhou *et al.* [6] reported that even following UV irradiation of 12 h and 24 h, CQDs display a very minute reduction in fluorescent intensity.

pH-dependent Features

Liu *et al.* [7] observed that different fluorescence could be achieved by varying the excitation. Due to the sheer surface areas and dispersion of CQDs, it was observed that the maximal emission spike expanded between 450 to 550 nm whenever the excitation wavelength increased between 340 to 480 nm. The pH-dependent property of CQDs is caused by the deprotonation and protonation of functional groups upon their surface. Shi and co-workers [8] developed a fluorescence-based pH sensor that displayed linear behavior between pH and fluorescent intensity in the pH range of 3.5 to 10.00. pH-dependent properties of CQDs play a crucial role and control man mechanisms such as multicolor imaging and pH detection [9 - 11].

Phosphorescence

Fluorescence is not the only feature of the CQDs that enables them to be employed for sensing applications. Phosphorescent properties of CQDs have also been discovered recently [12 - 15]. Deng and co-workers [16] obtained a pure organic phosphorescent (RTP) substance at ambient temperature. When UV light was used to excite the CQDs embedded within the polyvinyl alcohol (PVA) matrix, phosphorescence was seen. The phosphorescence mechanism of aromatic carbonyls on the surface of CQDs was discovered to be caused by their triple excited modes. The phosphorescent lifespan was enhanced to a sub-second array of ~ 380 milliseconds.

Chemiluminescence

Chemiluminescence (CL) is a phenomenon of the emission of light during a chemical reaction. Lin and co-workers [17] observed the CL features of CQDs. These characteristics of CQDs were observed in the presence of potassium permanganate and cerium (IV), which are classical oxidants.

The general proposed CL mechanism of the conducted experiment is shown in Fig. (**4**) [18]. The thermal equilibrium of the electron dispersion within CQDs causes the CL emission to increase with temperature. Modifying the surface groups can change the CL properties [19]. Zhao *et al.* [20] reported that in the existence of superior alkaline (KOH and NaOH) solution, an interesting CL behavior of CQDs occurred. Similar findings were reported by Dou and co-workers [21], where CL features were investigated in a concentrated NaOH solution. The CL characteristics were due to the CQDs ability to donate electrons to dissolved oxygen, culminating in the development of a superoxide anion radical. CQDs ability to operate as both an electron acceptor and the donor has a wide range of uses in optronics and catalysis. Li *et al.* illustrated the CL mechanism for Monitoring Hydroxylated Intermediates Generated in Persulfate-Based AOPs (Advanced Oxidation Processes), as shown in Fig. (**5**) [22].

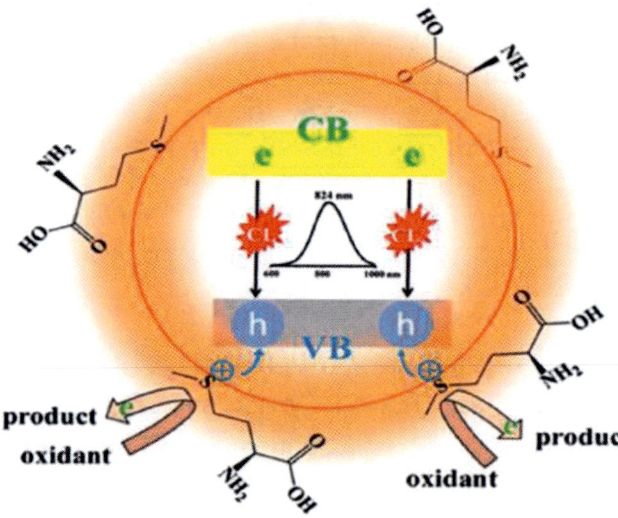

Fig. (4). Proposed chemiluminescence mechanism [18] .

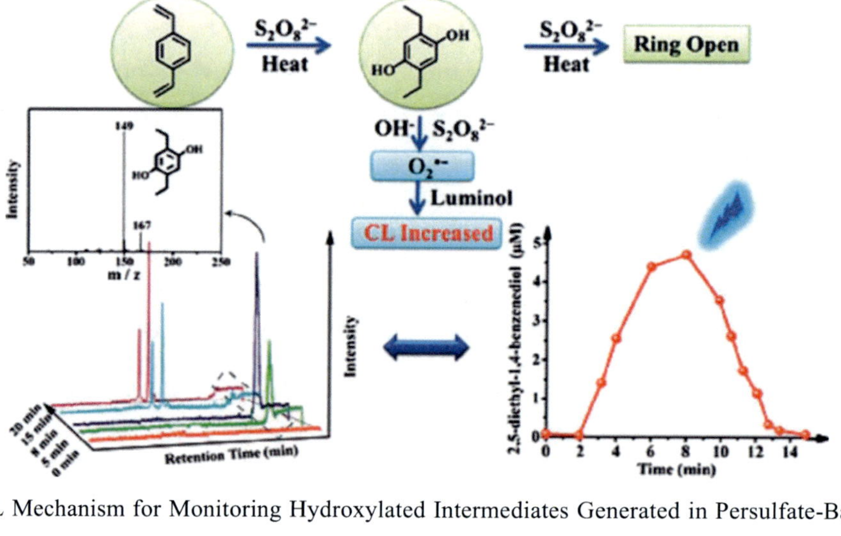

Fig. (5). CL Mechanism for Monitoring Hydroxylated Intermediates Generated in Persulfate-Based AOPs [22].

Adsorption

Studies revealed that CQDs had been demonstrated to have excellent optical characteristics in terms of absorption capabilities. CQDs typical absorption spectra show maximum bands in the UV band (230-320nm), featuring tails extension to the visible area. Usually, the bands are contributed to π→π* transition and n→π* of C=C and C=O bonds, correspondingly as shown in Fig. (**6**) [29].

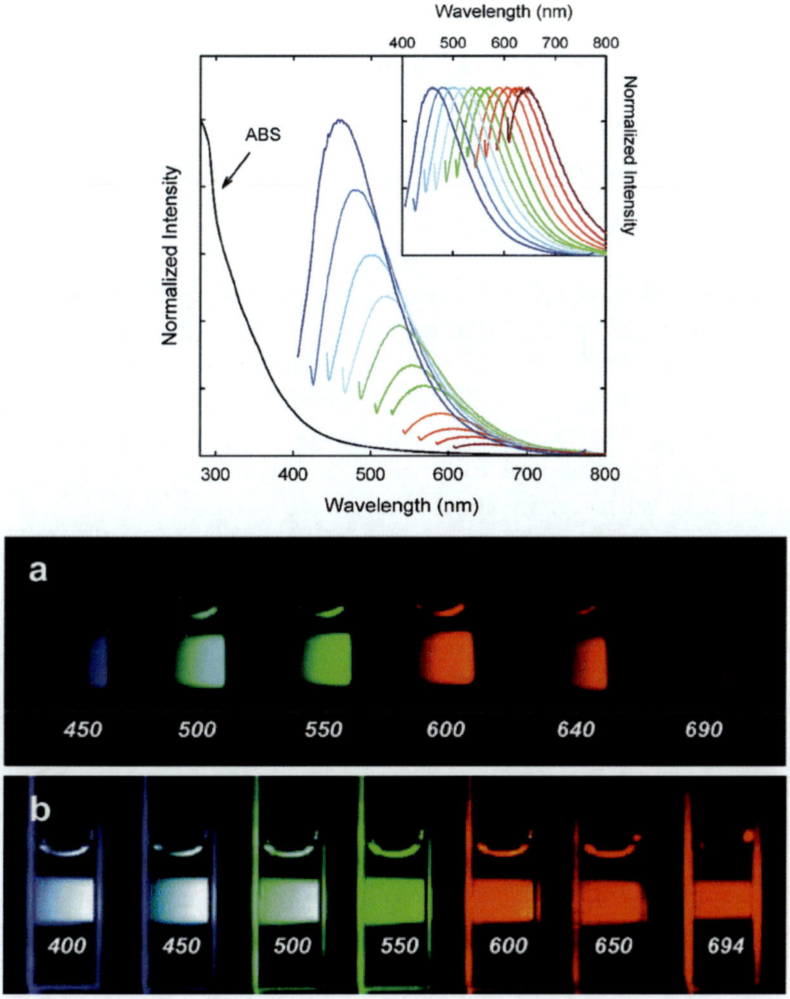

Fig. (6). The absorbance (ABS) and luminescent emission spectrum of PPEI-EI carbon dots in an aqueous phase (having successively larger excitation wavelengths commencing 400 nm on the left in 20 nm increments). Quantum outputs are used to standardize the emission spectrum maxima. The aqueous phase of the PEG1500N-attached carbon dots was stimulated at 400 nm in both (**a**) and (**b**) and imaged immediately after passing over the relevant band-pass filtration and wavelengths [23].

Electrical Properties

The electrical properties of CQDs have not been explored much. Very little research has been done on this aspect. Generally, the interaction between functional groups, doped heteroatom and the carbon core is accountable for the electrical properties [24]. The CQDs and graphene oxide sheets both have oxygenated functional groups, due to which their electrochemical nature is similar. CQDs are appropriate for oxygen reduction reactions (ORR) due to their tiny structure, good resilience, and conductivity. Li and colleagues recently found that doping nitrogen atoms with pyridinium units significantly affects the electrocatalytic properties of ORR. Li *et al.* found comparable results to Yan *et al.* [25] and Liu and associates [26]. Zhu and coworkers [27] explored the electro-catalytic impact of CQDs produced by biological biomass-soymilk. N-CQDs and the electrochemical reduction properties of the oxygen generated were identical.

Electrochemical Luminescence (ECL)

ECL is an extended branch of chemiluminescence where luminescence is produced by an electrochemical reaction. In order to create exciting forms which generate light, the electrodes create moieties that go through an electron transfer process. A report by Zhang *et al.* [28] highlighted that when CQDs were dispersed in ethanol having 0.1 M tetrabutyl ammonium bromide (TBAB), dual ECL peaks were observed. Based on the dual ECL of CQDs, iron ions were detected up to the range of micromoles with sensitivity and accuracy.

In recent research by Xu and associates [29], the impact of CQD surface structure, morphology, and configuration on PL and ECL in a variety of applications was studied. The oxidized CQDs diffused onto the electrode surface and controlled the electrochemical response. At 1.30 V, ECL achieved its maximum peak. By using different precursors, two kinds of CQDs were prepared. One of them was prepared by the carbonization-extraction method, and the other by the carbonization-oxidation strategy. Carbonization-oxidation CQDs (o-CQDs) feature a carbon core and a loose shell containing oxygen. On the contrary, CQDs prepared by carbonization-extraction (r-CQDs) mainly contain the carbon core as the same procedure has been depicted in Fig. (**7**) [36].

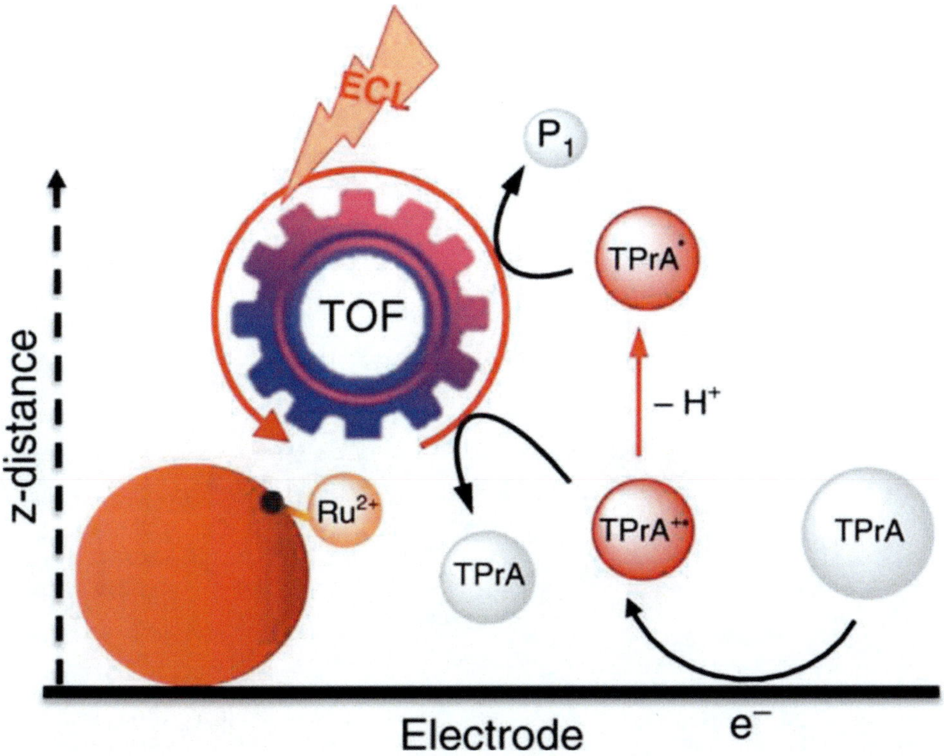

Fig. (7). Schematic representation of the remote Electrochemiluminescence (ECL) mechanism [30].

Photo-induced Electron Transfer

The photo-induced electron transfer (PET), photo-induced charge separation, and photo response have all been explored in order to use photoluminescent substances in areas such as light energy conversion. In this regard, Wang *et al.* [31] deliberated that the addition of electron acceptors, including 2,4-dinitrotoluene and 4-nitrotoluene, quenched the internal fluorescence of CQDs. The photoluminescence of CQDs solution was also quenched by the addition of electron donors, as shown in Fig. (**8**) [38].

The doping of CQDs with metals was also found to quench the fluorescence efficiently *via* the excited state redox process. Yu *et al.* [33] studied the nanocomposites of CQDs, namely CQDs-GO, CQDs-MWNTs and CQDs-TiO_2, lacking any linkers and investigated the electron transfer process in them. The authors reported that in the system of CQDs-GO, PL quenching was observed, and a very fast transfer of electrons occurred from CQDs to GO with 400 fs of the time constant. However, CQDs-TiO_2 and CQDs-MWNTs nanocomposites displayed no charge transfer. These features entitle CQDS to open new applications in catalysis, light-energy conversion and mechanistic interpretation.

Cui *et al.* showed the PL spectra of NCD solution in the presence of Fe^{+3}, as shown in Fig. (**9**) [34].

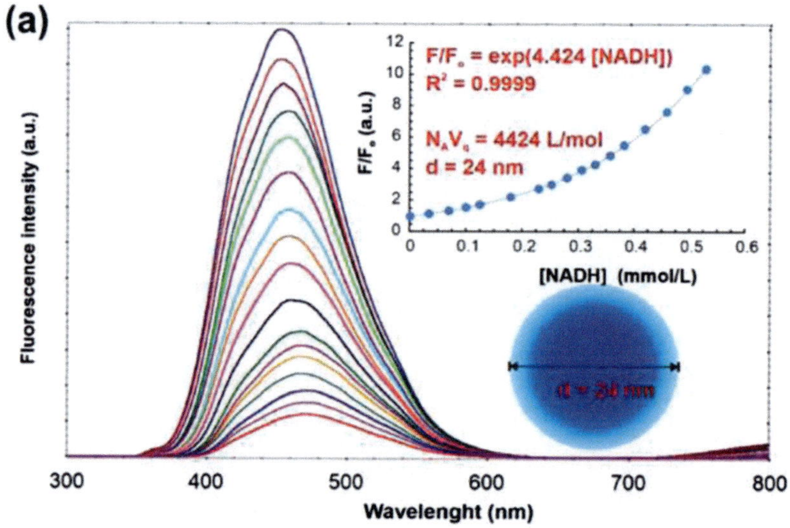

Fig. (8). Fluorescence emission spectra of N-CQD and their quenching by sequential addition of a reduced form of nicotinamide adenine dinucleotide (NADH) aliquots. Inset shows Stern–Volmer plots with quenching with various concentrations of NADH. F_0 is the fluorescence intensity of the initial sample, where F means observed intensities [32].

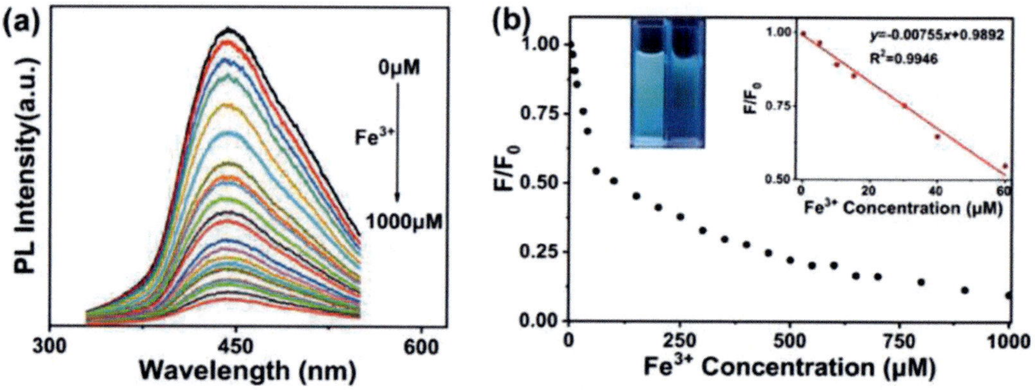

Fig. (9). (**a**) PL spectra of the NCD solution in the presence of Fe^{3+}. (**b**) PL intensity of the NCD solution in the presence of Fe^{3+}. Inset: images of the NCD solution under a UV lamp without (left) and with Fe^{3+} (right, 500 μM) [34].

Biological Properties

Good progress has been made by CQDs in the area of biosensors with good stability. However, biocompatibility is still a critical issue for the suitability of functionalized CQDs in live cells. Both raw and passivated CQDs were investigated for systematic cytotoxicity during the last few years. For the cytotoxicity assay, Wang *et al.* [35] utilized CQDs synthesized through arc-discharge of graphite rods. The authors concluded that even if a high concentration (0.4 mg/L) of bare CQDs were used, they were non-toxic to cells. Similarly, CQDs manufactured by an electrochemical graphite process were used in a cytotoxicity study utilizing a human kidney line. Cell viability was found to be unaffected by luminescent CQDs [36].

Moreover, Ray and co-workers [37] investigated the cell viability of CQDs synthesized by an improved soot-based approach. At a concentration adequate for fluorescent bio-imaging, the experimental results demonstrated negligible cytotoxicity. Furthermore, the hazardous impact of CQDs functionalized with groups, including PAA (poly (acrylic acid) [38], PPEI-EI [37], PEG (polyethylene glycol) [39] and BPEI (branched poly-(ethylenimine)) [40]. It was found that at low concentrations of 50 mg/L and exposure time of 24 h, both free PAA and PAA functionalized CQDs were found to be toxic to cells. The experimental evidence indicated that although at high concentration, PEI was non-toxic to HT-29 cells, PEI-passivated CQDs have more toxicity in contrast to the PPEI-EI functionalized CQDs. It was concluded that more ethylenimine units in PEI could be related to the cytotoxicity of CQDs. The PEG-functionalized CQDs were found to be non-toxic when injected in mice and observed for 28 days. Overall, it can be inferred that PEG and PPEI passivated CQDs are appropriate and non-toxic for *in vivo* and bioimaging. The CQDs prepared after surface functionalization using PAA and BPEI, although toxic at higher concentrations, can be used for *in vivo* and bioimaging at low concentrations.

Toxicity

Toxicity is a parameter to measure the potential of the material to modify and manipulate the environment. The toxicity of CQDs has been analyzed by multiple groups using 3-(4,5-Dimethyl thiazol-2-yl)-2,5 diphenyltetrazoliumbromide (MTT) assay for *in-vitro* toxic effects measurement [39, 41, 42]. Yang *et al.* [43] discovered that at a high concentration of 40 mg/kg (CQDs/mice body wt.), the toxicity of PEG-coated CQDs in male CD-1 mice is in accordance with serum biochemistry and histopathological analyses. Their results suggested that CQDs were biocompatible. In every case where CQDs were tested *in vivo*, their toxicity was shown to be significantly lower than that of CdSe-based QDs. In another research by Hsu and co-workers [44], the cytotoxicity of CQDs fabricated from fresh green tea was estimated. It was discovered that it inhibited MCF-7 and MDA-MB-231 cells more effectively than catechin-based CQDs. Moreover, easy excretion through mice's urinary systems also indicated minimal toxicity of CQDs. Further, Li *et al.* [45] revealed that fluorescent CQDs fabricated from ginger juice led to minimal toxicity to normal liver cells (FL83B) and normal mammary epithelial cells (MCF-10A) and suppressed the growth of human hepatocellular carcinoma cells (HepG2The inhibition was selective to HepG2 over other cancer cells with 50% inhibiting concentration (IC_{50}) value of the C-dots on HepG2 cells is 0.35 mg mL^{-1}. Therefore, based on the literature reports, CQDs can be treated as an efficient way in terms of sustainability and toxicity. Hence, CQDs toxicity is evaluated to be low when the concentration of CQDs is not so high.

CONCLUSION

In concluding lines, this chapter summarizes current advances in the domain of CQDs, focusing on the features of CQDs. The remarkable features of CQDs make them applicable in diverse streams ranging from energy storage to bio-analytics and sensing. The intriguing qualities of CQDs are highlighted in their applications for bioimaging, radiotherapy, chemotherapy, drug delivery, *etc.* It is also emphasized that with the combination of theoretical and experimental evaluation, further efforts should be made on the development of biocompatible CQDs and exploring their medicinal applications.

REFERENCES

[1] Wen, X.; Yu, P.; Toh, Y.R.; Ma, X.; Tang, J. On the upconversion fluorescence in carbon nanodots and graphene quantum dots. *Chem. Commun.,* **2014**, *50*(36), 4703-4706. [http://dx.doi.org/10.1039/C4CC01213E] [PMID: 24675809]

[2] Hu, J.; Bian, X.; Wang, R.; Liu, L.; Ilyas, N.; Wang, F.; Song, Z.; Fu, H. Giant Enhancement in Upconversion Luminescence of β-Ba_2ScAlO_5 :Yb^{3+} /Er^{3+} Phosphor by the Intermediate Band through Ca^{2+} Doping. *Chem. Mater.,* **2022**, *34*(7), 3089-3098. [http://dx.doi.org/10.1021/acs.chemmater.1c04142]

[3] Zeng, Y.W.; Ma, D.K.; Wang, W.; Chen, J.J.; Zhou, L.; Zheng, Y.Z.; Yu, K.; Huang, S.M. N, S co-doped carbon dots with orange luminescence synthesized through polymerization and carbonization reaction of amino acids. *Appl. Surf. Sci.,* **2015**, *342*, 136-143. [http://dx.doi.org/10.1016/j.apsusc.2015.03.029]

[4] Li, H.; He, X.; Liu, Y.; Huang, H.; Lian, S.; Lee, S.T.; Kang, Z. One-step ultrasonic synthesis of water-soluble carbon nanoparticles with excellent photoluminescent properties. *Carbon,* **2011**, *49*(2), 605-609. [http://dx.doi.org/10.1016/j.carbon.2010.10.004]

[5] Wang, B.; Tang, W.; Lu, H.; Huang, Z. Hydrothermal synthesis of ionic liquid-capped carbon quantum dots with high thermal stability and anion responsiveness. *J. Mater. Sci.,* **2015**, *50*(16), 5411-5418. [http://dx.doi.org/10.1007/s10853-015-9085-y]

[6] Zhuo, Y.; Miao, H.; Zhong, D.; Zhu, S.; Yang, X. One-step synthesis of high quantum-yield and excitation-independent emission carbon dots for cell imaging. *Mater. Lett.,* **2015**, *139*, 197-200. [http://dx.doi.org/10.1016/j.matlet.2014.10.048]

[7] Liu, C.; Zhang, P.; Tian, F.; Li, W.; Li, F.; Liu, W. One-step synthesis of surface passivated carbon nanodots by microwave assisted pyrolysis for enhanced multicolor photoluminescence and bioimaging. *J. Mater. Chem.,* **2011**, *21*(35), 13163-13167.

[http://dx.doi.org/10.1039/c1jm12744f]

[8] Shi, L.; Li, Y.; Li, X.; Zhao, B.; Wen, X.; Zhang, G.; Dong, C.; Shuang, S. Controllable synthesis of green and blue fluorescent carbon nanodots for pH and Cu 2+ sensing in living cells. *Biosens. Bioelectron.,* **2016**, *77*, 598-602. [http://dx.doi.org/10.1016/j.bios.2015.10.031] [PMID: 26485174]

[9] Jin, X.; Sun, X.; Chen, G.; Ding, L.; Li, Y.; Liu, Z.; Wang, Z.; Pan, W.; Hu, C.; Wang, J. pH-sensitive carbon dots for the visualization of regulation of intracellular pH inside living pathogenic fungal cells. *Carbon,* **2015**, *81*, 388-395. [http://dx.doi.org/10.1016/j.carbon.2014.09.071]

[10] Wang, C.; Xu, Z.; Cheng, H.; Lin, H.; Humphrey, M.G.; Zhang, C. A hydrothermal route to water-stable luminescent carbon dots as nanosensors for pH and temperature. *Carbon,* **2015**, *82*, 87-95. [http://dx.doi.org/10.1016/j.carbon.2014.10.035]

[11] Jia, X.; Li, J.; Wang, E. One-pot green synthesis of optically pH-sensitive carbon dots with upconversion luminescence. *Nanoscale,* **2012**, *4*(18), 5572-5575. [http://dx.doi.org/10.1039/c2nr31319g] [PMID: 22786671]

[12] Chen, Y.; He, J.; Hu, C.; Zhang, H.; Lei, B.; Liu, Y. Room temperature phosphorescence from moisture-resistant and oxygen-barred carbon dot aggregates. *J. Mater. Chem. C Mater. Opt. Electron. Devices,* **2017**, *5*(25), 6243-6250. [http://dx.doi.org/10.1039/C7TC01615H]

[13] Kang, X.; Wang, J.; Wu, H.; Aksay, I.A.; Liu, J.; Lin, Y. Glucose Oxidase–graphene–chitosan modified electrode for direct electrochemistry and glucose sensing. *Biosens. Bioelectron.,* **2009**, *25*(4), 901-905. [http://dx.doi.org/10.1016/j.bios.2009.09.004] [PMID: 19800781]

[14] Li, Q.; Zhou, M.; Yang, M.; Yang, Q.; Zhang, Z.; Shi, J. Induction of long-lived room temperature phosphorescence of carbon dots by water in hydrogen-bonded matrices. *Nat. Commun.,* **2018**, *9*(1), 734. [http://dx.doi.org/10.1038/s41467-018-03144-9] [PMID: 29467414]

[15] Joseph, J.; Anappara, A.A. Long life-time room-temperature phosphorescence of carbon dots in aluminum sulfate. *Chem. Sel.,* **2017**, *2*(14), 4058-4062. [http://dx.doi.org/10.1002/slct.201700786]

[16] Deng, Y.; Zhao, D.; Chen, X.; Wang, F.; Song, H.; Shen, D. Long lifetime pure organic phosphorescence based on water soluble carbon dots. *Chem. Commun.,* **2013**, *49*(51), 5751-5753. [http://dx.doi.org/10.1039/c3cc42600a] [PMID: 23685471]

[17] Lin, Z.; Xue, W.; Chen, H.; Lin, J.M. Classical oxidant induced chemiluminescence of fluorescent carbon dots. *Chem. Commun.,* **2012**, *48*(7), 1051-1053. [http://dx.doi.org/10.1039/C1CC15290D] [PMID: 22158939]

[18] Fu, L.; Zhang, B.; Gao, X.; Dong, S.; Wang, D.; Zou, G. A general route for chemiluminescence of n-type au nanocrystals. *Anal. Chem.,* **2022**, *94*(24), 8811-8817. [http://dx.doi.org/10.1021/acs.analchem.2c01770] [PMID: 35675670]

[19] Teng, P.; Xie, J.; Long, Y.; Huang, X.; Zhu, R.; Wang, X.; Liang, L.; Huang, Y.; Zheng, H. Chemiluminescence behavior of the carbon dots and the reduced state carbon dots. *J. Lumin.,* **2014**, *146*, 464-469. [http://dx.doi.org/10.1016/j.jlumin.2013.09.036]

[20] Zhao, L.; Di, F.; Wang, D.; Guo, L.H.; Yang, Y.; Wan, B.; Zhang, H. Chemiluminescence of carbon dots under strong alkaline solutions: A novel insight into carbon dot optical properties. *Nanoscale,* **2013**, *5*(7), 2655-2658. [http://dx.doi.org/10.1039/c3nr00358b] [PMID: 23456202]

[21] Dou, X.; Lin, Z.; Chen, H.; Zheng, Y.; Lu, C.; Lin, J.M. Production of superoxide anion radicals as

evidence for carbon nanodots acting as electron donors by the chemiluminescence method. *Chem. Commun.,* **2013**, *49*(52), 5871-5873.
[http://dx.doi.org/10.1039/c3cc41145a] [PMID: 23703092]

[22] Li, Z.; Chen, X.; Teng, X.; Lu, C. Chemiluminescence as a new indicator for monitoring hydroxylated intermediates in persulfate-based advanced oxidation processes. *J. Phys. Chem. C,* **2019**, *123*(35), 21704-21712.
[http://dx.doi.org/10.1021/acs.jpcc.9b07058]

[23] Sun, Y.P.; Zhou, B.; Lin, Y.; Wang, W.; Fernando, K.A.S.; Pathak, P.; Meziani, M.J.; Harruff, B.A.; Wang, X.; Wang, H.; Luo, P.G.; Yang, H.; Kose, M.E.; Chen, B.; Veca, L.M.; Xie, S.Y. Quantum-sized carbon dots for bright and colorful photoluminescence. *J. Am. Chem. Soc.,* **2006**, *128*(24), 7756-7757.
[http://dx.doi.org/10.1021/ja062677d] [PMID: 16771487]

[24] Ambrosi, A.; Chua, C.K.; Bonanni, A.; Pumera, M. Electrochemistry of graphene and related materials. *Chem. Rev.,* **2014**, *114*(14), 7150-7188.
[http://dx.doi.org/10.1021/cr500023c] [PMID: 24895834]

[25] Ye, R.; Xiang, C.; Lin, J.; Peng, Z.; Huang, K.; Yan, Z.; Cook, N.P.; Samuel, E.L.G.; Hwang, C.C.; Ruan, G.; Ceriotti, G.; Raji, A.R.O.; Martí, A.A.; Tour, J.M. Coal as an abundant source of graphene quantum dots. *Nat. Commun.,* **2013**, *4*(1), 2943.
[http://dx.doi.org/10.1038/ncomms3943] [PMID: 24309588]

[26] Liu, Y.; Wu, P. Graphene quantum dot hybrids as efficient metal-free electrocatalyst for the oxygen reduction reaction. *ACS Appl. Mater. Interfaces,* **2013**, *5*(8), 3362-3369.
[http://dx.doi.org/10.1021/am400415t] [PMID: 23532722]

[27] Zhu, C.; Zhai, J.; Dong, S. Bifunctional fluorescent carbon nanodots: Green synthesis *via* soy milk and application as metal-free electrocatalysts for oxygen reduction. *Chem. Commun.,* **2012**, *48*(75), 9367-9369.
[http://dx.doi.org/10.1039/c2cc33844k] [PMID: 22911246]

[28] Zhang, P.; Xue, Z.; Luo, D.; Yu, W.; Guo, Z.; Wang, T. Dual-peak electrogenerated chemiluminescence of carbon dots for iron ions detection. *Anal. Chem.,* **2014**, *86*(12), 5620-5623.
[http://dx.doi.org/10.1021/ac5011734] [PMID: 24855934]

[29] Xu, Y.; Wu, M.; Liu, Y.; Feng, X.Z.; Yin, X.B.; He, X.W.; Zhang, Y.K. Nitrogen-doped carbon dots: A facile and general preparation method, photoluminescence investigation, and imaging applications. *Chemistry,* **2013**, *19*(7), 2276-2283.
[http://dx.doi.org/10.1002/chem.201203641] [PMID: 23322649]

[30] Zanut, A.; Fiorani, A.; Canola, S.; Saito, T.; Ziebart, N.; Rapino, S.; Rebeccani, S.; Barbon, A.; Irie, T.; Josel, H.P.; Negri, F.; Marcaccio, M.; Windfuhr, M.; Imai, K.; Valenti, G.; Paolucci, F. Insights into the mechanism of coreactant electrochemiluminescence facilitating enhanced bioanalytical performance. *Nat. Commun.,* **2020**, *11*(1), 2668.
[http://dx.doi.org/10.1038/s41467-020-16476-2] [PMID: 32472057]

[31] Wang, X.; Cao, L.; Lu, F.; Meziani, M.J.; Li, H.; Qi, G.; Zhou, B.; Harruff, B.A.; Kermarrec, F.; Sun, Y.P. Photoinduced electron transfers with carbon dots. *Chem. Commun.,* **2009**, (25), 3774-3776.
[http://dx.doi.org/10.1039/b906252a] [PMID: 19557278]

[32] Wiśniewski, M.; Czarnecka, J.; Bolibok, P.; Świdziński, M.; Roszek, K. New insight into the fluorescence quenching of nitrogen-containing carbonaceous quantum dots—From surface chemistry to biomedical applications. *Materials,* **2021**, *14*(9), 2454.
[http://dx.doi.org/10.3390/ma14092454] [PMID: 34065161]

[33] Yu, P.; Wen, X.; Toh, Y.R.; Lee, Y.C.; Huang, K.Y.; Huang, S.; Shrestha, S.; Conibeer, G.; Tang, J. Efficient electron transfer in carbon nanodot–graphene oxide nanocomposites. *J. Mater. Chem. C Mater. Opt. Electron. Devices,* **2014**, *2*(16), 2894-2901.
[http://dx.doi.org/10.1039/c3tc32395a]

[34] Cui, J.; Zhu, X.; Liu, Y.; Liang, L.; Peng, Y.; Wu, S.; Zhao, Y. N-Doped Carbon Dots as Fluorescent "Turn-Off" Nanosensors for Ascorbic Acid and Fe^{3+} Detection. *ACS Appl. Nano Mater.*, **2022**, *5*(5), 7268-7277.
[http://dx.doi.org/10.1021/acsanm.2c01170]

[35] Wang, Y.; Anilkumar, P.; Cao, L.; Liu, J.H.; Luo, P.G.; Tackett, K.N., II; Sahu, S.; Wang, P.; Wang, X.; Sun, Y.P. Carbon dots of different composition and surface functionalization: cytotoxicity issues relevant to fluorescence cell imaging. *Exp. Biol. Med.*, **2011**, *236*(11), 1231-1238.
[http://dx.doi.org/10.1258/ebm.2011.011132] [PMID: 22036734]

[36] Zhao, Q.L.; Zhang, Z.L.; Huang, B.H.; Peng, J.; Zhang, M.; Pang, D.W. Facile preparation of low cytotoxicity fluorescent carbon nanocrystals by electrooxidation of graphite. *Chem. Commun.*, **2008**, (41), 5116-5118.
[http://dx.doi.org/10.1039/b812420e] [PMID: 18956040]

[37] Ray, S.C.; Saha, A.; Jana, N.R.; Sarkar, R. Fluorescent carbon nanoparticles: Synthesis, characterization, and bioimaging application. *J. Phys. Chem. C,* **2009**, *113*(43), 18546-18551.
[http://dx.doi.org/10.1021/jp905912n]

[38] Wang, Y.; Bao, L.; Liu, Z.; Pang, D.W. Aptamer biosensor based on fluorescence resonance energy transfer from upconverting phosphors to carbon nanoparticles for thrombin detection in human plasma. *Anal. Chem.*, **2011**, *83*(21), 8130-8137.
[http://dx.doi.org/10.1021/ac201631b] [PMID: 21923110]

[39] Yang, S.T.; Cao, L.; Luo, P.G.; Lu, F.; Wang, X.; Wang, H.; Meziani, M.J.; Liu, Y.; Qi, G.; Sun, Y.P. Carbon dots for optical imaging *in vivo*. *J. Am. Chem. Soc.*, **2009**, *131*(32), 11308-11309.
[http://dx.doi.org/10.1021/ja904843x] [PMID: 19722643]

[40] Dong, Y.; Wang, R.; Li, G.; Chen, C.; Chi, Y.; Chen, G. Polyamine-functionalized carbon quantum dots as fluorescent probes for selective and sensitive detection of copper ions. *Anal. Chem.*, **2012**, *84*(14), 6220-6224.
[http://dx.doi.org/10.1021/ac3012126] [PMID: 22686413]

[41] Anand, A.; Unnikrishnan, B.; Wei, S.C.; Chou, C.P.; Zhang, L.Z.; Huang, C.C. Graphene oxide and carbon dots as broad-spectrum antimicrobial agents :A minireview. *Nanoscale Horiz.*, **2019**, *4*(1), 117-137.
[http://dx.doi.org/10.1039/C8NH00174J] [PMID: 32254148]

[42] Zhang, H.; Chen, Y.; Liang, M.; Xu, L.; Qi, S.; Chen, H.; Chen, X. Solid-phase synthesis of highly fluorescent nitrogen-doped carbon dots for sensitive and selective probing ferric ions in living cells. *Anal. Chem.*, **2014**, *86*(19), 9846-9852.
[http://dx.doi.org/10.1021/ac502446m] [PMID: 25211236]

[43] Yang, S.T.; Wang, X.; Wang, H.; Lu, F.; Luo, P.G.; Cao, L.; Meziani, M.J.; Liu, J.H.; Liu, Y.; Chen, M.; Huang, Y.; Sun, Y.P. Carbon dots as nontoxic and high-performance fluorescence imaging agents. *J. Phys. Chem. C,* **2009**, *113*(42), 18110-18114.
[http://dx.doi.org/10.1021/jp9085969] [PMID: 20357893]

[44] Hsu, P.C.; Chen, P.C.; Ou, C.M.; Chang, H.Y.; Chang, H.T. Extremely high inhibition activity of photoluminescent carbon nanodots toward cancer cells. *J. Mater. Chem. B Mater. Biol. Med.*, **2013**, *1*(13), 1774-1781.
[http://dx.doi.org/10.1039/c3tb00545c] [PMID: 32261141]

[45] Li, C.L.; Ou, C.M.; Huang, C.C.; Wu, W.C.; Chen, Y.P.; Lin, T.E.; Ho, L.C.; Wang, C.W.; Shih, C.C.; Zhou, H.C.; Lee, Y.C.; Tzeng, W.F.; Chiou, T.J.; Chu, S.T.; Cang, J.; Chang, H.T. Carbon dots prepared from ginger exhibiting efficient inhibition of human hepatocellular carcinoma cells. *J. Mater. Chem. B Mater. Biol. Med.*, **2014**, *2*(28), 4564-4571.
[http://dx.doi.org/10.1039/c4tb00216d] [PMID: 32261557]

CHAPTER 4

Characterization of Carbonaceous Quantum Dots

Abhinay Thakur[1], **Harpreet Kaur**[1], **Ashish Kumar**[2,*] and **Sumayah Bashir**[3]

[1] *Department of Chemistry, School of Chemical Engineering and Physical Sciences, Lovely Professional University, Phagwara, Punjab, India*

[2] *NCE, Department of Science and Technology, Government of Bihar, India*

[3] *Department of Chemistry, Central University of Kashmir, Kashmir, India*

Abstract: Carbonaceous quantum dots (CQDs), a prominent figure of carbon materials, offer remarkable impetus in a variety of sectors, including biosensors, biomedical imaging, drug delivery, photonics, photovoltaics, and electrocatalysis, due to their distinctive physicochemical, optical, and electrical capabilities. This chapter attempts to show current advances in CQD characterization, with an emphasis on the essential multifarious function of CQDs using various techniques, such as photoluminescence and fluorescence emission spectroscopy. Additionally, with the aim of developing highly efficient and long-term sustainable CQD-based components, we explore the obstacles and potential directions of CQD-based substances in this developing research field.

Keywords: CQDs, Characterization, Energy storage, Nanomaterials, Photoluminescence, TEM.

INTRODUCTION

Carbonaceous Quantum Dots (CQDs)

Due to their intriguing properties, carbonaceous quantum dots have gained considerable attention presently. These compounds, in general, have great biocompatibility, low cytotoxicity, considerable mechanical and thermal properties, and are simple to functionalize. CQDs are also notable for their great stability, low hazardous activity, solubility in water, and transesterification accessibility [1 - 4]. Heteroatoms (functional groups) bound to a carbonised core constitute the bulk of them. Among other nanomaterials, carbon dots include carbon nanodots, polymer dots and graphene quantum dots. Particles having a small size (less than 10 nm) composed of sp^2 hybrid carbon core-shell connected among carbon (core) and organic functional groups (shell), including –C = O,

* **Corresponding author Ashish Kumar:** NCE, Department of Science and Technology, Government of Bihar, India; Tel: +91- 94784-78088; E-mail: drashishchemlpu@gmail.com

Kulvinder Singh & Vineet Kumar (Eds.)

N–H, –OH, COOH, C–N and C–O, or polymer agglomeration, are described. According to numerous studies, carbonaceous quantum dots with diverse configurations are manufactured using a variety of techniques and carbon sources. CQDs have photoluminescence (PL) qualities that are equivalent to semiconductor quantum dots, but they surpass their counterpart in terms of low cytotoxicity, environmentally friendly nature, relatively inexpensive, and a wide range of simple synthetic pathways, and a proclivity for surface functionalization. CQDs sparked massive research on their synthesis, characteristics, and fabrication techniques in the disciplines of optoelectronics, electrical and chemical sensors, photocatalysis, and electrocatalysis due to their outstanding physicochemical features [5 - 7]. In 2010, the emphasis was on finding viable synthetic routes for CQDs. To summarise, there are two types of synthetic techniques for CQDs: “top-down” and “bottom-up.” The latter deals with the oxidation of bulk carbon materials such as graphite, CNTs, carbon soot, and activated carbon using laser ablation, arc discharge, chemical reactions, and electrochemical reactions [8 - 12]. The “bottom-up” techniques use hydrothermal, microwave, and calcination to synthesize CQDs using molecular precursors like citrate, carbohydrates, and a variety of feedstock. CQDs have amorphous to nanocrystalline cores and diverse surface functional groups due to the diversity of raw materials and processing processes. As the manufacturing of CQDs entails oxidation in particular, substantial quantities of oxygen-containing moieties, including carbonyl, carboxyl and hydroxyl, are added [13 - 19]. The distinctive properties of CQDs, such as their innocuous chemical structure, controllable fluorescence particulates, ease of functionalization, and outstanding physicochemical and photochemical consistency (non-photobleaching or non-photo blinking), make them extremely appealing for advanced applications when contrasted to traditional semiconductor quantum dots [20 - 24].

This chapter explores the advances made in the evolution of CQDs and their technological applications in terms of characterization approaches. In light of multiple outstanding CQD characterization techniques, such as FTIR, TEM, AFM, UV-vis, XRD, Photoluminescence, and so on, it is desired that this chapter would offer an extensive understanding of recent CQD research and expose different insights toward the research and innovation of CQDs with significantly enhanced physicochemical characteristics [25 - 27]. Taking into the consideration of the superior quality of CQDs, Vercelli *et al.* [28] gave a brief overview of the use of CQDs in organic photovoltaic devices (OPVs) in an attempt to investigate the function that nanomaterials serve in improving the devices' efficiency. They have piqued the research society's attention ever since due to their unique optical features (*e.g.*, excitation wavelength-dependent fluorescence and size-dependent), that render CQDs extremely comparable to well-recognized semiconductor QDs and appropriate for utilization in photovoltaic systems (PVs). In essence, with the suitable architectural design, it is feasible to manage the photoluminescence characteristics, energy levels, and band gap of CQDs, allowing for perfect charge carrier directional passage inside the PV device structure where they were embedded. Fig. (**1**) depicts the several green quantum dot synthesis attributes for CQDs and GQDs along with their extinctive potential applications.

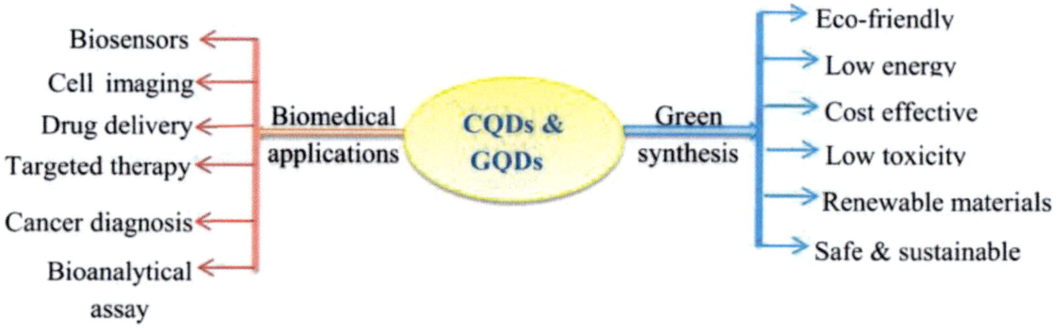

Fig. (1). Green quantum dot synthesis attributes for CQDs and GQDs along with their potential applications. Adapted from Ref [29]. Distributed under attribution-based CCBY license 4.0.

In research, Mehmandoust *et al.* [30] devised an electrochemical sensor for determining the azo dye tartrazine (TRT). 3D porous reduced graphene oxide embellished containing CQDs and platinum nanoparticles (Pt/CQDs@rGO/SPCE) was deposited on a screen-printed carbon electrode (SPCE). The quantity of TRT produced was measured using differential pulse voltammetry. The sensor had two main linearities under ideal circumstances, spanning from 0.01 to 1.57 M and 1.57–9.3 M, with corresponding reliability coefficients of determination of 0.991 and 0.992. 7.93 nM was also determined as the detection limit (LOD). Furthermore, in the vicinity of numerous disrupting reagents and azo dye complexes with comparable structures, the Pt/CQDs@rGO/SPCE indicated excellent selectivity. Furthermore, the Pt/CQDs@rGO/SPCE demonstrated exceptional recovery rates of 96.5–101.6 percent for candy, 96.0–101.2 percent for jelly powder, 99.7–103.5 percent for soft drinks and 98.0–103.0% for water specimens. In addition, the constructed sensor has outstanding selection, durability, reproducibility, and consistency, suggesting that it has a promising future in TRT measurement. As a result, the suggested electrode might be used to accurately quantify TRT in food specimens. Similarly, Shen *et al.* [31] established a simple one-step hydrothermal of m-phenylenediamine and glucose to produce nitrogen-doped CQDs (N-CQDs). The chemical structure, crystal structure and exterior functional groups of N-CQDs produced were all studied in detail. The CQDs were chemically doped with nitrogen, and the N-CQDs crystallised in a graphene configuration, according to the results. Photoluminescence (PL) studies revealed that the N-CQDs emit a significant blue light when treated with UV light. The emission was excitation-dependent, resisted photobleaching and had considerable ionic strength, and dropped somewhat as the temperature rose. They have a quantum efficiency of roughly 17.5 percent. The PL strength of N-CQDs gradually decreases as Fe^{3+} (0.5–1.0 mM) and CrO_4^{2-} (0.3–0.6 mM) levels raise, making them a good fluorescent probe for Fe^{3+} and CrO_4^{2-} detection. The quenching process of Fe^{3+} and CrO_4^{2-} is confirmed to be a static inner filter impact quenching mechanism. N-CQDs have also been discovered to be an effective cell imaging reagent in Hela cells.

In a case [32], due to the efficiency of dialysis in ultrafiltration and the formation of compact and size-regulated CQDs, which may readily pass thru the microbial cell membrane, triggering eventual cell damage, CQDs after dialysis demonstrated dramatically improved antibacterial efficacy. The findings were consistent with previous research. However, as illustrated in Fig. (**2**), the efficiency of as-fabricated CQDs as powerful antibacterial agents was mostly due to their content of ornamental hydroxyl function groups, which were primarily accountable for microbial cell demise through the formation of reactive oxygen species (ROS).

CHARACTERIZATION OF CQDS

To characterize CQDs, a multitude of techniques or analyses are routinely utilized. For instance, Fourier Transform Infrared Spectroscopy could be used to investigate functional groups on CQD surfaces (FTIR). CQD surface morphology, microstructures, and particle size could all be studied using transmission electron microscopy (TEM) [33 - 35]. Aside from that, UV-V absorption spectra containing wavelengths ranging from 200 to 800 nm could be utilized to evaluate the optical activity of CQDs, while photoluminescence (PL) could be utilized to evaluate emission and excitation light with wavelengths ranging from 200 to 1000 nm. This section discusses a number of effective characterization approaches for evaluating CQD surface morphology, emission spectra, purity assay, and absorption, among other things.

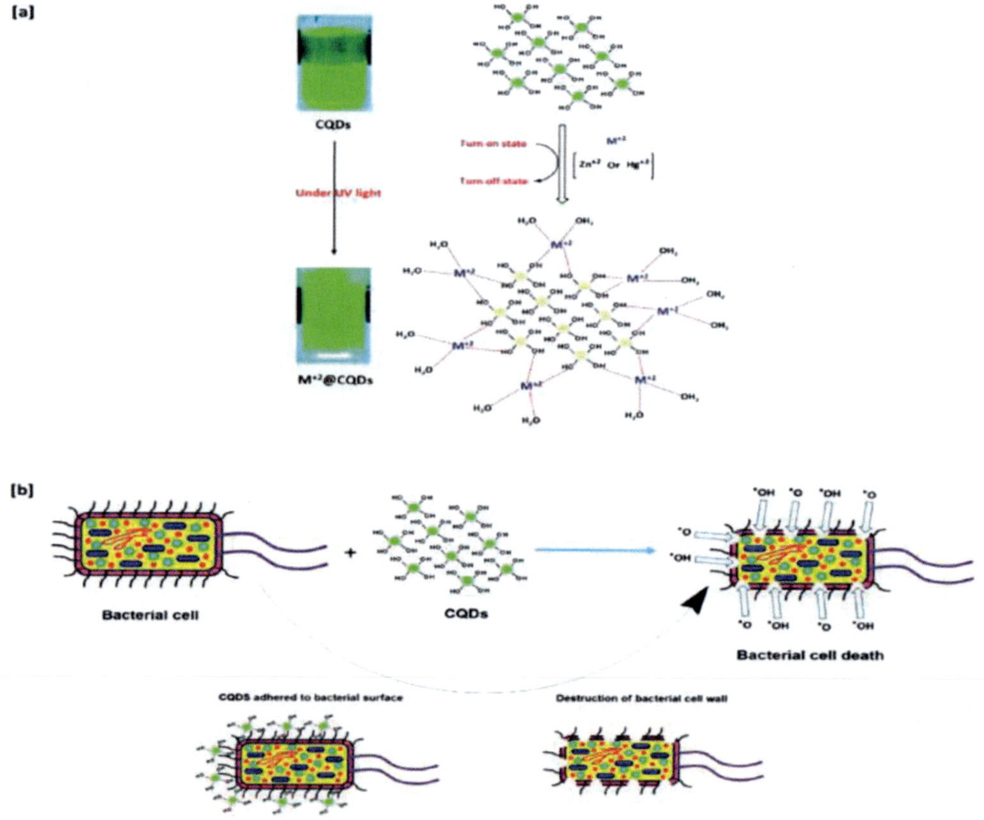

Fig. (2). Mechanism for [**a**] metal detection and [**b**] antibacterial effect of CQDs. Adapted from Ref [32]. Distributed under attribution-based CCBY license 4.0.

Fluorescence Spectroscopy

Fluorescence spectroscopy investigates a molecule's fluorescence in terms of its fluorescent features. A molecule is energized by photons, which causes it to enter an electrically exciting phase and produce fluorescence. Fluorescence spectroscopy is the process of employing a laser pulse to excite the electrons in a compound's molecules, leading them to produce light. For detection and evaluation of the molecule or modifications in the molecule, that beam is transmitted toward that filter and onto a detector. In a study, Wi´sniewski *et al.* [36] researched at nitrogen-containing carbonaceous quantum dots (N-CQDs), which are simple to produce, biocompatible, and have steady fluorescence characteristics which are resistant to vast pH variations. Following 24 hours of culture in the vicinity of 50, 100, 200, and 300 g/mL N-CQDs, the quantity of N-CQD endocytosed by cells was measured. Multiple freeze/thaw cycles were used to lyse the cell monolayer. A 360 nm excitation wavelength and a 450 nm emission wavelength were used to quantify the fluorescence intensities in cell lysates and the medium retrieved from culture plates. By analyzing changes in N-CQD fluorescence among control specimens (without cells) and post-culture media, the quantity of internalized material was estimated and subsequently normalized to the cell count. Unexpectedly, the intensity of N-CQD fluorescence in cells (as measured by micrographs or lysates) was lesser than the CQD number predicted based on N-CQD reduction in the extracellular environment. Following a 24-hour fluorescence evaluation of the extracellular milieu (measured in culture media), all cell types show a considerable drop in N-CQD concentration of more than 100 g/mL. The underestimation of endocytosed N-CQDs in the concentration range utilized cannot be explained by self-quenching or pH responsiveness. They looked at a number of drugs and conditions that affect fluorescence quenching within the intracellular milieu to better understand this manifestation. They discovered that reducing substances, such as a reduced variant of -nicotinamide adenine dinucleotide (NADH), are excellent N-CQD fluorescence quenchers (Fig. **3a**). The significant bands of surface –NH_2 and C–C functionalities occurred at 1600 and 1400 cm^{-1} along with the procedure, suggesting the formation of surface primary amines (Fig. **3b**). Also evident are distinctive bands owing to adsorbed NAD^+. However, the decrease of the –C=O signal is obscured by the 1700 cm^{-1} signal from NAD^+. The presence of multiple quenching operations running at the same time in this condition could indicate a collisional quenching event, resulting in the formation of non-fluorescent CQDs. The Perrin paradigm, as seen in the foreground of (Fig. **3a**), provides an accurate explanation of energy transmission from a donor to adjacent acceptors *via* the electron exchange channel within the sphere. Considering the redox mechanism, which involves the reduction of N-CQDs to nano-sized spheres and the oxidation of NADH see Fig. (**3b,c**). As a result, the radius of effective quenching calculated using Perrin's equation is bigger than the width predicted using HRTEM or AFM analysis, revealing that the reactive redox zone extends 5 nm beyond the N-CQD surface.

The photoluminescent (PL) spectra of the lemon-PEI CDs, while exposed at excitation wavelengths in the region of 250 to 400 nm, are illustrated below in Fig. (**4a**) [37]. When excited at 340 nm, the highest emission spectra were centered at 470 nm. Fig. (**4b**) shows a fluorescence mapping showing a limited specified area where the fluorescence is contained within a rather brief window of emission wavelengths. Regarding the usage of MoCo, this property permits fluorescence to be readily distinguished from any interferents existing inside the communication channel. Fig. (4**c**) shows the qualitative impression of the aqueous suspensions of lemon-PEI and lemon-only CQDs. Lemon-only CQDs have a wider fluorescence emission spectrum (Fig. **4d**) than PEI-treated CQDs, despite the fact that their intensity is equal. Fig. (**4e**) shows the excitation-emission layouts of aqueous suspensions of lemon-PEI CQDs and lemon-only CQDs, respectively. The inclusion of suspended particle debris in the lemon-only CQD solution causes it to resemble "cloudy," as seen in Fig. (**4f**), that could be isolated by centrifugation (up to 15,000 rpm) or filtration (0.2 m pore size).

Fourier Transform Infrared Spectroscopy (FTIR)

The functional moieties of CQDs could be affirmed by FTIR analysis. Altajer *et al.* [38] used a hydrothermal strategy to manufacture CQDs from green precursors. FT-IR spectroscopy was utilized to characterize the produced products. The functional moieties connected to the interface of CQDs are important in determining their properties. The surface functional groups have a big influence on the optical characteristics and solubility of CQDs. For the characterization of functional groups, FTIR analysis was used. The O-H functional moieties were responsible for the broad peak at 3000-3400 cm^{-1}. The stretching vibration of C=C and C=O bonds caused the mild peak at 1638 $cm^{-1,}$ as shown in Fig. (**5**) [50]. The C-C bonds might well be responsible for the weak absorption peaks between 1200 and 1500. The C-O bond can be linked to the peak at 110 cm^{-1}. According to FTIR research, a large number of COOH and O-H functional moieties adhered to the surface of the generated CQDs. The high solubility of generated CQDs could support these observations.

Velmurugan *et al.* [39] employed a fresh guava leaf extract (Psidium guajava) as a green carbon precursor in a hydrothermal procedure to make blue luminous CQDs (GCQDs). In UV light, the GCQDs emit a strong blue fluorescence having an excitation wavelength of 350 nm and an emission wavelength of 450 nm. FTIR was utilized to determine the physical composition of GCQDs (FT-IR). The absorption peaks in the FTIR spectra at 1400 and 1633 cm^{-1} could correspond to C=O bond stretching, demonstrating the inclusion of carboxyl group dots. As aliphatic hydrocarbons in the guava leaf extract contained methyl or methylene groups, considerable absorption at 2077 and 3459 cm^{-1} might be attributed to C–H stretching. O–H stretching was attributable to the low absorption at 3858 cm^{-1}. The –C = C– the extension is shown by the hint at 1636 cm^{-1}. This functional group's presence clarifies the functionalization of C-specks comprising $-CH_3$ and –OH.

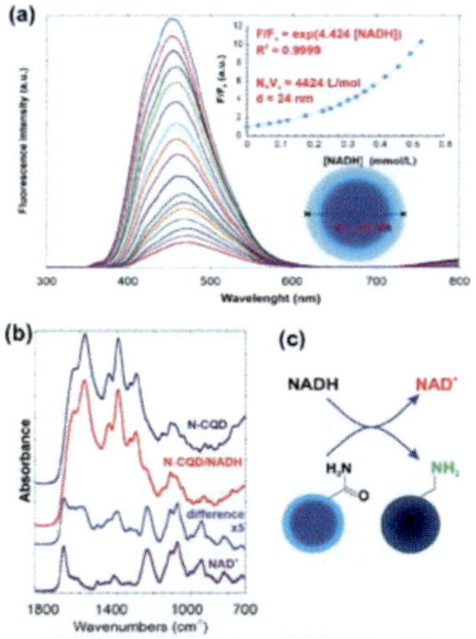

Fig. (3). (a) N-CQD fluorescence emission spectra and associated quenching by adding nicotinamide adenine dinucleotide (NADH) aliquots sequentially. Stern–Volmer graphs for quenching using varying NADH concentrations are shown in the inset. **(b)** FTIR spectrum of NAD+ and N-CQD/NADH, with the blue spectrum representing the differential; **(c)** conceptual depiction of the mechanisms underlying phenomena. Figure adapted from Ref [36]. Distributed under attribution-based CCBY license 4.0.

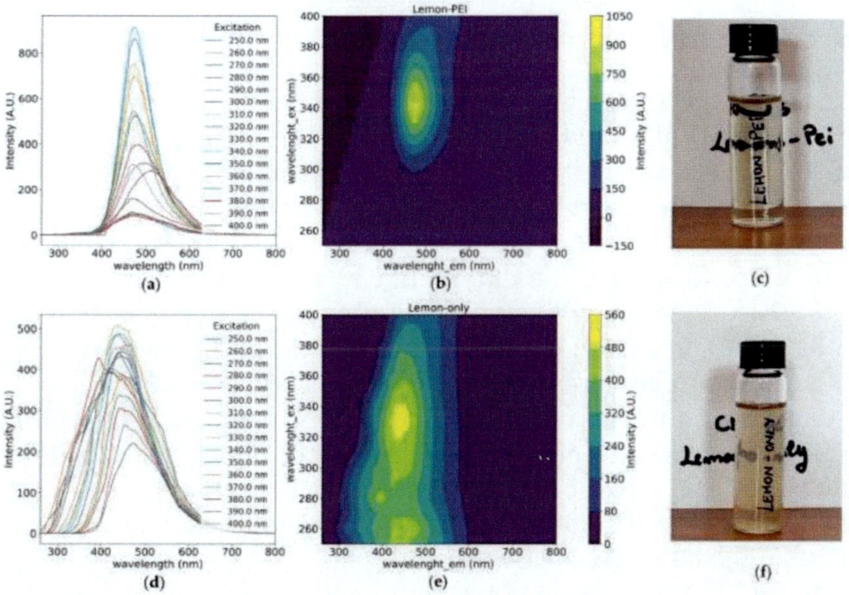

Fig. (4). (a,d) Fluorescence spectra of aqueous systems of lemon-PEI CQDs and lemon-only CQDs, procured at various excitation wavelengths as disclosed in the markings; **(b,e)** excitation-emission layouts of aqueous suspensions of lemon-PEI CQDs and lemon-only CQDs, respectively; **(c,f)** photos showing the qualitative impression of the aqueous suspensions of lemon-PEI and lemon-only CQDs. Adapted from Ref [37]. Distributed under attribution-based CCBY license 4.0.

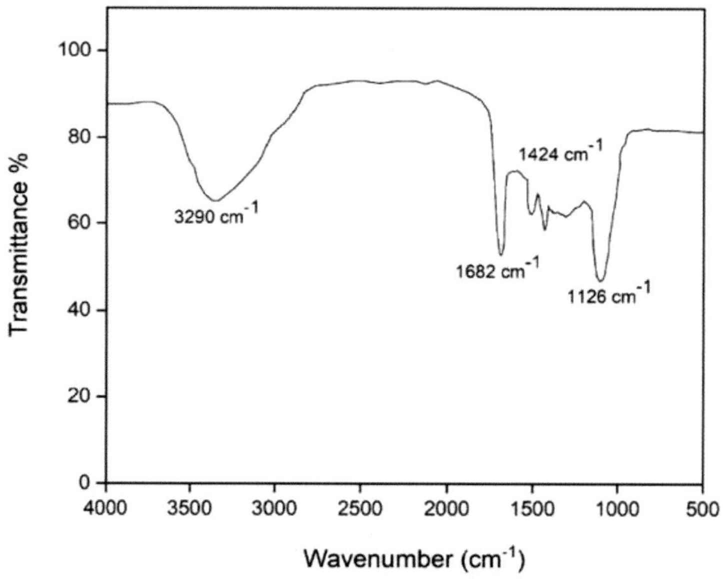

Fig. (5). FTIR spectra of CQDs that have been synthesized. Adapted from Ref [38]. Distributed under attribution-based CCBY license 4.0.

Transmission Electron Microscopy (TEM)

The nanostructures of the manufactured CQDs were investigated using TEM, as well as the sizes of CQDs down to the quantum range. The average size of CQDs is 2–12 nm. Mujica *et al.* [40] developed CQDs having size dispersion in the region of 1–5 nm. The interior composition of CQDs was studied using high-resolution TEM images. The CQDs formed by SADW possess approximately sizes of 1 to 5 nm, according to a thorough analysis of numerous RTEM pictures. The FFT of the extreme magnification images was used to calculate the interlayer lengths in the (002) and (100) crystallographic planes. The (002) distance, which corresponds to the range across graphene planes, was discovered to be around 0.34 and 0.35 nm, somewhat higher than the 0.335 nm seen in well-mannered macroscopic graphite. As per the average 002 inter-layer width and the approximate sizes of the particles, the CQDs have between 5 and 14 layers. In all of the analyzed images, the 100 range was calculated to be 0.21 nm based on the fringe pattern in the HRTEM images.

Ahmed *et al.* [32] used a basic and environmentally friendly method to manufacture biocide/fluorescent metal marker CQDs having hydrophilic properties. In a severe alkaline environment, the natural biopolymer employed in this research for the manufacture of CQDs fragments. The carbonic nanostructures were then developed and aggregated in hydrothermal treatment, which included aromatization, additional oxidation and re-polymerization. The transmission electron microscopy (TEM) micrographs demonstrate the topographical characteristics and geometric characteristics of the manufactured CQDs during hydrothermal treatment for 3 hr (Fig. **6a**), 6 hr (Fig. **6b**), and 12 hr (Fig. **6c**), from which their size dispersion was also determined and depicted. All of the as-manufactured CQDs were spherical, having a width lower than 10 nm, irrespective of the hydrothermal period. CQDs of smaller diameters and wider size dispersion were created by extending the hydrothermal process from 3 to 12 hours. CQDs synthesized at hydrothermal treatment for 3 hours, 6 hours, and 12 hours had particle size diameters of 5.1 nm (1.7–10.2 nm), 2.2 nm (1.1–3.5 nm), and 1.7 nm (1.0–2.5 nm). CQDs generated under hydrothermal treatment for 12 hours were found to be homogenous, well-distributed, and reproducibly aggregated in the reaction mixture, with relatively compact sizes. Fig. (**6d**) shows the particle size analyzer findings for the produced CQDs. Despite utilizing a different method to measure particle size (dynamic light scattering), comparable particle sizes (6.5 nm for 3 hr, 3.1 nm for 6 hr, and 1.5 nm for 12 hr) were found. Because of the prolonged hydrothermal duration of 3 h, 6 h, and 12 h, the poly-dispersity index (pdI) of the generated CQDs reduced from 1.00 to 0.75 and 0.59. This might correspond to the extended hydrothermal time allowed for greater re-polymerization and aromatization, resulting in CQDs with regulated sizes. As a result, the zetasizer results corroborated the TEM findings for sequential seeding of tiny and regulated CQDs. Finally, the results suggest that alkali hydrolyzed guar gum is capable of forming hydrophilic CQDs after hydrothermal treatment; nevertheless, running the reaction for 12 hr was favored for developing size-regulated and well-dispersed/homogeneous CQDs.

Atomic Force Microscopy (AFM)

Size, morphology, surface appearance, and roughness are just a few of the physical attributes that AFM can provide both qualitative and quantitative insight on. Statistical data such as dimension, surface area, and volumetric ratios could also be derived. From 1 nm to 8 m, a wide variety of particle sizes could be characterized in a single scan. AFM provides great resolution and visibility in three-dimensional pictures from tip motion. Cali *et al.* [37] created two types of CQDs: one with only lemon juice as a carbonaceous source (lemon-only CQDs) and another with polyethyleneimine added to the lemon juice (polyethyleneimine-only CQDs) (named lemon-PEI CQDs). It is well known that synthesizing CQDs in the vicinity of PEI improves the grade of the CQDs produced in two ways. For starters, the PEI works as an extra reservoir of nitrogen throughout the hydrothermal breakdown process, although to a smaller extent than formerly described due to stability concerns in the spectrum up to 200 °C. As a consequence, the quantum output of CQDs is improved due to the increased doping of the carbonaceous core. Furthermore, because of its polyamine composition, it functions as a passivating reagent, functionalizing the CQDs' surface. The impact of surface passivation creates emissive energy traps that facilitate electron–carbon pair interaction, culminating in higher quantum efficiency. CQDs as informational NPs for MoCo required not only distinctive optical qualities but also appropriate fluid dynamic properties to limit information dispersion during transport. As a result, investigating CQD dimensions is critical since they are linked to fluid dynamic features. In this situation, a small size dispersion normalizes self-diffusivity impacts, reducing chemical signal widening throughout communication channel travel. The magnitude of these NPs' dispersion is recognized to be dependent on the synthesis process used.

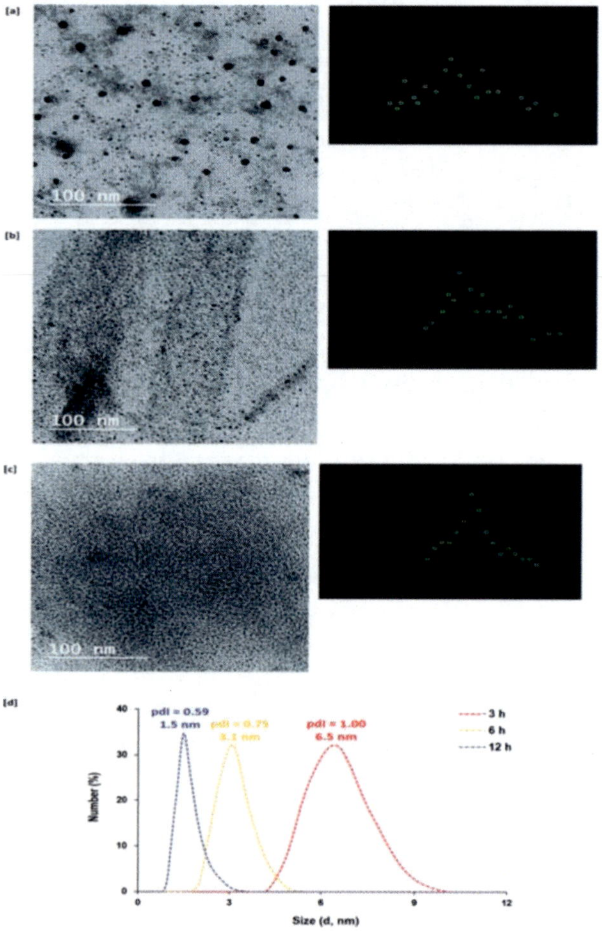

Fig. (6). (a–c) TEM micrographs of CQDs generated at three distinct hydrothermal periods: **(a)** three hours, **(b)** six hours, and **(c)** twelve hours. **(d)** Data from the Zetasizer analyzer. Adapted from Ref [32]. Distributed under attribution-based CCBY license 4.0.

Fig. (**7a**) illustrates the characteristic dispersion of lemon-PEI CQDs achieved using the chosen synthesis technique using AFM characterization. In cross-sectional form, a segment assessment was conducted by examining the relative height of each NP (example in Fig. (**7b**). A size dispersion of roughly 0.5 nm can be seen in the histogram, which is centered around 1.5 nm. In terms of ingredients and solvents, this dimensional homogeneity was achieved using a relatively simple and cheap synthesis technique. The utilization of PEI in the carbonization stage of the active ingredients in lemon juice, as well as the dialysis procedure, resulted in an effective synthesis. A size divergence of roughly 0.5 nm was achieved for lemon-only CQDs, however, this was time-centered about 2 nm (Fig. **7c, d**). However, somewhat bigger particles of greater than 3nm were found, which were most likely generated due to the lack of PEI as a capping agent.

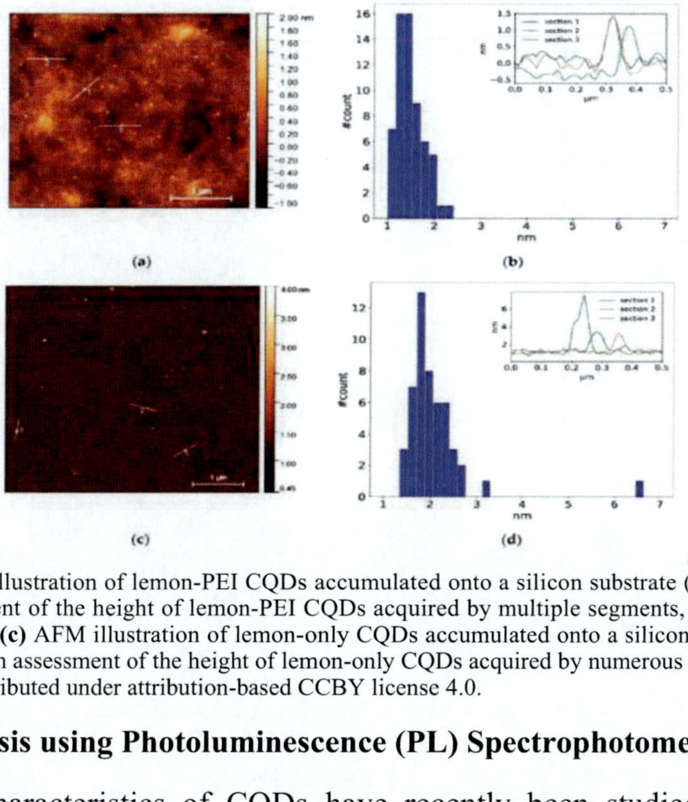

Fig. (7). (a) AFM illustration of lemon-PEI CQDs accumulated onto a silicon substrate (z scale: 0–20); **(b)** histogram assessment of the height of lemon-PEI CQDs acquired by multiple segments, three of which are shown in the inset; **(c)** AFM illustration of lemon-only CQDs accumulated onto a silicon substrate (z scale: 0–20); **(d)** histogram assessment of the height of lemon-only CQDs acquired by numerous segments. Adapted from Ref [37]. Distributed under attribution-based CCBY license 4.0.

Optical Analysis using Photoluminescence (PL) Spectrophotometer

The optical characteristics of CQDs have recently been studied using a PL spectrophotometer. With a growing emission peak between 300 to 480 nm and a progressive reduction in intensity, the excitation wavelengths could be changed. Altajer *et al.* [38] used a hydrothermal approach to manufacture carbonaceous quantum dots from green precursors. Photoluminescence (PL) spectroscopy was used to characterize the produced products. 1 g soy flour was disseminated in 50 mL deionized H_2O for 3 hours with vigorous shaking. The resulting mixture was then introduced in a stainless-steel autoclave and heated to 200°C for 10 hours. To eliminate any insoluble elements, the dark brown specimen was centrifuged at 10000 rpm for 40 minutes. For 20 days, the liquid was maintained at 4°C and centrifuged each day to eliminate residing sediments. For additional characterization and sensor experiments, the resultant carbonaceous quantum dots were stored at 4°C. The PL spectrum produced by activating the specimen at various excitation levels is shown in Fig. (**8**). (360, 400, and 440 nm). Excitation-dependent fluorescence intensity is observed in the CQDs that have been synthesized. It is essential for the development of antibodies and iron absorption, as well as a range of other critical human processes. As previously stated, the CQDs surface functional groups are particularly efficient at generating photoluminescence features. The PL intensity of CQDs was mildly reduced when Cr(VI) was introduced into the pre-prepared CQDs mixture, which could be attributed to the interactions between (Cr(VI) ions and surface functional groups (OH and COOH). This could be performed as a sophisticated configuration. It is self-evident that raising Cr(VI) ion concentration reduces PL intensity. With the addition of ascorbic acid, the procedure can be reverted. At mild reaction conditions, ascorbic acid may convert Cr(VI) to Cr(III), resulting in the eradication of the Cr(VI)-CQDs interaction. As a result, the PL intensity of CQDs begins to proceed once more.

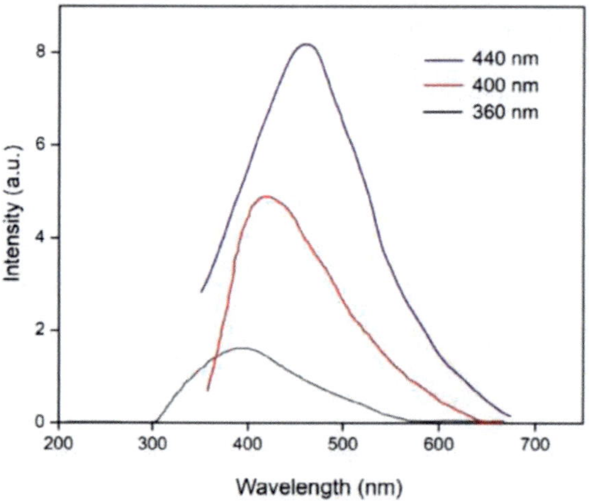

Fig. (8). The photoluminescence spectra of CQDs generated by excitation of the sample at various levels. Adapted with permission from Ref [38]. Distributed under attribution-based CCBY license 4.0.

Zou *et al.* [41] thoroughly investigated the luminescent characteristics of CQDs@Ln (TFA)3 at ambient temperature. CQDs@Eu $(TFA)_3$ (a) and CQDs@Tb $(TFA)^3$ (b) emission spectra were acquired primarily to highlight the matching distinctive peaks in (Fig. **9**). The emission spectrum of CQDs@Eu $(TFA)_3$ Fig. (**9a**) reveals four strong peaks at 580, 591, 613, and 652 nm, that correspond to Eu^{3+} 5D07FJ (J = 0–3) transitions, accordingly. As could be observed, the powerful 5D07 F2 transition at 613 nm dominates the emission spectra of CQDs@Eu $(TFA)_3$, whereas the peak centered at 544 nm arises from Tb^{3+} 5D47F5 in CQDs@Tb $(TFA)_3$ as shown in Fig. (**9 b**). CQDs are responsible for the peak in the 380–500 nm range, and the emission band of CQDs in the CQDs@Ln $(TFA)_3$ is concentrated at around 435 nm. The dual-emitting fluorescent composite not only represents a shift from "one" to "two," but it also improves CQD proportional monitoring layout and increases related applications.

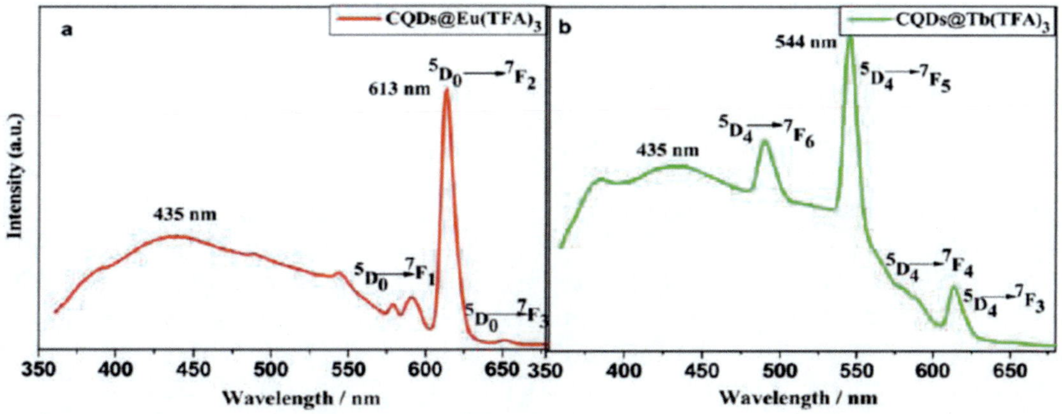

Fig. (9). The emission spectra of CQDs@Eu(TFA)3 **(a)** and CQDs@Tb(TFA)3 **(b)**. Adapted from Ref [41]. Distributed under attribution-based CCBY license 4.0.

MnO_4 is a disinfectant that is extensively employed in agriculture and has a significant function in health and medicine. MnO_4 in large concentrations, on the other hand, increases the environmental hazard. As a result, identifying these solutions and ions is critical in everyday life. The quenching ability of CQDs@Tb $(TFA)_3$ was tested in the influence of multiple anion ions with a quantity of 103 M at 365 nm excitation wavelength to detect the specific fluorescence sensing of anion ions. Fig. (**10**) illustrates the PL emission spectra of CQDs@Tb $(TFA)_3$ in the vicinity of several anion ions such as NO_3, I, Br, Cl, CrO4 2, and MnO_4 at 365 nm, as well as the associated anion bar graph (ICQDs/ITb). The luminous intensity of Tb^{3+} characterization could help to elucidate the chemical-sensing impact. The green emission is reduced with various anions, and the peak centered at 545 nm is attributed to the 5D47F5 of Tb^{3+} ions (Fig. **10**). Moreover, it is clear that MnO_4 significantly reduces the intensity. When the fluorescence intensity ratio (ICQDs/ITb) of the dual emission peaks of each anion was analyzed, it was found that the value of ICQDs/ITb was minimal when MnO_4 was involved. As a result, the CQDs@Tb $(TFA)_3$ fluorescence sensor was adequate for MnO_4 over the other anion ions, as demonstrated by the results.

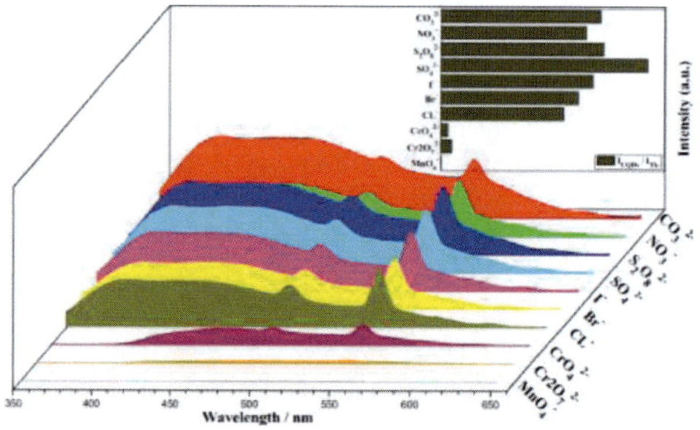

Fig. (10). CQDs@Tb $(TFA)_3$ PL emission spectrum in the influence of several anions (ICQDs/ITb histogram). Adapted from Ref [41]. Distributed under attribution-based CCBY license 4.0.

Light Absorption by UV-Vis Spectrophotometer

UV–Vis light absorption investigations could be used to evaluate the linear optical absorption characteristic of CQDs. For instance, Teymoorian *et al.* [42] synthesized non-metal elements for doping CQDs using nitrogen and sulphur (N, S-CQDs), which were then put within hexagonal mesoporous silicon (HMS) to eliminate methylene blue dye (MB) out of an aqueous phase. The hydrothermal approach was used to manufacture N, S-CQDs utilizing cysteine amino acids and histidine as the supply, as depicted in Fig. (**11**). Various microscopy and spectroscopic approaches were used to characterize the morphology and composition of the N, S-CQDs, and adsorbent (N, S-CQDs/HMS). The effects of the amount of adsorbent (0.25–1 g/L), pH (2–10), interaction period (15–75 min), and baseline MB dye intensity (20–300 mg/L) on adsorption were studied. Under ideal circumstances at 303 K, the maximal adsorption capability and elimination effectiveness of MB were estimated to be 370.4 mg/g and 97 percent, correspondingly.

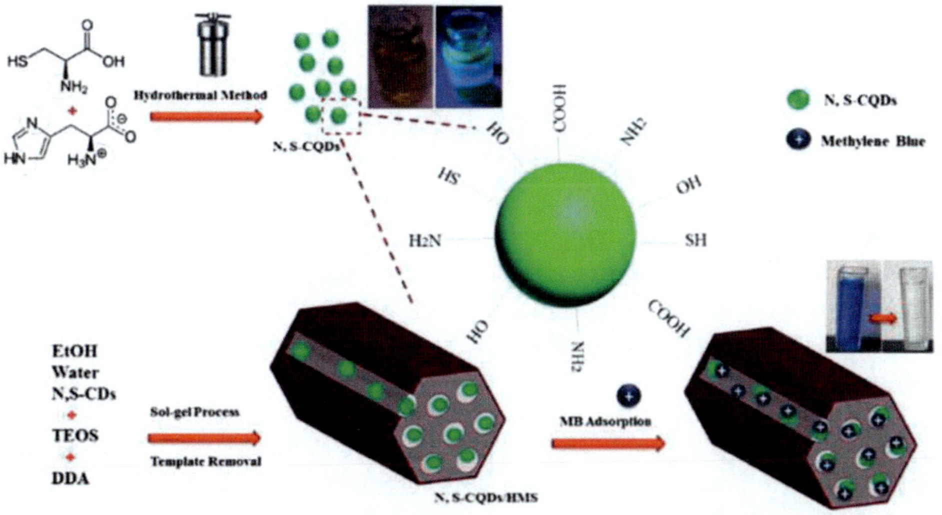

Fig. (11). N, S-CQDs/HMS synthesis routes and Methylene blue dye adsorption. Adapted from Ref [42]. Distributed under attribution-based CCBY license 4.0.

UV-Vis and photoluminescence (PL) spectroscopy were utilized to explore the optical characteristics of N, N-CQDs and S-CQDs. The UV–Vis absorption of CQD solutions is depicted in Fig. (**12**) [42]. The π-π* transition of electrons in the conjugated C=C band has a particular absorption peak at 268 nm, whereas the n-π* transition of C=N, C–N, C=S, and C=O bonds has an adsorption peak at 300–350 nm.

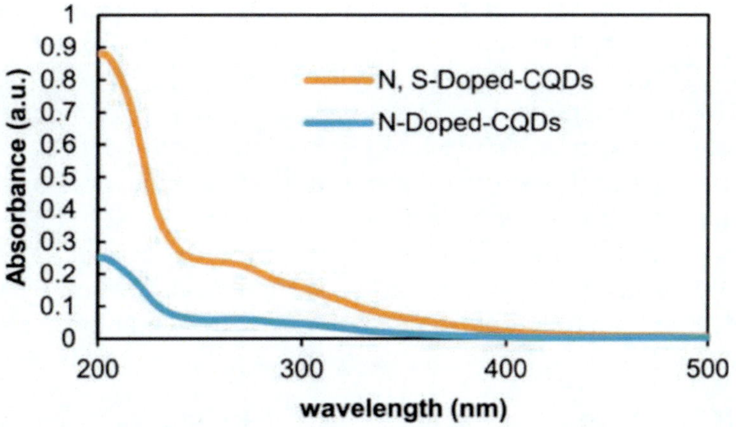

Fig. (12). UV–Vis absorption spectrum of N-CQDs and N, S-CQDs. Adapted from Ref [42]. Distributed under attribution-based CCBY license 4.0.

Marouzi *et al.* [43] used a hydrothermal method to explore the production of the nutrient Salvia hispanica L. (Chia) as a future prospective carbon source, along with the biological dispersion and toxicity of CQDs. The generated CQDs were examined using a variety of physicochemical techniques, and they have been found to have a mean dimension of 5.4 nm, a spherical morphology, acceptable fluorescence, and a quantum yield (QY) of 17.8% in ideal conditions (180°C for 4 h). The optimum absorption of CQDs produced *via* ultraviolet light was noticed at 280 nm owing to the transmitting of π-π* and the existence of CC (aromatic carbon) functional moieties, and the presence of n-π* transmits that were affiliated with the CO (carbonyl) bonds induced in the pattern of a slight shoulder among the spectrum of 350–370 nm, that suggests the creation of CQDs, as per the executed electronic excitation. The CQDs were found to have substantial optical absorption in the ultraviolet (200–400 nm) region, with a tail that prolonged into the visual spectrum (600 nm). The achieved dimension and composition of quantum nanoparticles' surface culminated in a distinctive property recognized as fluorescence as a response to the synthesizing process used. Thus, it could be believed that the strength and dispersion of fluorescent rays are based on the wavelength of excitation.

Powder X-Ray Diffraction Analysis (XRD)

X-ray powder diffraction (XRD) is a rapid analytical technique that could provide insight into unit cell measurements and is mostly utilized for the phase determination of crystalline materials. The research material is homogenized and ground into powder, and the mass content is determined on an average basis. Rajamanikandan *et al.* [44] used a simple hydrothermal procedure to manufacture CQDs using agro-waste Ananas comosus. CQD fluorescence characteristics reveal significant blue emission irradiation with a QY of 10.65%. CQDs were tested for DPPH (2, 2-diphenyl-1-picrylhydrazyl), Superoxide anion radical scavenging, hydroxyl radical scavenging, and hydrogen peroxide radical scavenging properties in an antioxidant experiment. Powder X-ray diffraction (PXRD) was utilized to investigate the crystallinity of pineapple peel CQDs. The CQDs' XRD pattern shows a large diffraction peak at $2\theta = 22°$ to 25°, indicating that the CQDs are unorganized. The aspect relates to the crystal lattice (002), that is an important characteristic of carbon-based materials. Carbon dots' amorphous character is also demonstrated by the quantity of nitrogen and oxygen-containing groups on them.

Rezaei *et al.* [45] sonicated for 5 minutes after dissolving chitosan and Arginine-CQD in 200 mM acetate buffer (pH = 5.5). The solution was then sterilized using a 0.22 mm filter. The carrier/pDNA complex was created by mixing carrier with 200 ng pDNA at various mass ratios (carrier/pDNA, W/W) ranging from 30:1 to

70:1, as the whole synthesis has been presented in Fig. (**13**). At ambient temperature, the mixture was fermented for 30 minutes. The complexes were electrophoresed on a 1 percent agarose gel at 100 V for 60 minutes in 0.5 TAE (TrisacetateEDTA) buffer after 2 liters of loading buffer were introduced to 8 liters of the mixture. Furthermore, after hydrothermal carbonization, the crystalline peak of chitosan at 19.96° in the XRD spectra was decreased, and a wide peak at $2\theta = 23$ ascribed to the amorphous graphitic-carbon phase of the tiny size specimen emerged.

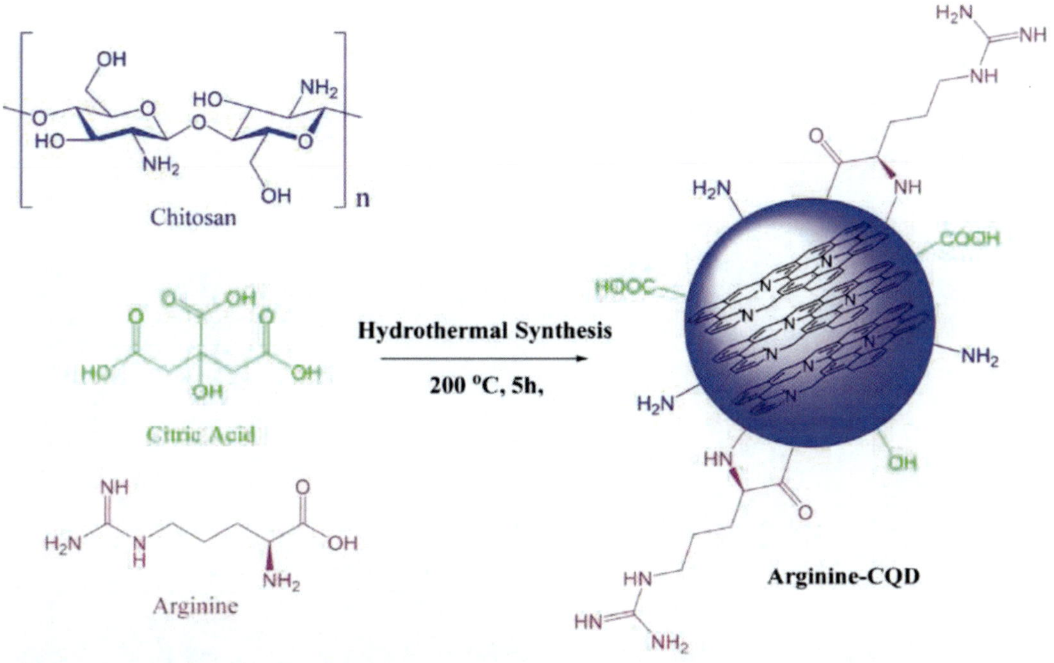

Fig. (13). The Arginine-CQDs are synthesized hydrothermally. Adapted from Ref [45]. Distributed under attribution-based CCBY license 4.0.

Charges on the surface of nanoparticles are determined *via* zeta potential analysis, which is advantageous for DNA condensation. In this context, zeta potential is used to analyze bare chitosan, DNA and Arginine-CQDs (Fig. **14b**). Due to the elevated density of amine groups in the chitosan precursors, it has a positive charge. As a result, Arginine-CQDs produced from chitosan have a positive charge as well. In addition, arginine is involved in the hydrothermal pyrolysis of citric acid and chitosan to form a graphitic core and passivate the surface of CQDs. Fig. (**14a**) shows the arginine-CQDs XRD patterns, while Fig. (**14d**) depicts the chitosan gel retardation analysis, and Fig. (**14e**) depicts the arginine-CQDs gel retardation test, at several weight ratios. As a result, the suggested nanocarrier has a positive charge, making it easier to build stable complexes with negatively charged DNA. Rapid release of the nanocarrier/DNA complex within endosomes is a crucial phase for effective gene delivery, which is regulated by the nanovector's buffering capability. According to the proton sponge hypothesis, if indeed the nanocarrier has the ability to bind protons and buffer lysosome medium, the lysosomal breakdown is avoided by complex enlargement and vesicle rupture into the cell. The buffering capability of chitosan and Arginine-CQDs is determined by titration using 0.01 N HCl or NaOH in this case. The buffering capability of substances in this region could aid to burst endosomal vesicles by osmosis and swelling even though the pH of the endosomal media is in the region of 5.5 to 7. The buffering capability of chitosan and Arginine-CQDs is shown in Fig. (**14c**). Chitosan amine groups possess a pKa value of roughly 6, and the results show that chitosan has a limited buffering capability.

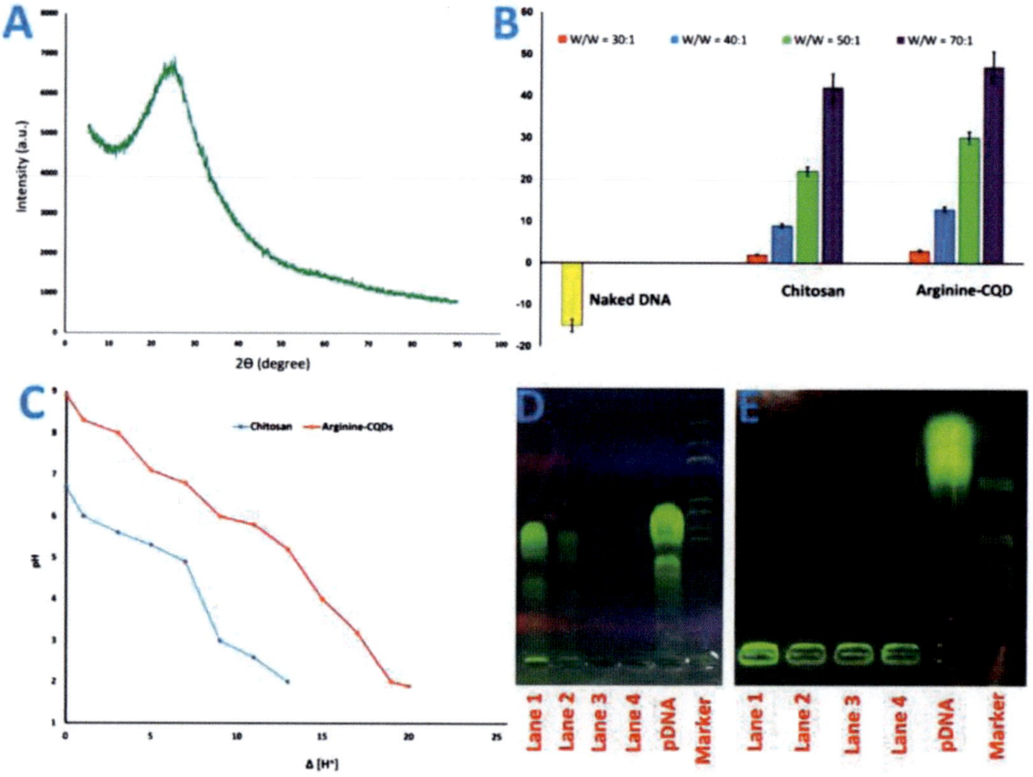

Fig. (14). **(a)** Arginine-CQDs XRD patterns, **(b)** bare DNA, chitosan, and Arginine-CQDs zeta potential, **(c)** chitosan and Arginine-CQDs buffering capability, **(d)** chitosan gel retardation analysis, **(e)** Arginine-CQDs gel retardation test, at several weight ratios. Lane 1 has a W/W of 30; Lane 2 has a W/W of 40; Lane 3 has a W/W of 50; Lane 4 has a W/W of 70; Lane 5 has pDNA; Lane 6 has a DNA marker. Adapted from Ref [45]. Distributed under attribution-based CCBY license 4.0.

Raman Spectroscopy

Raman scattering is a vibrational molecular spectroscopy technique, which may be used to investigate alterations in the molecular structure of an element and can also be used to develop CQDs. Shen *et al.* [31] devised a simple one-step hydrothermal of m-phenylenediamine and glucose to produce nitrogen-doped CQDs (N-CQDs). Raman spectroscopy is used to determine the geometry and crystalline structure of N-CQDs. To validate the crystalline structure and volume of the N-CQDs, a Raman spectrum was initially acquired. The spectrum shows the typical Raman bands at 1357 and 1565 cm^{-1}, which are commonly referred to as D and G bands, correspondingly. The D band is concerned with crystalline size and flaws, while the G band is concerned with the sample's graphite phase. The lack of 2D band and the significant amplitude of G band above D band show that the CQD crystallized successfully in the graphene framework without stacking. The ratio ID/IG is proportional to 1/La (ID symbolizes D band intensity, IG indicates G band intensity, and La indicates crystalline lateral size), and the crystalline lateral size estimated using this relation is ca. 8 nanometers.

CHALLENGES AND FUTURE OUTLOOKS

CQDs have been utilized in a variety of applications in the last three years, covering detectors, cell imaging, drug/gene delivery, Semiconductors, and photocatalysis. CQDs are used in photocatalysis and solar energy transmission because of their capacity to sensitize, operate as an electron sink, reduce electron-void pair combination, generate electron-void pairs, and absorb sunlight over a wide spectrum range. CQDs have various benefits over their heavy metal-containing cousins and could be utilized as a substitute for semiconductor QDs. The optical characteristics of CQDs are greatly influenced by the synthesis technique, functionalization, and doping with heteroatoms. Except in a few rare circumstances, CQDs possess low QY values, necessitating additional study to address this scarcity. The influence of varied synthesis parameters, synthesis methods, and doping on the photocatalytic capabilities of CQDs must be investigated. CQDs have the capacity to absorb light across the entire spectrum, from UV to IR. Still, there are still numerous obstacles to overcome. The most important difficulties are **(1)** thoroughly comprehending and managing the functioning of CQDs, and **(2)** developing and synthesizing CQDs utilizing traditional methods with low-cost precursors with high yields and a range of functions for a number of applications. It would be too unclear to estimate when CQD will be commercialized. The factors that enhance the CQDs' light absorption are still unknown. The impact of width and dimension on the optical characteristics of CQDs is still unknown. The photocatalytic efficiency of photocatalyst substances is influenced by charge separation and charge transport.

It is necessary to explore the effects of size, functionalization, manufacturing parameters and doping on charge separation and transport properties. It appears that by addressing these ambiguous places in the evolution of CQDs, these new nanostructures will be able to find a marketplace for use in cost-effective applications.

CONCLUSION

Owing to its distinctive and customizable physicochemical, photonic, and electrical characteristics, there has been significant research on the manufacturing, alterations, characteristics, and uses of CQDs since their finding in 2004. It's encouraging to see how CQD research is shifting away from conventional domains like biological imaging, bio-sensing, and medication delivery and toward alternative energy storage and conversion disciplines like synthetic photocatalysis. As a result, we've sought to provide a broad overview of current breakthroughs in CQD characterization techniques, concentrating on their synthesis, characteristics, and various applications. Varied basic, less expensive, size-controllable, and large-scale strategies for the production of CQDs with several components and topologies have been discovered too far. CQDs, specifically, are capable of acting as electron mediators, photosensitizers, spectral converters, and single photocatalysts in photocatalysis. Despite the premise that several concerns related to CQDs are still up in the air, a plethora of exciting opportunities and challenges await us in the nearest term as we investigate the terrific possibilities of CQDs in a variety of technical applications. In order to improve the rapidly developing characterization approaches for CQDs, it is anticipated that this chapter will offer more details on their different functions. To validate the existence of CQDs, scientists generally depend on morphology, optical properties, and compositional elements, which is insufficient and inaccurate, particularly since the finding of graphene quantum dots, therefore, a description that can identify CQDs from other substances is desperately required. Because of their superior crystal organization and surface characteristics, CQD nanocomposites often helped to reduce charge transfer paths and maintain the cycle durability of electrode materials. Further investigation is anticipated in building cost-effective and environmentally acceptable nanocomposites utilizing CQDs for supplying high energy density and stable electrodes for a broad array of applications, bearing in mind the above-mentioned strengths and limits.

REFERENCES

[1] Nair, A.; Haponiuk, J.T.; Thomas, S.; Gopi, S. Natural carbon-based quantum dots and their applications in drug delivery: A review. *Biomed. Pharmacother.,* **2020**, *132*, 110834. [http://dx.doi.org/10.1016/j.biopha.2020.110834] [PMID: 33035830]

[2] Surendran, P.; Lakshmanan, A.; Priya, S.S.; Balakrishnan, K.; Rameshkumar, P.; Kannan, K.; Geetha, P.; Hegde, T.A.; Vinitha, G. Bioinspired fluorescence carbon quantum dots extracted from natural

honey: Efficient material for photonic and antibacterial applications. *Nano-Structures & Nano-Objects,* **2020**, *24*, 100589.
[http://dx.doi.org/10.1016/j.nanoso.2020.100589]

[3] Iravani, S.; Varma, R.S. Green synthesis, biomedical and biotechnological applications of carbon and graphene quantum dots. A review. *Environ. Chem. Lett.,* **2020**, *18*(3), 703-727.
[http://dx.doi.org/10.1007/s10311-020-00984-0] [PMID: 32206050]

[4] Saikia, M.; Das, T.; Dihingia, N.; Fan, X.; Silva, L.F.O.; Saikia, B.K. Formation of carbon quantum dots and graphene nanosheets from different abundant carbonaceous materials. *Diamond Related Materials,* **2020**, *106*, 107813.
[http://dx.doi.org/10.1016/j.diamond.2020.107813]

[5] Tan, C.; Zuo, S.; Zhao, Y.; Shen, B. Preparation of multicolored carbon quantum dots using HNO_3 /$HClO_4$ oxidation of graphitized carbon. *J. Mater. Res.,* **2019**, *34*(20), 3428-3438.
[http://dx.doi.org/10.1557/jmr.2019.261]

[6] Shokry, A.; Khalil, M.M.A.; Ibrahim, H.; Soliman, M.; Ebrahim, S. Highly Luminescent Ternary Nanocomposite of Polyaniline, Silver Nanoparticles and Graphene Oxide Quantum Dots. *Sci. Rep.,* **2019**, *9*(1), 16984.
[http://dx.doi.org/10.1038/s41598-019-53584-6] [PMID: 31740719]

[7] Serhan, M.; Sprowls, M.; Jackemeyer, D.; Long, M.; Perez, I.D.; Maret, W.; Tao, N.; Forzani, E. Total iron measurement in human serum with a smartphone. *AIChE Annu. Meet. Conf. Proc.,* **2019**.

[8] Bashir, S.; Thakur, A.; Lgaz, H.; Chung, I.M.; Kumar, A. Corrosion inhibition efficiency of bronopol on aluminium in 0.5 M HCl solution: Insights from experimental and quantum chemical studies. *Surf. Interfaces,* **2020**, *20*, 100542.
[http://dx.doi.org/10.1016/j.surfin.2020.100542]

[9] Bashir, S.; Lgaz, H.; Chung, I.M.; Kumar, A. Effective green corrosion inhibition of aluminium using analgin in acidic medium: an experimental and theoretical study. *Chem. Eng. Commun.,* **2020**, *0*, 1-10.
[http://dx.doi.org/10.1080/00986445.2020.1752680]

[10] Bashir, S.; Thakur, A.; Lgaz, H.; Chung, I.M.; Kumar, A. Corrosion inhibition performance of acarbose on mild steel corrosion in acidic medium: An experimental and computational study. *Arab. J. Sci. Eng.,* **2020**, *45*(6), 4773-4783.
[http://dx.doi.org/10.1007/s13369-020-04514-6]

[11] Parveen, G.; Bashir, S.; Thakur, A.; Saha, S.K.; Banerjee, P.; Kumar, A. Experimental and computational studies of imidazolium based ionic liquid 1-methyl- 3-propylimidazolium iodide on mild steel corrosion in acidic solution. *Mater. Res. Express,* **2020**, *7*(1), 016510.
[http://dx.doi.org/10.1088/2053-1591/ab5c6a]

[12] Bashir, S.; Thakur, A.; Lgaz, H.; Chung, I.M.; Kumar, A. Computational and experimental studies on phenylephrine as anti-corrosion substance of mild steel in acidic medium. *J. Mol. Liq.,* **2019**, *293*, 111539.
[http://dx.doi.org/10.1016/j.molliq.2019.111539]

[13] Thakur, A.; Kumar, A. Recent trends in nanostructured carbon-based electrochemical sensors for the detection and remediation of persistent toxic substances in real-time analysis. *Mater. Res. Express,* **2023**, *10*(3), 034001.
[http://dx.doi.org/10.1088/2053-1591/acbd1a]

[14] Thakur, A.; Kaya, S.; Kumar, A. Recent trends in the characterization and application progress of nano-modified coatings in corrosion mitigation of metals and alloys. *Appl. Sci.,* **2023**, *13*(2), 730.
[http://dx.doi.org/10.3390/app13020730]

[15] Thakur, A.; Kaya, S.; Kumar, A. Recent innovations in nano container-based self-healing coatings in the construction industry. *Curr. Nanosci.,* **2022**, *18*(2), 203-216.
[http://dx.doi.org/10.2174/1573413717666210216120741]

[16] Thakur, A.; Kumar, A.; Kaya, S.; Marzouki, R.; Zhang, F.; Guo, L. Recent advancements in surface modification, characterization and functionalization for enhancing the biocompatibility and corrosion resistance of biomedical implants. *Coatings,* **2022**, *12*(10), 1459. Available from:https://doi.org/https://doi.org/10.3390/coatings12101459

[17] Thakur, A.; Kaya, S.; Abousalem, A.S.; Kumar, A.; Experimental, D.F.T. Experimental, DFT and MC simulation analysis of vicia sativa weed aerial extract as sustainable and eco-benign corrosion inhibitor for mild steel in acidic environment. *Sustain. Chem. Pharm.,* **2022**, *29*, 100785. [http://dx.doi.org/10.1016/j.scp.2022.100785]

[18] Thakur, A.; Kumar, A. Recent advances on rapid detection and remediation of environmental pollutants utilizing nanomaterials-based (bio)sensors. *Sci. Total Environ.,* **2022**, *834*, 155219. [http://dx.doi.org/10.1016/j.scitotenv.2022.155219] [PMID: 35421493]

[19] Thakur, A.; Kumar, A.; Sharma, S.; Ganjoo, R.; Assad, H. Computational and experimental studies on the efficiency of Sonchus arvensis as green corrosion inhibitor for mild steel in 0.5 M HCl solution. *Mater. Today Proc.,* **2022**, *66*, 609-621. [http://dx.doi.org/10.1016/j.matpr.2022.06.479]

[20] Thakur, A.; Sharma, S.; Ganjoo, R.; Assad, H.; Kumar, A. Anti-corrosive potential of the sustainable corrosion inhibitors based on biomass waste: A review on preceding and perspective research. *J. Phys. Conf. Ser.,* **2022**, *2267*(1), 012079. [http://dx.doi.org/10.1088/1742-6596/2267/1/012079]

[21] Thakur, A.; Kaya, S.; Abousalem, A.S.; Sharma, S.; Ganjoo, R.; Assad, H.; Kumar, A. Computational and experimental studies on the corrosion inhibition performance of an aerial extract of Cnicus Benedictus weed on the acidic corrosion of mild steel. *Process Saf. Environ. Prot.,* **2022**, *161*, 801-818. [http://dx.doi.org/10.1016/j.psep.2022.03.082]

[22] Kumar, A.; Thakur, A. Encapsulated nanoparticles in organic polymers for corrosion inhibition. In: *Corrosion Protection at the Nanoscale*; Elsevier Inc., **2020**; pp. 345-362. [http://dx.doi.org/10.1016/B978-0-12-819359-4.00018-0]

[23] Thakur, A.; Kumar, A.; Kaya, S.; Vo, D.V.N.; Sharma, A. Suppressing inhibitory compounds by nanomaterials for highly efficient biofuel production: A review. *Fuel,* **2022**, *312*, 122934. [http://dx.doi.org/10.1016/j.fuel.2021.122934]

[24] Thakur, A.; Kumar, A. Sustainable inhibitors for corrosion mitigation in aggressive corrosive media: A comprehensive study. *J Bio Tribo Corros,* **2021**, *7*, 67. [http://dx.doi.org/10.1007/s40735-021-00501-y]

[25] Molaei, M.J. Carbon quantum dots and their biomedical and therapeutic applications: A review. *RSC Advances,* **2019**, *9*(12), 6460-6481. [http://dx.doi.org/10.1039/C8RA08088G] [PMID: 35518468]

[26] Mintz, K.J.; Zhou, Y.; Leblanc, R.M. Recent development of carbon quantum dots regarding their optical properties, photoluminescence mechanism, and core structure. *Nanoscale,* **2019**, *11*(11), 4634-4652. [http://dx.doi.org/10.1039/C8NR10059D] [PMID: 30834912]

[27] Khan, Z.M.S.H.; Rahman, R.S.; Shumaila, S.; Islam, S.; Zulfequar, M. Hydrothermal treatment of red lentils for the synthesis of fluorescent carbon quantum dots and its application for sensing Fe3+. *Opt. Mater.,* **2019**, *91*, 386-395. [http://dx.doi.org/10.1016/j.optmat.2019.03.054]

[28] Vercelli, B. The role of carbon quantum dots in organic photovoltaics: A short overview. *Coatings,* **2021**, *11*(2), 232. [http://dx.doi.org/10.3390/coatings11020232]

[29] El-Shabasy, R.M.; Farouk Elsadek, M.; Mohamed Ahmed, B.; Fawzy Farahat, M.; Mosleh, K.N.;

Taher, M.M. Recent developments in carbon quantum dots: Properties, fabrication techniques, and bio-applications. *Processes,* **2021**, *9*(2), 388.
[http://dx.doi.org/10.3390/pr9020388]

[30] Mehmandoust, M.; Erk, N.; Karaman, O.; Karimi, F.; Bijad, M.; Karaman, C. Three-dimensional porous reduced graphene oxide decorated with carbon quantum dots and platinum nanoparticles for highly selective determination of azo dye compound tartrazine. *Food Chem. Toxicol.,* **2021**, *158*, 112698.
[http://dx.doi.org/10.1016/j.fct.2021.112698] [PMID: 34838678]

[31] Shen, T.Y.; Jia, P.Y.; Chen, D.S.; Wang, L.N. Hydrothermal synthesis of N-doped carbon quantum dots and their application in ion-detection and cell-imaging. *Spectrochim. Acta A Mol. Biomol. Spectrosc.,* **2021**, *248*, 119282.
[http://dx.doi.org/10.1016/j.saa.2020.119282] [PMID: 33316652]

[32] Ahmed, H.B.; Emam, H.E. Environmentally exploitable biocide/fluorescent metal marker carbon quantum dots. *RSC Advances,* **2020**, *10*(70), 42916-42929.
[http://dx.doi.org/10.1039/D0RA06383E] [PMID: 35514886]

[33] Zhao, D.L.; Chung, T.S. Applications of carbon quantum dots (CQDs) in membrane technologies: A review. *Water Res.,* **2018**, *147*, 43-49.
[http://dx.doi.org/10.1016/j.watres.2018.09.040] [PMID: 30296608]

[34] Haque, E.; Kim, J.; Malgras, V.; Reddy, K.R.; Ward, A.C.; You, J.; Bando, Y.; Hossain, M.S.A.; Yamauchi, Y. Recent advances in graphene quantum dots: Synthesis, properties, and applications. *Small Methods,* **2018**, *2*(10), 1800050.
[http://dx.doi.org/10.1002/smtd.201800050]

[35] Kaur, M.; Kaur, M.; Sharma, V.K. Nitrogen-doped graphene and graphene quantum dots: A review onsynthesis and applications in energy, sensors and environment. *Adv. Colloid Interface Sci.,* **2018**, *259*, 44-64.
[http://dx.doi.org/10.1016/j.cis.2018.07.001] [PMID: 30032930]

[36] Wiśniewski, M.; Czarnecka, J.; Bolibok, P.; Świdziński, M.; Roszek, K. New insight into the fluorescence quenching of nitrogen-containing carbonaceous quantum dots—From surface chemistry to biomedical applications. *Materials,* **2021**, *14*(9), 2454.
[http://dx.doi.org/10.3390/ma14092454] [PMID: 34065161]

[37] Calì, F.; Cantaro, V.; Fichera, L.; Ruffino, R.; Trusso Sfrazzetto, G.; Li-Destri, G.; Tuccitto, N. Carbon quantum dots from lemon waste enable communication among biodevices. *Chemosensors,* **2021**, *9*(8), 202.
[http://dx.doi.org/10.3390/chemosensors9080202]

[38] Altajer, A.H.; Tanjung, F.A.; Jabbar, A.H.; Mezan, S.O.; Ihangavelu, L.; Kadhim, M.M.; Alkaim, A.R. Novel carbon quantum dots: Green and facile synthesis, characterization and its application in on-of--on fluorescent probes for ascorbic acid. *J. Nanostructures.,* **2021**, *11*, 236-242.
[http://dx.doi.org/10.22052/JNS.2021.02.004]

[39] Velmurugan, P.; Kumar, R.V.; Sivakumar, S.; Ravi, A.V. *Fabrication of blue fluorescent carbon quantum dots using green carbon precursor psidium guajava leaf extract and its application in water treatment*; *Carbon Lett*; Springer, **2021**, 32, pp. 119-129.
[http://dx.doi.org/10.1007/s42823-021-00259-w]

[40] Chao-Mujica, F.J.; Garcia-Hernández, L.; Camacho-López, S.; Camacho-López, M.; Camacho-López, M.A.; Reyes Contreras, D.; Pérez-Rodríguez, A.; Peña-Caravaca, J.P.; Páez-Rodríguez, A.; Darias-Gonzalez, J.G.; Hernandez-Tabares, L.; Arias de Fuentes, O.; Prokhorov, E.; Torres-Figueredo, N.; Reguera, E.; Desdin-García, L.F. Carbon quantum dots by submerged arc discharge in water: Synthesis, characterization, and mechanism of formation. *J. Appl. Phys.,* **2021**, *129*(16), 163301.
[http://dx.doi.org/10.1063/5.0040322]

[41] Zou, F.R.; Wang, S.N.; Wang, F.F.; Liu, D.; Li, Y. Synthesis of lanthanide-functionalized carbon

quantum dots for chemical sensing and photocatalytic application. *Catalysts,* **2020**, *10*(8), 833. [http://dx.doi.org/10.3390/catal10080833]

[42] Teymoorian, T.; Hashemi, N.; Mousazadeh, M.H.; Entezarian, Z. N, S doped carbon quantum dots inside mesoporous silica for effective adsorption of methylene blue dye. *SN Applied Sciences,* **2021**, *3*(3), 305. [http://dx.doi.org/10.1007/s42452-021-04287-z]

[43] Marouzi, S.; Darroudi, M.; Hekmat, A.; Sadri, K.; Kazemi Oskuee, R. One-pot hydrothermal synthesis of carbon quantum dots from Salvia hispanica L. seeds and investigation of their biodistribution, and cytotoxicity effects. *J. Environ. Chem. Eng.,* **2021**, *9*(4), 105461. [http://dx.doi.org/10.1016/j.jece.2021.105461]

[44] Rajamanikandan, S.; Biruntha, M.; Ramalingam, G. Blue emissive carbon quantum dots (cqds) from bio-waste peels and its antioxidant activity. *J. Cluster Sci.,* **2021**, *0* [http://dx.doi.org/10.1007/s10876-021-02029-0]

[45] Rezaei, A.; Hashemi, E. A pseudohomogeneous nanocarrier based on carbon quantum dots decorated with arginine as an efficient gene delivery vehicle. *Sci. Rep.,* **2021**, *11*(1), 13790. [http://dx.doi.org/10.1038/s41598-021-93153-4] [PMID: 34215792]

CHAPTER 5

Application of Carbonaceous Quantum Dots in Biomedical

Yarima S. García[1], **Adelmo M. Estrada**[1], **Eneida R. López**[1], **Ivan H. Herrera**[1] and **Jorge L. Neves**[1,*]

[1] *Center for Advanced Studies, HAV, Cuba and Federal University of Pernambuco, PE, Brazil*

Abstract: Numerous research fields, including chemistry, electronics, and medical sciences, have concentrated on the production and use of novel functional nanomaterials. Carbon, a component of all organic life forms, is essential for the creation of nanomaterials. The modern carbon-based family component known as carbonaceous quantum dots (CQD) was unintentionally discovered in 2004 while single-walled carbon nanotubes were being purified. Additionally, CQDs have exceptional qualities like outstanding photoluminescence and minimal toxic effects. Outstanding *in vitro* andin vivo biomedical implications of CQDs include drug/gene delivery, biosensor biotherapy, and theragnostic evolution. Also, CQDs can pass through specific body sites of endothelial inflammation (epithelium of the intestinal tract, liver, for example), tumors or penetrate capillaries due to their small size. For the same reason, nanoparticles are more suitable for intravenous administration than microparticles and also prevent particle aggregation and bypass emboli or thrombi formation. This chapter describes the most contemporary applications of CQDs in diverse biomedical fields. We hope it will provide incalculable insights to inspire discoveries on CQD and delineate a road map toward a broader range of bio applications.

Keywords: Biomedical application, Biosensor, Biotherapy, CQDs, Contrast agent, Drug/gene delivery, Theragnostic.

INTRODUCTION

Throughout history, medicine and science have advanced together to improve people's life quality. However, the advancement of nanoscience and nanotechnology at the beginning of the 21st century has brought many innovations in this field [1]. Over the years, nanotechnology has focused on developing the electronic industry and the quest for new materials showing excellent attributes and new strategies for synthesizing nanomaterials [2]. The

* **Corresponding author Jorge L. Neves:** Center for Advanced Studies, HAV, Cuba and Federal University of Pernambuco, PE, Brazil, Email: jorge.lneves@ufpe.br

Kulvinder Singh & Vineet Kumar (Eds.)

new discipline of Nanomedicine materializes due to the growth of nanotechnology in the health sciences. In recent decades, this nanotechnology discipline has emerged with great force from the scientific sector in search of an enormous impact on today's society. That is why medicine is presented as a potential market in the nanoworld that could lead to the cure of diseases such as cancer nanotechnology is defined as the scientific-technological activity involved in the conception, synthesis, characterization, and application of materials and devices (presenting a minimum of one dimension) at the nanoscale (generally 1-100 nanometers) [2].

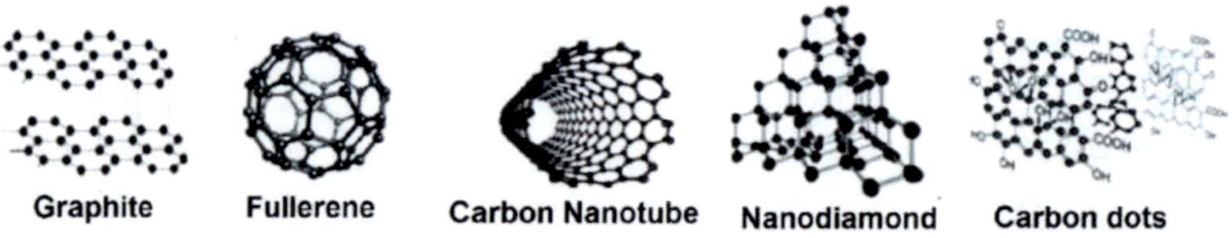

Fig. (1). Carbon nanoforms, Reproduced with permission from [3], under an open access Creative Common CC BY license.

Among the different types of nanomaterials studied for these purposes is carbon, which can take various forms see Fig. (**1**), the best-known being: fullerenes, graphene, nanodiamonds, carbon nanotubes and carbon quantum dots (CQD), which in the last two decades have attracted much interest in biomedicine [4, 5] with excellent physical-chemical properties that are of great value. Carbon Quantum Dots (CQD) commonly refer to charming zero-dimensional (0D) carbon nanomaterials showing a diameter ≤ 10 nm. However, Scrivens and colleagues accidentally uncovered CQDs (4) in 2004 by filtration of single-walled carbon nanotubes. Later on, Sun and colleagues [6] stamped the term carbon dots in 2006 for these quantum-sized nanoparticles exhibiting strong photoluminescence in solid-state and solution. Since then, CQDs' biomedical applications have achieved notable interest owing to their great advantages in optical and chemical characteristics, biocompatibility, and low toxicity [7]. Likewise, they have great aqueous stability and unique electronic features that allow them to be electron donors and receptor agents, which could be advantageous in numerous biomedical applications [8].

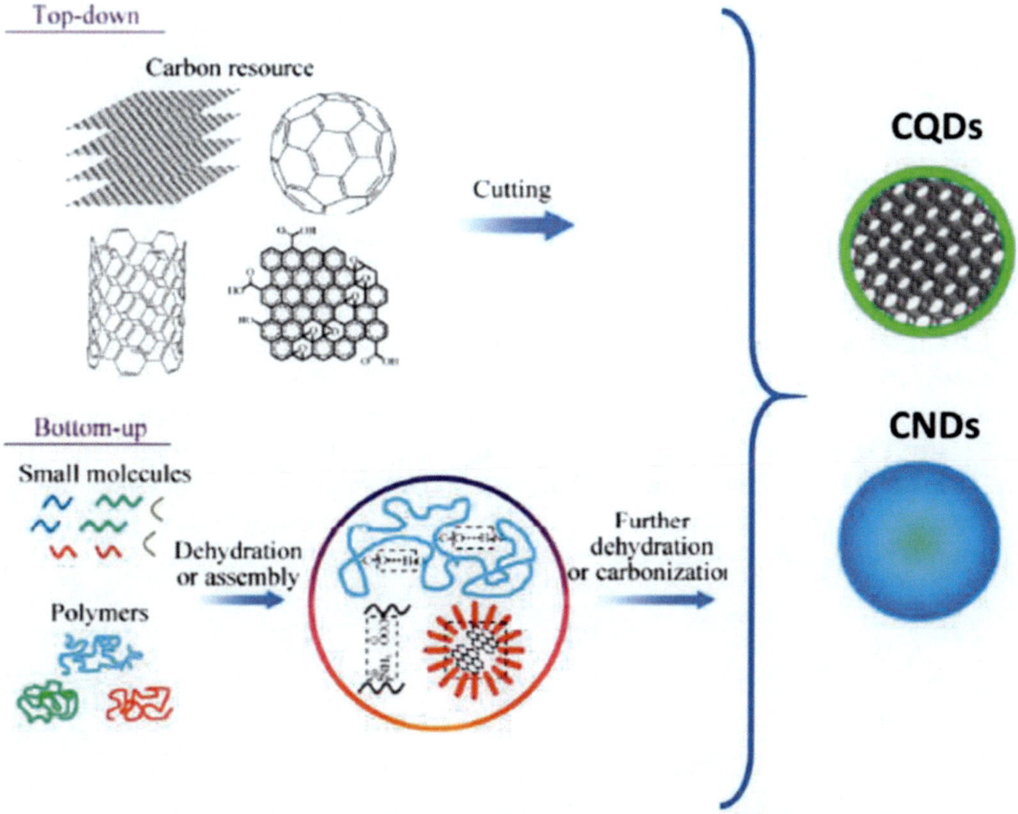

Fig. (2). Top-down and bottom-up approaches for CQD synthesis. Reproduced with permission from [9], under an open-access Creative Common CC BY license.

Two main approaches have been applied to CQD synthesis see Fig. (**2**). On one side, the "top-down" approach comprises breaking down long carbon structures and includes methods like electrochemical oxidation, laser ablation, chemical oxidation, and ultrasound fabrication. On another side, CQD synthesis starts with molecular precursors in the "bottom-up" approach and the advantages of hydrothermal and plasma treatment, thermal decomposition and microwave synthesis [10, 11]. Hydrothermal and solvothermal approaches provided CQDs functionalized with various superficial groups (mostly carboxyl and hydroxyl groups attached to oxygen), which improves water solubility without additional surface chemical transformations. In addition, reactive groups on the border of CQDs greatly simplify passivation and functionalization of the surface, making them suitable for drug delivery.

As shown in Fig. (**3**), CQDs are suitable alternatives for theranostics applications (bioimaging [12, 13] and *in vivo* and *in vitro* drug delivery [14, 15] owing to their optical and surface chemical attributes. Additionally, CQDs may help detect tumors in the premature stages (using fluorescence molecular bioimaging techniques), enabling remarkable observation of pathological and physiological occurrences in a non-invasive, highly sensitive manner. Furthermore, CQDs have been widely used in developing and guiding the evolution of new optical biosensors, particularly due to early reported advantages [16 - 19] allied to low cost, high sensitivity, and utility. Besides the common elemental and biomarker analysis, CDQs has allowed investigations at the single-cell level by following fluorescence changes [17]. The fluorescent properties of CQDs are sensitive to the medium in which they are found. Passivation of its surface allows its optical properties to be unaffected or reduced. Similarly, the fluorescence activity of the nanomaterial and the functionalization of the interface of CQDs perform a significant role in biological applications. In such a way, it alters their capacity to engage with drugs, ions, organic molecules, and living organisms [20]. In this scenario, the presented review intends to update the biological applications of CQDs and their recent scientific developments.

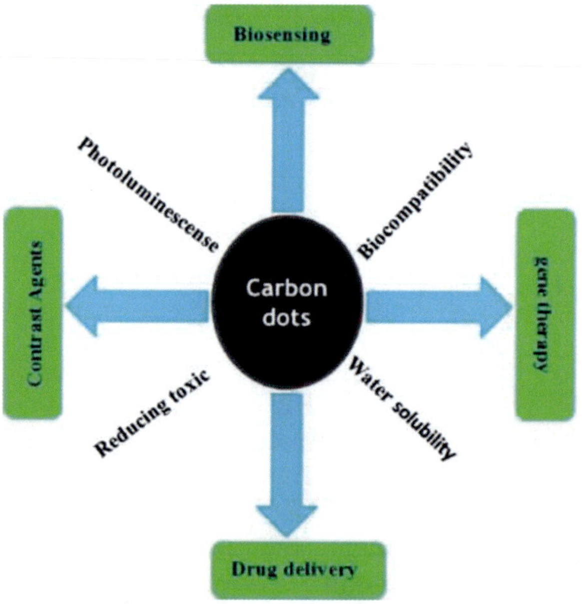

Fig. (3). Biomedical applications and relevant properties of the CQDs.

Carbon Quantum Dots for Theranostic Development

Recent years have seen a lot of attention and work put into the development of nano-theranostics, which combines bioimaging diagnostics with treatment to maximize performance. Theranostics is not limited to treatment or cinematography but instead enables accurate and concurrent diagnosis as well as the administration of effective medicines using a drug delivery system (DDS)(19). Thus, potential theranostic suitability for personalized medicine is among the most encouraging challenges in nanomedicine. However, the dissimilarity between theranostic and DDS is not clear. By fusing therapeutics and traditional diagnosis into a single system, Theranostic may be regarded as a special DDS type. Theranostic applications established on Quantum Dots, silica nanoparticles, polymers and especially carbon nanomaterials (CQDs) have been registered [21, 22]. In these theranostic approaches, diverse therapy procedures (*i.e.*, photo-/chemo-/radio-therapy) are merged with imaging techniques to diagnose and treat diseases simultaneously. Performing multiple functions is challenging for theranostic agents [23 - 25] and demands the integration of composite modalities and functions (*i.e.*, carrier, therapy and diagnosis).

Theranostic nanoparticles based on CQDs have attracted raising interest in areas like cancer therapy [26 - 31]. The possible explanation is that theranostic CQDs can achieve high sensitivity, low interference (owing to their physical-chemical characteristics), and more detailed and precise information than conventional cancer treatment. In addition, integration can lead to elevated therapeutic effectiveness by controlling tumor metastasis and regression. Therefore, a wide miscellany of nano-theranostic agents (based on nanostructures such as CQDs) merge many imaging features like magnetic resonance imaging (MRI), and gene therapy (GT), and have applications as a biosensor and in drug-controlled release [32 - 37]. The growing nature of the field towards this sensitive and intelligent theranostic particle can be used as tools for researchers in life sciences, particularly in the characterization of cells and diseases and the identification of markers throughout their useful life. Many clinical imaging technologies have limitations in resolution, but theranostic nanoparticles CQDs can be used in combination to offer early detection and therapy of diseases. In addition, they have the potential to improve platforms for superior performance.

Carbon Quantum Dots and their Application in Drug Delivery

Among the most important problems that affect humanity is the confrontation with many infectious diseases influenced by the growing resistance of pathogenic microorganisms to antibiotics and cancer in its various forms and manifestations. The fight against these diseases begins once the diagnosis is made. Medicines play an outstanding role in this fight, but the research to find new effective drugs is difficult, and much remains to be done. Although chemotherapy has many drawbacks (it lacks specificity and causes multi-drug resistance and toxicity [38]), it has been widely used to battle cancer and other localized diseases. Therefore, rehabilitation is essential for a better quality of life for people who survive these diseases. Hence, the development of new routes of administration, treatments and novel materials is required to treat diseases that are currently devastating [39].

Targeted drug delivery is an option for drug efficacy and reduced side effects [40 - 42]. Nevertheless, this alternative method impacts the targeted agents by early drug leakage before contacting the target site [43, 44]. Hence, a key step is developing new efficacious management agents with many distinguishable advantages. Nanotechnology offers highly novel alternatives like drug delivery systems that are selectively targeted at their site of action. For example, drug-coupled nanostructures can enhance DDS concerning drug absorption, distribution, and elimination and lead to several advantages. In this approach, the drug is targeted to a distinct tissue/cell, delivery is enhanced for minimal water-soluble drugs, different therapies (combination of different drugs or therapeutic modalities) are administered concomitantly, large macromolecules are transferred, and the drug site can be monitored by optical nanoparticle properties feasible to bioimaging [45, 46].

Due to the intrinsic capacity of its components to address dispersion and durability issues, direct medication delivery to particular areas, and diagnose *in vivo* through imaging, nanomedicine has grown into a multimillion thriving industry. Along with genomics, nanomedicine may soon generate long-awaited individualized medicine. Future nanomedicine innovations could produce multifunctional entities qualified for coexisting diagnosing, treating, and monitoring [47]. Nanotechnology enables the drug release to be minimally invasive since it allows the synthesis of CQDs at the nanoscale, with sizes that enable these devices to cross pores and cell membranes, increasing drug effectiveness by optimizing the required dose and size, morphology and surface properties of the compounds. On one side, if CQDs can associate with busted organs, tissues or cells and then unleash the active principle *in situ*, the associated drug toxicity should be lowered. On the other side, the gradual drug release is viable according to the patient's necessities, reducing the possible adverse events resulting from the direct uncontrolled drug supply [48 - 51].

Many investigations have employed CQD in DDS [50]. For instance, researchers covalently conjugated a drug (derived chlorambucil called Qucbl) to nitrogen-containing CQDs [52]. The Qucbl-loaded CQDs were characterized by FTIR spectroscopy, and their aggregation in the cell cytoplasm and nucleus is confirmed by *In vitro* experiments. Furthermore, irradiation from the Qucbl-loaded CQDs has manipulated drug liberation to destroy cancer cells. Other investigations (53) showed that doxorubicin (DOX) loaded CQDs (CQD-DOX) were quickly absorbed by cells, being localized primarily in the A549 cell cytoplasm and nucleus (according to the Fluorescence microscopy) and have pH-controlled drug liberation. Additionally, CQDs have been passivated using polyamine-containing organosilane molecules to deliver DOX [53], and organophilic CQDs were utilized for curcumin anti-malignant drugs and fast internalization by the HeLa cells [54].

Other cell organelles have also been targeted by exploring advancements in CQD research for drug delivery. Due to the paramount role of mitochondria in cancer, CQDs have been obtained (utilizing various media and a concise one-step hydrothermal strategy), which hold mitochondrial targeting features [55]. Such specialized CQDs could be utilized for drug delivery and mitochondrially imaging, presenting diverse benefits concerning commercial mitochondrial probes [56]. Besides organelle-target nanoprobes, the brain is an essential organ that can benefit from advancements in targeted CQD research. Nevertheless, the brain displays various challenges to be overcome to make targeted DDS useful to treat brain diseases (neurodegenerative disorders and tumors). More specifically, the brain decreased permeability, quick drug retention and the inclusion of the blood-brain barrier are considered the main difficulties to be tackled. In this scenario, CQDs have been integrated into tumor-penetrating peptides to attack brain tumors, providing a proper diagnosis and prognosis tool for such diseases.

Carbon Quantum Dots as Contrast Agents

Currently, new compounds capable of acting as contrast agents (ACs) are subjected to the most intense investigations in medicine. The most widely used ACs are Gd^{3+} complexes [57, 58], capable of altering T_1 and T_2 relaxation tempos, and despite being quite effective, their high toxicity in patients with kidney disease has led to new studies to replace these classic ACs with more efficient and less toxic compounds to the organism [59]. In parallel to imaging techniques, a new generation of tracers and contrast agents has been created to increase sensitivity and provide greater contrast in imaging techniques from synthetic and biological nanoparticles. Owing to the physical-chemical characteristics that carbon dots present [60, 61], the production of nanodevices as contrast agents in medical imaging has clear advantages over traditional agents used in the diagnosis of diseases, such as better optical dispersion, increase in biocompatibility, decrease in the probability of denaturation and especially, their ability to bind ligands, which turns them into devices with multiple functions.

Therefore, CQDs as contrast agents can be designed to be directed to some relevant biological target to determine diseases, achieving an earlier and more precise diagnosis. CQDs were introduced as a fluorescent contrast agent in mice for the first time in 2009 [62]. Subsequently, a miscellany of CQDs has been synthesized and employed for *In vivo* imaging [63]. Due to more in-depth light penetration into the tissue samples, lengthier wavelength excitation is favorably desired for *in vivo* imaging. Combining magnetic resonance imaging (MRI) and fluorescence imaging would enable remarkable tissue infiltration, superior spatial accuracy, and straightforward microscopic tissue inspection. For instance, magnetic fluorescent CQDs have been fabricated by merging ions like Eu^{3+}, Fe^{3+}, Mn^{2+} and Gd^{3+} in the nanoparticle structure. These types of carbon dots with different heteroatoms have the ideal size. The ability to participate and interact with molecules of biological interest serves as a tool in health and biomedicine

since they have proved to be capable of crossing pores and cell membranes without causing damage.

The usage of metal-free and metal-incorporated CQDs [64] is the principal strategy to embrace CQDs for MRI applications. Examples of such usages include the work of Chen *et al.* [65]. They prepared relatively monodisperse and crystalline Gd@GCN (high Gd content of 37.6%), and the material displayed outstanding T_1/T_2 MRI contrast. Also, tumor T_1-weighted images (7T device) of an SCC-7 tumor-bearing mouse enhance greatly after intravenous injection of Gd@GCN. Other studies [66] obtained Gd^{3+} and Yb^{3+} dopped CQDs, which act as a single multimodal contrast agent. The prepared material reveals an intense and stable Fluorescence emission and high MR contrast capabilities, even superior longitudinal relaxation (r_1) than commercial Magnevist (a Gd-DTPA contrast agent). Also, because of the high X-ray Yb absorption [67], Yb^{3+} dopped CQDs could be employed as CT contrast agents.

However, due to safety concerns, Gd- or Yb-based material applications are hindered for renal failure patients. In order to defeat such shortcomings, CQDs can be prepared with nontoxic Mn and be applied as a T1-weighted MRI contrast. Mn^{2+}-based CQDs deliver positive image contrasts and comparable pharmacokinetics and pharmacodynamics features to Gd^{3+} materials, with less toxicity [68 - 70]. Furthermore, the increase in Mn relaxation is similar to Gd-based complexes, and the potent concentration is similar between Mn and Gd (68). In 2016, Cuiping Han *et al.* presented the first instance of hybrid manganese-carbon quantum dots (Mn-CQD) targeting ovarian cancer cells and revealing exceptional magnetic-fluorescent imaging features. The obtained Mn-CQDs demonstrated a high relaxation r_1=3.26 ± 0.11 $mM^{-1}s^{-1}$ and a high fluorescence quantum productivity of 90.76% (69). In addition, the nanoparticle superficial amino and carboxyl groups also allow target agent functionalization against ovarian cancer biomarkers HE4 and CA-125.

Carbon Quantum Dots in Gene Therapy

The goal of gene therapy is to repair a person's genome by inserting functional parts that are lacking. To treat a disease or carry out a marking, it occurs in the cells and tissue. Since this method is still being developed, it is currently only used in controlled studies to treat severe inherited or acquired disorders. It was developed initially mainly to treat hereditary illnesses, but it is now being studied for practically any ailment. Notable efforts have been dedicated to gene therapy, and it is attracted substantial attention in the biomedical and biomedical areas. Control over the disease origins can be performed by expressing and supplying an exogenous encoded DNA with the absent or faulty gene product. Hence, appropriate genetic vectors [70] are an essential element for gene therapy, being distinct classes of nanoparticles [71 - 73] and QD [74 - 78] employed for this task. In this respect, CQDs have an excellent capability to pack plasmid DNA with remarkable transfection efficacy providing a fascinating gene delivery platform [79]. Also, complexes of cationic CQDs and negatively charged siRNAs can be assembled and deliver siRNA in a reducing environment [80].

The function of proteins in cellular environments can be dissected by the transfection of exogenous proteins within cells. Protein genes are conventionally cloned with a protein tag (GFP) before the protein transfection. However, these systems suffer from signal loss and lower transfection efficiency [81], making it critical to mark them with protein/gene dyes. Also, the self-image attribute is absent in several of the current non-viral gene vectors [82 - 84]. In recent investigations, sodium alginate-derived CQDs were employed to act simultaneously as gene delivery and fluorescent marker in human cell lines. CQD transfection efficiency was comparable to lipofectamine, with the advantage of CQD's very low cytotoxic and high biocompatible cell imaging [85]. It was also reported that cationic PEI functionalized CQDs (CQD-PEI) aided DNA condensation on CQD through electrostatic interactions (through the surface protonated amino groups), which indicates that CQDs functionalized with cationic groups are more suitable for gene / DNA delivery applications. Furthermore, besides protonated amino groups in CQD-PEI, non-protonated ones are also featured in cytoplasm gene release by buffering endolysosomal pH [86]. Newly, CQDs synthesized with glucose and PEI display 104 times greater transfection rate than the conventional technique in models of human cells. Also, nanomaterials resulting from gold nanoparticle conjugation with CQDs and PEI-pDNA have been reported for cell DNA delivery. Zhang *et al.* synthesized hyaluronic acid (HA) / PEI (HP-CQD) functionalized CQDs (utilizing a bottom-up strategy) for tumor selection and gene therapy. As a result, HP-CD spots exhibited excellent biocompatibility, gene condensing capability, and water solubility, efficiently undergoing cancer cell internalization by endocytosis [87].

CONCLUSION

This chapter has summarized contemporary advancements in CQD research and its biological applications. CQD development for biomedical applications has experienced dramatic advances in this scenario. However, coming investigations will still face elemental knowledge and practical challenges. CQDs will advance their theranostic bioimaging applications, gene therapy, and the bio-sensorial drug controlled release. Carrying out combinations of experimental and theoretical studies that require more effort to continue developing biocompatible CQDs and explore their clinical applications.

REFERENCES

[1] Puvvada, N.; Kumar, B.N.P.; Konar, S.; Kalita, H.; Mandal, M.; Pathak, A. Synthesis of biocompatible multicolor luminescent carbon dots for bioimaging applications. *Sci. Technol. Adv. Mater.,* **2012**, *13*(4), 045008.
[http://dx.doi.org/10.1088/1468-6996/13/4/045008] [PMID: 27877508]

[2] Hola, K.; Zhang, Y.; Wang, Y.; Giannelis, E.P.; Zboril, R.; Rogach, A.L. Carbon dots—Emerging light emitters for bioimaging, cancer therapy and optoelectronics. *Nano Today,* **2014**, *9*(5), 590-603.
[http://dx.doi.org/10.1016/j.nantod.2014.09.004]

[3] Yan, Q.L.; Gozin, M.; Zhao, F.Q.; Cohen, A.; Pang, S.P. Highly energetic compositions based on functionalized carbon nanomaterials. *Nanoscale,* **2016**, *8*(9), 4799-4851.
[http://dx.doi.org/10.1039/C5NR07855E] [PMID: 26880518]

[4] Xu, X; Ray, R; Gu, Y; Ploehn, HJ; Gearheart, L; Raker, K; Scrivens, WA Electrophoretic analysis and purification of fluorescent single-walled carbon nanotube fragments. *J Am Chem Soc.,* **2004**, *126*(410), 12736-12737.

[5] Baker, S.N.; Baker, G.A. Luminescent carbon nanodots: emergent nanolights. *Angew. Chem. Int. Ed.,* **2010**, *49*(38), 6726-6744.
[http://dx.doi.org/10.1002/anie.200906623] [PMID: 20687055]

[6] Sun, Y.P.; Zhou, B.; Lin, Y.; Wang, W.; Fernando, K.A.S.; Pathak, P.; Meziani, M.J.; Harruff, B.A.; Wang, X.; Wang, H.; Luo, P.G.; Yang, H.; Kose, M.E.; Chen, B.; Veca, L.M.; Xie, S.Y. Quantum-sized carbon dots for bright and colorful photoluminescence. *J. Am. Chem. Soc.,* **2006**, *128*(24), 7756-7757.
[http://dx.doi.org/10.1021/ja062677d] [PMID: 16771487]

[7] Yuan, F.; Li, S.; Fan, Z.; Meng, X.; Fan, L.; Yang, S. Shining carbon dots: Synthesis and biomedical and optoelectronic applications. *Nano Today,* **2016**, *11*(5), 565-586.
[http://dx.doi.org/10.1016/j.nantod.2016.08.006]

[8] Li, Y.; Zhao, Y.; Cheng, H.; Hu, Y.; Shi, G.; Dai, L.; Qu, L. Nitrogen-doped graphene quantum dots with oxygen-rich functional groups. *J. Am. Chem. Soc.,* **2012**, *134*(1), 15-18.
[http://dx.doi.org/10.1021/ja206030c] [PMID: 22136359]

[9] Carbonaro, C.; Corpino, S.; Mocci, T. On the emission properties of carbon dots: reviewing data and discussing models. *J. Carbon Research,* **2019**, *5*, 60.
[http://dx.doi.org/10.3390/c5040060]

[10] Guo, Q.; Yuan, F.; Zhang, B.; Zhou, S.; Zhang, J.; Bai, Y.; Fan, L.; Hayat, T.; Alsaedi, A.; Tan, Z. Passivation of the grain boundaries of $CH_3NH_3PbI_3$ using carbon quantum dots for highly efficient perovskite solar cells with excellent environmental stability. *Nanoscale,* **2019**, *11*(1), 115-124.
[http://dx.doi.org/10.1039/C8NR08295B] [PMID: 30525161]

[11] Cao, L.; Wang, X.; Meziani, M.J.; Lu, F.; Wang, H.; Luo, P.G.; Lin, Y.; Harruff, B.A.; Veca, L.M.; Murray, D.; Xie, S.Y.; Sun, Y.P. Carbon dots for multiphoton bioimaging. *J. Am. Chem. Soc.,* **2007**, *129*(37), 11318-11319.
[http://dx.doi.org/10.1021/ja073527l] [PMID: 17722926]

[12] Antaris, A.L.; Robinson, J.T.; Yaghi, O.K.; Hong, G.; Diao, S.; Luong, R.; Dai, H. Ultra-low doses of chirality sorted (6,5) carbon nanotubes for simultaneous tumor imaging and photothermal therapy. *ACS Nano,* **2013**, *7*(4), 3644-3652.
[http://dx.doi.org/10.1021/nn4006472] [PMID: 23521224]

[13] Zhu, S.; Meng, Q.; Wang, L.; Zhang, J.; Song, Y.; Jin, H. Highly photoluminescent carbon dots for multicolor patterning, sensors, and bioimaging. In: *Angewandte Chemie*; wiley, **2013**. ; 52, 3953-3957.
[http://dx.doi.org/10.1002/anie.201300519]

[14] Wang, R.; Lu, K.Q.; Tang, Z.R.; Xu, Y.J. Recent progress in carbon quantum dots: Synthesis, properties and applications in photocatalysis. *J. Mater. Chem. A Mater. Energy Sustain.,* **2017**, *5*(8), 3717-3734.
[http://dx.doi.org/10.1039/C6TA08660H]

[15] Peng, J.; Zhao, Q.; Zheng, W.; Li, W.; Li, P.; Zhu, L.; Liu, X.; Shao, B.; Li, H.; Wang, C.; Yang, Y. Peptide-functionalized nanomaterials for the efficient isolation of her2-positive circulating tumor cells. *ACS Appl. Mater. Interfaces,* **2017**, *9*(22), 18423-18428.
[http://dx.doi.org/10.1021/acsami.7b03905] [PMID: 28523908]

[16] Pandit, S.; Behera, P.; Sahoo, J.; De, M. *In Situ* synthesis of amino acid functionalized carbon dots with tunable properties and their biological applications. *ACS Appl. Bio Mater.,* **2019**, *2*(8), 3393-3403.
[http://dx.doi.org/10.1021/acsabm.9b00374] [PMID: 35030781]

[17] Malina, T.; Poláková, K.; Skopalík, J.; Milotová, V.; Holá, K.; Havrdová, M.; Tománková, K.B.; Čmiel, V.; Šefc, L.; Zbořil, R. Carbon dots for *in vitro* fluorescence imaging of adipose tissue-derived mesenchymal stromal cells. *Carbon,* **2019**, *152*, 434-443.
[http://dx.doi.org/10.1016/j.carbon.2019.05.061]

[18] Fasbender, S.; Zimmermann, L.; Cadeddu, R.P.; Luysberg, M.; Moll, B.; Janiak, C.; Heinzel, T.; Haas, R. The low toxicity of graphene quantum dots is reflected by marginal gene expression changes of primary human hematopoietic stem cells. *Sci. Rep.,* **2019**, *9*(1), 12028.
[http://dx.doi.org/10.1038/s41598-019-48567-6] [PMID: 31427693]

[19] Wu, L.; Luderer, M.; Yang, X.; Swain, C.; Zhang, H.; Nelson, K.; Stacy, A.J.; Shen, B.; Lanza, G.M.; Pan, D. Surface passivation of carbon nanoparticles with branched macromolecules influences near infrared bioimaging. *Theranostics,* **2013**, *3*(9), 677-686.
[http://dx.doi.org/10.7150/thno.6535] [PMID: 24019852]

[20] Interaction of carbon dots with endothelial cells: Implications for biomedical applications. **2016**.

[21] Wu, L.; Cai, X.; Nelson, K.; Xing, W.; Xia, J.; Zhang, R.; Stacy, A.J.; Luderer, M.; Lanza, G.M.; Wang, L.V.; Shen, B.; Pan, D. A green synthesis of carbon nanoparticles from honey and their use in real-time photoacoustic imaging. *Nano Res.,* **2013**, *6*(5), 312-325.
[http://dx.doi.org/10.1007/s12274-013-0308-8] [PMID: 23824757]

[22] Huang, P.; Lin, J.; Wang, X.; Wang, Z.; Zhang, C.; He, M.; Wang, K.; Chen, F.; Li, Z.; Shen, G.; Cui, D.; Chen, X. Light-triggered theranostics based on photosensitizer-conjugated carbon dots for simultaneous enhanced-fluorescence imaging and photodynamic therapy. *Adv. Mater.,* **2012**, *24*(37), 5104-5110.
[http://dx.doi.org/10.1002/adma.201200650] [PMID: 22718562]

[23] Demir, B.; Lemberger, M.M.; Panagiotopoulou, M.; Medina Rangel, P.X.; Timur, S.; Hirsch, T.; Tse Sum Bui, B.; Wegener, J.; Haupt, K. Tracking hyaluronan: Molecularly imprinted polymer coated carbon dots for cancer cell targeting and imaging. *ACS Appl. Mater. Interfaces,* **2018**, *10*(4), 3305-3313.

[http://dx.doi.org/10.1021/acsami.7b16225] [PMID: 29299913]

[24] Wang, S.J.; Wang, B.B.; Bai, F.W.; Ma, X.J. Tumor cell responses to carbon dots derived from chondroitin sulfate. *RSC Advances,* **2015**, *5*(99), 81388-81394.
[http://dx.doi.org/10.1039/C5RA14585F]

[25] Wang, H.; Wang, K.; Tian, B.; Revia, R.; Mu, Q.; Jeon, M.; Chang, F.C.; Zhang, M. Preloading of hydrophobic anticancer drug into multifunctional nanocarrier for multimodal imaging, nir-responsive drug release, and synergistic therapy. *Small,* **2016**, *12*(46), 6388-6397.
[http://dx.doi.org/10.1002/smll.201602263] [PMID: 27671114]

[26] Chen, H.; Zhang, W.; Zhu, G.; Xie, J.; Chen, X. Rethinking cancer nanotheranostics. *Nat. Rev. Mater.,* **2017**, *2*(7), 17024.
[http://dx.doi.org/10.1038/natrevmats.2017.24] [PMID: 29075517]

[27] Liu, Y.; Bhattarai, P.; Dai, Z.; Chen, X. Photothermal therapy and photoacoustic imaging *via* nanotheranostics in fighting cancer. *Chem. Soc. Rev.,* **2019**, *48*(7), 2053-2108.
[http://dx.doi.org/10.1039/C8CS00618K] [PMID: 30259015]

[28] Nabil, G.; Bhise, K.; Sau, S.; Atef, M.; El-Banna, H.A.; Iyer, A.K. Nano-engineered delivery systems for cancer imaging and therapy: Recent advances, future direction and patent evaluation. *Drug Discov. Today,* **2019**, *24*(2), 462-491.
[http://dx.doi.org/10.1016/j.drudis.2018.08.009] [PMID: 30121330]

[29] Liu, J.; Lécuyer, T.; Seguin, J.; Mignet, N.; Scherman, D.; viana, B.; Richard, C. Imaging and therapeutic applications of persistent luminescence nanomaterials. *Adv. Drug Deliv. Rev.,* **2019**, *138*, 193-210.
[http://dx.doi.org/10.1016/j.addr.2018.10.015] [PMID: 30414492]

[30] Silva, C.O.; Pinho, J.O.; Lopes, J.M.; Almeida, A.J.; Gaspar, M.M.; Reis, C. Current trends in cancer nanotheranostics: Metallic, polymeric, and lipid-based systems. *Pharmaceutics,* **2019**, *11*(1), 22.
[http://dx.doi.org/10.3390/pharmaceutics11010022] [PMID: 30625999]

[31] Singh, R.K.; Patel, K.D.; Leong, K.W.; Kim, H.W. Progress in nanotheranostics based on mesoporous silica nanomaterial platforms. *ACS Appl. Mater. Interfaces,* **2017**, *9*(12), 10309-10337.
[http://dx.doi.org/10.1021/acsami.6b16505] [PMID: 28274115]

[32] Sonali, S.; Viswanadh, MK.; Singh, RP; Agrawal, P; Mehata, AK. Nanotheranostics: Emerging strategies for early diagnosis and therapy of brain cancer. *Nanotheranostics,* **2018**, *2*, 70-86.
[http://dx.doi.org/10.7150/ntno.21638]

[33] Knežević, N.Ž.; Kaluđerović, G.N. Silicon-based nanotheranostics. *Nanoscale,* **2017**, *9*(35), 12821-12829.
[http://dx.doi.org/10.1039/C7NR04445C] [PMID: 28853473]

[34] Chen, Y.; Wu, Y.; Sun, B.; Liu, S.; Liu, H. Two-dimensional nanomaterials for cancer nanotheranostics. *Small,* **2017**, *13*(10), 1603446.
[http://dx.doi.org/10.1002/smll.201603446] [PMID: 28075057]

[35] Li, H.; Huang, J.; Lu, F.; Liu, Y.; Song, Y.; Sun, Y.; Zhong, J.; Huang, H.; Wang, Y.; Li, S.; Lifshitz, Y.; Lee, S.T.; Kang, Z. Impacts of carbon dots on rice plants: Boosting the growth and improving the disease resistance. *ACS Appl. Bio Mater.,* **2018**, *1*(3), 663-672.
[http://dx.doi.org/10.1021/acsabm.8b00345] [PMID: 34996197]

[36] Sharma, A.; Das, J. Small molecules derived carbon dots: synthesis and applications in sensing, catalysis, imaging, and biomedicine. *J. Nanobiotechnology,* **2019**, *17*(1), 92.
[http://dx.doi.org/10.1186/s12951-019-0525-8] [PMID: 31451110]

[37] Chen, Y.; Zhang, Y. Application of the CRISPR/Cas9 system to drug resistance in breast cancer. *Adv. Sci.,* **2018**, *5*(6), 1700964.
[http://dx.doi.org/10.1002/advs.201700964] [PMID: 29938175]

[38] Inoue-Yamauchi, A.; Jeng, P.S.; Kim, K.; Chen, H.C.; Han, S.; Ganesan, Y.T.; Ishizawa, K.; Jebiwott,

S.; Dong, Y.; Pietanza, M.C.; Hellmann, M.D.; Kris, M.G.; Hsieh, J.J.; Cheng, E.H. Targeting the differential addiction to anti-apoptotic BCL-2 family for cancer therapy. *Nat. Commun.,* **2017**, *8*(1), 16078.
[http://dx.doi.org/10.1038/ncomms16078] [PMID: 28714472]

[39] Liu, Y.; Xiao, N.; Gong, N.; Wang, H.; Shi, X.; Gu, W.; Ye, L. One-step microwave-assisted polyol synthesis of green luminescent carbon dots as optical nanoprobes. *Carbon,* **2014**, *68*, 258-264.
[http://dx.doi.org/10.1016/j.carbon.2013.10.086]

[40] Rosenholm, J.M.; Sahlgren, C.; Lindén, M. Towards multifunctional, targeted drug delivery systems using mesoporous silica nanoparticles – opportunities & challenges. *Nanoscale,* **2010**, *2*(10), 1870-1883.
[http://dx.doi.org/10.1039/c0nr00156b] [PMID: 20730166]

[41] Sharma, G.; Anabousi, S.; Ehrhardt, C.; Ravi Kumar, M.N.V. Liposomes as targeted drug delivery systems in the treatment of breast cancer. *J. Drug Target.,* **2006**, *14*(5), 301-310.
[http://dx.doi.org/10.1080/10611860600809112] [PMID: 16882550]

[42] Roy, D.; Cambre, J.N.; Sumerlin, B.S. Future perspectives and recent advances in stimuli-responsive materials. *Prog. Polym. Sci.,* **2010**, *35*(1-2), 278-301.
[http://dx.doi.org/10.1016/j.progpolymsci.2009.10.008]

[43] Wang, L.; Zhou, H.S. Green synthesis of luminescent nitrogen-doped carbon dots from milk and its imaging application. *Anal. Chem.,* **2014**, *86*(18), 8902-8905.
[http://dx.doi.org/10.1021/ac502646x] [PMID: 25181643]

[44] Khan, M.S.; Pandey, S.; Talib, A.; Bhaisare, M.L.; Wu, H.F.; Wu, H-F. Controlled delivery of dopamine hydrochloride using surface modified carbon dots for neuro diseases. *Colloids Surf. B Biointerfaces,* **2015**, *134*, 140-146.
[http://dx.doi.org/10.1016/j.colsurfb.2015.06.006] [PMID: 26186107]

[45] Farokhzad, O.C.; Langer, R. Impact of nanotechnology on drug delivery. *ACS Nano,* **2009**, *3*(1), 16-20.
[http://dx.doi.org/10.1021/nn900002m] [PMID: 19206243]

[46] Molaei, M.J. A review on nanostructured carbon quantum dots and their applications in biotechnology, sensors, and chemiluminescence. *Talanta,* **2019**, *196*, 456-478.
[http://dx.doi.org/10.1016/j.talanta.2018.12.042] [PMID: 30683392]

[47] Pan, D.; Zhang, J.; Li, Z.; Wu, M. Hydrothermal route for cutting graphene sheets into blue-luminescent graphene quantum dots. *Adv. Mater.,* **2010**, *22*(6), 734-738.
[http://dx.doi.org/10.1002/adma.200902825] [PMID: 20217780]

[48] Liu, Z.; Chen, X.; Zhang, X.; Gooding, J.J.; Zhou, Y. Targeted drug delivery: Carbon□quantum□dots□loaded mesoporous silica nanocarriers with ph□switchable zwitterionic surface and enzyme□responsive pore□cap for targeted imaging and drug delivery to tumor. *Adv. Healthc. Mater.,* **2016**, *5*(12), 1380-1380.
[http://dx.doi.org/10.1002/adhm.201670058]

[49] Feng, T.; Ai, X.; Ong, H.; Zhao, Y. Dual-responsive carbon dots for tumor extracellular microenvironment triggered targeting and enhanced anticancer drug delivery. *ACS Appl. Mater. Interfaces,* **2016**, *8*(29), 18732-18740.
[http://dx.doi.org/10.1021/acsami.6b06695] [PMID: 27367152]

[50] Chowdhuri, A.R.; Singh, T.; Ghosh, S.K.; Sahu, S.K. Carbon dots embedded magnetic nanoparticles @chitosan @metal organic framework as a nanoprobe for ph sensitive targeted anticancer drug delivery. *ACS Appl. Mater. Interfaces,* **2016**, *8*(26), 16573-16583.
[http://dx.doi.org/10.1021/acsami.6b03988] [PMID: 27305490]

[51] Peng, Z.; Miyanji, E.H.; Zhou, Y.; Pardo, J.; Hettiarachchi, S.D.; Li, S.; Blackwelder, P.L.; Skromne, I.; Leblanc, R.M. Carbon dots: Promising biomaterials for bone-specific imaging and drug delivery. *Nanoscale,* **2017**, *9*(44), 17533-17543.

[http://dx.doi.org/10.1039/C7NR05731H] [PMID: 29110000]

[52] Karthik, S.; Saha, B.; Ghosh, S.K.; Pradeep Singh, N.D. Photoresponsive quinoline tethered fluorescent carbon dots for regulated anticancer drug delivery. *Chem. Commun.,* **2013**, *49*(89), 10471-10473.
[http://dx.doi.org/10.1039/c3cc46078a] [PMID: 24080791]

[53] Wang, Q.; Huang, X.; Long, Y.; Wang, X.; Zhang, H.; Zhu, R.; Liang, L.; Teng, P.; Zheng, H. Hollow luminescent carbon dots for drug delivery. *Carbon,* **2013**, *59*, 192-199.
[http://dx.doi.org/10.1016/j.carbon.2013.03.009]

[54] Yang, J.; Gao, G.; Zhang, X.; Ma, Y.H.; Jia, H.R.; Jiang, Y.W.; Wang, Z.; Wu, F.G. Ultrasmall and photostable nanotheranostic agents based on carbon quantum dots passivated with polyamine-containing organosilane molecules. *Nanoscale,* **2017**, *9*(40), 15441-15452.
[http://dx.doi.org/10.1039/C7NR05613C] [PMID: 28976508]

[55] Shu, Y.; Lu, J.; Mao, Q.X.; Song, R.S.; Wang, X.Y.; Chen, X.W.; Wang, J-H. Ionic liquid mediated organophilic carbon dots for drug delivery and bioimaging. *Carbon,* **2017**, *114*, 324-333.
[http://dx.doi.org/10.1016/j.carbon.2016.12.038]

[56] Hu, Y.; Zhang, L.; Li, X.; Liu, R.; Lin, L.; Zhao, S. Green preparation of s and n co-doped carbon dots from water chestnut and onion as well as their use as an off–on fluorescent probe for the quantification and imaging of coenzyme A. *ACS Sustain. Chem.& Eng.,* **2017**, *5*(6), 4992-5000.
[http://dx.doi.org/10.1021/acssuschemeng.7b00393]

[57] Wu, X.; Sun, S.; Wang, Y.; Zhu, J.; Jiang, K.; Leng, Y.; Shu, Q.; Lin, H. A fluorescent carbon-dot--based mitochondria-targetable nanoprobe for peroxynitrite sensing in living cells. *Biosens. Bioelectron.,* **2017**, *90*, 501-507.
[http://dx.doi.org/10.1016/j.bios.2016.10.060] [PMID: 27825883]

[58] Cao, X.; Deng, W.; Qu, R.; Yu, Q.; Li, J.; Yang, Y.; Cao, Y.; Gao, X.; Xu, X.; Yu, J. Non-viral co-delivery of the four yamanaka factors for generation of human induced pluripotent stem cells *via* calcium phosphate nanocomposite particles. *Adv. Funct. Mater.,* **2013**, *23*(43), 5403-5411.
[http://dx.doi.org/10.1002/adfm.201203646]

[59] Agarwal, S.; Sane, R.; Oberoi, R.; Ohlfest, J.R.; Elmquist, W.F. Delivery of molecularly targeted therapy to malignant glioma, a disease of the whole brain. *Expert Rev. Mol. Med.,* **2011**, *13*, e17.
[http://dx.doi.org/10.1017/S1462399411001888] [PMID: 21676290]

[60] Li, X.; Robinson, S.M.; Gupta, A.; Saha, K.; Jiang, Z.; Moyano, D.F.; Sahar, A.; Riley, M.A.; Rotello, V.M. Functional gold nanoparticles as potent antimicrobial agents against multi-drug-resistant bacteria. *ACS Nano,* **2014**, *8*(10), 10682-10686.
[http://dx.doi.org/10.1021/nn5042625] [PMID: 25232643]

[61] Acharya, A. Luminescent magnetic quantum dots for *in vitro in vitro* imaging and applications in therapeutics. *J. Nanosci. Nanotechnol.,* **2013**, *13*(6), 3753-3768.
[http://dx.doi.org/10.1166/jnn.2013.7460] [PMID: 23862405]

[62] Bourlinos, A.B.; Bakandritsos, A.; Kouloumpis, A.; Gournis, D.; Krysmann, M.; Giannelis, E.P.; Polakova, K.; Safarova, K.; Hola, K.; Zboril, R. Gd(iii)-doped carbon dots as a dual fluorescent-MRI probe. *J. Mater. Chem.,* **2012**, *22*(44), 23327.
[http://dx.doi.org/10.1039/c2jm35592b]

[63] Gossuin, Y.; Gillis, P.; Hocq, A.; Vuong, Q.L.; Roch, A. Magnetic resonance relaxation properties of superparamagnetic particles. *Wiley Interdiscip. Rev. Nanomed. Nanobiotechnol.,* **2009**, *1*(3), 299-310.
[http://dx.doi.org/10.1002/wnan.36] [PMID: 20049798]

[64] Kokubo, K. Water-soluble single-nano carbon particles: Fullerenol and its derivatives. . INTECHOPEN, **2012**.
[http://dx.doi.org/10.5772/36352]

[65] Chen, H.; Wang, G.D.; Tang, W.; Todd, T.; Zhen, Z.; Tsang, C.; Hekmatyar, K.; Cowger, T.;

Hubbard, R.B.; Zhang, W.; Stickney, J.; Shen, B.; Xie, J. Gd-encapsulated carbonaceous dots with efficient renal clearance for magnetic resonance imaging. *Adv. Mater.*, **2014**, *26*(39), 6761-6766. [http://dx.doi.org/10.1002/adma.201402964] [PMID: 25178894]

[66] Biju, V. Chemical modifications and bioconjugate reactions of nanomaterials for sensing, imaging, drug delivery and therapy. *Chem. Soc. Rev.*, **2014**, *43*(3), 744-764. [http://dx.doi.org/10.1039/C3CS60273G] [PMID: 24220322]

[67] Yang, S.T.; Cao, L.; Luo, P.G.; Lu, F.; Wang, X.; Wang, H.; Meziani, M.J.; Liu, Y.; Qi, G.; Sun, Y.P. Carbon dots for optical imaging *in vitro*. *J. Am. Chem. Soc.*, **2009**, *131*(32), 11308-11309. [http://dx.doi.org/10.1021/ja904843x] [PMID: 19722643]

[68] Jia, Q.; Zhao, Z.; Liang, K.; Nan, F.; Li, Y.; Wang, J.; Ge, J.; Wang, P. Recent advances and prospects of carbon dots in cancer nanotheranostics. *Mater. Chem. Front.*, **2020**, *4*(2), 449-471. [http://dx.doi.org/10.1039/C9QM00667B]

[69] Rub Pakkath, S.A.; Chetty, S.S.; Selvarasu, P.; Vadivel Murugan, A.; Kumar, Y.; Periyasamy, L.; Santhakumar, M.; Sadras, S.R.; Santhakumar, K. Transition Metal Ion (Mn^{2+}, Fe^{2+}, Co^{2+}, and Ni^{2+})-Doped Carbon Dots Synthesized *via* Microwave-assisted pyrolysis: A potential nanoprobe for magneto-fluorescent dual-modality bioimaging. *ACS Biomater. Sci. Eng.*, **2018**, *4*(7), 2582-2596. [http://dx.doi.org/10.1021/acsbiomaterials.7b00943]

[70] Bouzas-Ramos, D.; Cigales Canga, J.; Mayo, J.C.; Sainz, R.M.; Ruiz Encinar, J.; Costa-Fernandez, J.M. Carbon quantum dots codoped with nitrogen and lanthanides for multimodal imaging. *Adv. Funct. Mater.*, **2019**, *29*(38), 1903884. [http://dx.doi.org/10.1002/adfm.201903884]

[71] Sui, J.; Liu, G.; Song, Y.; Li, D.; Dong, X.; Wang, J.; Yu, W. Integrating photoluminescence, magnetism and thermal conversion for potential photothermal therapy and dual-modal bioimaging. *J. Colloid Interface Sci.*, **2018**, *510*, 292-301. [http://dx.doi.org/10.1016/j.jcis.2017.09.085] [PMID: 28957745]

[72] Gale, E.M.; Atanasova, I.P.; Blasi, F.; Ay, I.; Caravan, P. A manganese alternative to gadolinium for mri contrast. *J. Am. Chem. Soc.*, **2015**, *137*(49), 15548-15557. [http://dx.doi.org/10.1021/jacs.5b10748] [PMID: 26588204]

[73] Han, C.; Xu, H.; Wang, R.; Wang, K.; Dai, Y.; Liu, Q.; Guo, M.; Li, J.; Xu, K. Synthesis of a multifunctional manganese(II)–carbon dots hybrid and its application as an efficient magnetic-fluorescent imaging probe for ovarian cancer cell imaging. *J. Mater. Chem. B Mater. Biol. Med.*, **2016**, *4*(35), 5798-5802. [http://dx.doi.org/10.1039/C6TB01250G] [PMID: 32263751]

[74] El-Sherbiny, I.M.; Elbaz, N.M.; Sedki, M.; Elgammal, A.; Yacoub, M.H. Magnetic nanoparticles-based drug and gene delivery systems for the treatment of pulmonary diseases. *Nanomedicine,* **2017**, *12*(4), 387-402. [http://dx.doi.org/10.2217/nnm-2016-0341] [PMID: 28078950]

[75] Fernandez-Piñeiro, I.; Pensado, A.; Badiola, I.; Sanchez, A. Development and characterisation of chondroitin sulfate- and hyaluronic acid-incorporated sorbitan ester nanoparticles as gene delivery systems. *Eur. J. Pharm. Biopharm.*, **2018**, *125*, 85-94. [http://dx.doi.org/10.1016/j.ejpb.2018.01.009] [PMID: 29355685]

[76] Huang, X.; Andina, D.; Ge, J.; Labarre, A.; Leroux, J.C.; Castagner, B. Characterization of calcium phosphate nanoparticles based on a pegylated chelator for gene delivery. *ACS Appl. Mater. Interfaces,* **2017**, *9*(12), 10435-10445. [http://dx.doi.org/10.1021/acsami.6b15925] [PMID: 28266206]

[77] Du, B.; Gu, X.; Han, X.; Ding, G.; Wang, Y.; Li, D.; Wang, E.; Wang, J. Lipid-coated gold nanoparticles functionalized by folic acid as gene vectors for targeted gene delivery *in vitro* and *in vitro*. *ChemMedChem,* **2017**, *12*(21), 1768-1775. [http://dx.doi.org/10.1002/cmdc.201700391] [PMID: 28967206]

[78] Kim, M.C.; Lin, M.M.; Sohn, Y.; Kim, J.J.; Kang, B.S.; Kim, D.K. Polyethyleneimine□associated polycaprolactone—Superparamagnetic iron oxide nanoparticles as a gene delivery vector. *J. Biomed. Mater. Res. B Appl. Biomater.,* **2017**, *105*(1), 145-154. [http://dx.doi.org/10.1002/jbm.b.33519] [PMID: 26443109]

[79] Zhang, P.; Li, B.; Du, J.; Wang, Y. Regulation the morphology of cationized gold nanoparticles for effective gene delivery. *Colloids Surf. B Biointerfaces,* **2017**, *157*, 18-25. [http://dx.doi.org/10.1016/j.colsurfb.2017.04.056] [PMID: 28558339]

[80] Yezhelyev, M.V.; Qi, L.; O'Regan, R.M.; Nie, S.; Gao, X. Proton-sponge coated quantum dots for siRNA delivery and intracellular imaging. *J. Am. Chem. Soc.,* **2008**, *130*(28), 9006-9012. [http://dx.doi.org/10.1021/ja800086u] [PMID: 18570415]

[81] Ghafary, S.M.; Nikkhah, M.; Hatamie, S.; Hosseinkhani, S. Simultaneous gene delivery and tracking through preparation of photo-luminescent nanoparticles based on graphene quantum dots and chimeric peptides. *Sci. Rep.,* **2017**, *7*(1), 9552. [http://dx.doi.org/10.1038/s41598-017-09890-y] [PMID: 28842617]

[82] Lin, G.; Chen, T.; Zou, J.; Wang, Y.; Wang, X.; Li, J.; Huang, Q.; Fu, Z.; Zhao, Y.; Lin, M.C.M.; Xu, G.; Yong, K.T. Quantum dots-sirna nanoplexes for gene silencing in central nervous system tumor cells. *Front. Pharmacol.,* **2017**, *8*, 182. [http://dx.doi.org/10.3389/fphar.2017.00182] [PMID: 28420995]

[83] Yin, F.; Hu, K.; Chen, S.; Wang, D.; Zhang, J.; Xie, M.; Yang, D.; Qiu, M.; Zhang, H.; Li, Z. Black phosphorus quantum dot based novel siRNA delivery systems in human pluripotent teratoma PA-1 cells. *J. Mater. Chem. B Mater. Biol. Med.,* **2017**, *5*(27), 5433-5440. [http://dx.doi.org/10.1039/C7TB01068K] [PMID: 32264082]

[84] Yang, X.; Wang, Y.; Shen, X.; Su, C.; Yang, J.; Piao, M.; Jia, F.; Gao, G.; Zhang, L.; Lin, Q. One-step synthesis of photoluminescent carbon dots with excitation-independent emission for selective bioimaging and gene delivery. *J. Colloid Interface Sci.,* **2017**, *492*, 1-7. [http://dx.doi.org/10.1016/j.jcis.2016.12.057] [PMID: 28068539]

[85] Zhou, J.; Deng, W.; Wang, Y.; Cao, X.; Chen, J.; Wang, Q.; Xu, W.; Du, P.; Yu, Q.; Chen, J.; Spector, M.; Yu, J.; Xu, X. Cationic carbon quantum dots derived from alginate for gene delivery: One-step synthesis and cellular uptake. *Acta Biomater.,* **2016**, *42*, 209-219. [http://dx.doi.org/10.1016/j.actbio.2016.06.021] [PMID: 27321673]

[86] Kim, T.K.; Eberwine, J.H. Mammalian cell transfection: The present and the future. *Anal. Bioanal. Chem.,* **2010**, *397*(8), 3173-3178. [http://dx.doi.org/10.1007/s00216-010-3821-6] [PMID: 20549496]

[87] Xu, X.; Cao, J.; Deng, W.; Cao, X.; Chen, J.; Wang, Y.; Wang, S.; Yu, J.; Gao, X.; Yu, Q.; Chen, B. Efficient gene delivery to human umbilical cord mesenchymal stem cells by cationized porphyra yezoensis polysaccharide nanoparticles. *Int. J. Nanomedicine,* **2015**, *10*, 7097-7107. [http://dx.doi.org/10.2147/IJN.S93122] [PMID: 26604758]

CHAPTER 6

Application of Carbonaceous Quantum Dots in Solar Cells

Ekta Sharma[1,*] and **Vaishali Thakur**[1]

[1] *Department of Chemistry, School of Basic and Applied Sciences, Maharaja Agrasen University, Baddi, India*

Abstract: Carbonaceous Quantum Dots (CQDs) are gaining the interest of various research groups owing to their significant features, including water stability, biocompatibility, minimal cytotoxicity, chemical inertness and fluorescence which make them a good candidate in solar cells applications such as photocatalysis, solar energy conversion, photovoltaic solar cells, and Photoelectrochemical cells. CQDs are used in photocatalytic reactions because they can be used as electron sinks to stop the coupling of electron void pairs. The high coefficient of absorption and the broad spectrum of absorption improve the photocatalytic activity. In solar cells, the CQDs are used as sensitizers. CQDs are employed in solar energy generation because they are non-toxic and affordable. This chapter discusses the use of CQDs in solar cells.

Keywords: Carbonaceous quantum dots, Photocatalysis, Photovoltaic solar cells, Photoelectrochemical water splitting.

INTRODUCTION

The massive usage of fossil fuels, combined with concern about worldwide warming produced by global industry, has compelled the search for clean, ecologically friendly energy sources [1 - 3]. Solar energy is among the best options for a potential source of energy [4]. With the sun contributing about 86PW of energy onto the earth's crust annually, the fundamental transition away from our inefficient and environmentally harmful petroleum-based capitalism is made possible [5]. Photovoltaic technologies such as thin-coating solar cells, silicon solar cells, dye-sensitized solar cells, perovskite solar cells and quantum dots sensitized solar cells are all common uses of solar energy [4, 6, 7]. Owing to the novel and adaptable properties of semiconductor QDs, such as extreme absorption coefficient and variable band gap, QDs solar cells are emerging as an excellent candidate among third-generation photovoltaics [8 - 11]. CdS, CdSe,

* **Corresponding author Ekta Sharma:** Department of Chemistry, School of Basic and Applied Sciences, Maharaja Agrasen University, Baddi, India; Email: ekta25515@gmail.com

Kulvinder Singh & Vineet Kumar (Eds.)

CdTe, PbS, PbSe, $CuInS_2$, and other low-band-gap metal chalcogenide QDs are commonly utilized as sensitizers in QDSCs [12 - 15].

In addition to semiconductor quantum dots, CQDs are attaining the interest of researchers nowadays. The CQDs were invented in 2004 by Xu *et al.* and further synthesized by Sun and Group in 2006, which resulted in better fluorescence emission [16, 17]. There are two approaches to synthesizing CQDs termed top-down and bottom-up techniques. The top-down approach includes acidic oxidation, chemical oxidation, electrochemical oxidation and laser ablation [18 - 21], whereas the bottom-up technique includes carbonization, thermal decomposition and hydrothermal synthesis [22 - 25]. CQDs are used in various applications owing to their superior attributes, including minimal cytotoxicity, high Quantum yield, easy functionalization, easy synthesis and chemical inertness [26].

Photocatalysts are supposed to ideally work in both the UV or near-infrared regions of the spectrum. There are many catalysts which fail to work in this region. Due to upconverted and downconverted PL, CQDs widely work in longer wavelength of light, which make CQDs the best candidate for photocatalysis [27]. Charge separation and photoexcitation could be the mechanism of photocatalysis in CQDs [28]. The holes and electrons would be confined at the surface locations. The fluorescent dispersion observed in CQDs is spurred on by the recombination of electrons and holes. The quantum containment effect in semiconductor QDs does not hold for CQDs, even though photoexcitation in both types of QDs is the same. Either interface imperfection entrapment regions or electron and hole interaction pathways are responsible for the wavelengths of emission in CQDs [29].

The CQDs can operate as electron donors as well as electron acceptors, allowing holes and electrons to be separated. CQDs could be utilized as the spectral converter, solitary photocatalysts, electron mediators and photosensitizers. In many circumstances, these effects can occur at the same time. Defects can scatter or trap free-charge carriers, increasing the probability of recombination. As a result, the segregation of photogenerated electrons and holes must be promoted in order to optimize photocatalytic performance. CQDs can delay recombination at junction interfaces due to their large electron storage capacity [30]. Due to their ability to down and upconversion of PL, CQDs can also operate as photosensitizers [31 - 33]. Due to upconversion PL, CQDs are also utilized as spectral converters. Surface oxygen-containing functional groups influence the photocatalytic activity of CQDs [30, 34].

Carbonaceous Quantum Dots in Photocatalysis

CQDs show superiority in features including minimal toxicity, water dispersion and chemical resilience, due to which CQDs are used in photocatalytic applications. Moreover, after particular surface modification, CQDs have outstanding and modified optical characteristics of absorbance and PL [35 - 41]. The term "UCPL with CQDs" refers to an optical phenomenon where substances produce lower wavelengths of light beyond the source of excitation and can considerably increase the absorption of wide band gap semiconductors in the visible and near-infrared. Additionally, photoinduced CQDs constitute effective electron acceptors and donors, enabling effective electron and hole separation. As a result, CQDs can be used in a variety of photocatalyst applications, including electron mediators, photosensitizers, spectrum converters, solitary photocatalysts, photo centres and catalytic centres [35 - 37, 39, 42]. In fact, in many circumstances, these many effects occur at the same time. In this section, we will highlight the many functions of CQDs in CQDs-based photocatalytic systems and their various applications.

Electron Mediator

For photocatalytic activity, photocatalysts' ability to transmit and separate photogenerated holes and electrons is crucial. Random flaws, however, can either capture or disperse free charge transfer (holes and electrons), which raises the possibility of recombination [43]. Enhancing the segregation of photogenerated voids and electrons will facilitate photocatalysis. It has been demonstrated that carbon nanostructures have a significant potential to store electrons [44]. The conductive matrix of CQDs can therefore be used to transport photo-stimulated electrons through semiconductors or similar photocatalysts, delaying photogenerated carrier splitting at the intersection junction. CQDs show substantial interest as an electron mediators due to their low toxicity and simplicity of production [45, 46].

Photosensitizer

CQDs exhibit UCPL and PL, making them suitable photosensitizers to be utilized in photocatalytic systems in addition to acting as electron mediators [31 - 33, 37, 47, 48]. CQDs, on the other hand, act as a photosensitizer in a variety of ways. To begin with, CQDs are commonly terminated at their surface by carboxylic acid groups, which provide them with great dispersion in an aqueous solution and allow for additional modification. Moreover, as organic dyes [49 - 53], CQDs can also be made on a wide scale by using eco-friendly processes at a low cost. CQDs might also be used to substitute CdS and CdSe QDs owing to their minimal cytotoxicity [54]. CQDs were used as a photosensitizer for photocatalytic CO_2

reduction using polyethylene glycol (PEG) that have been passivated and have noble metals Au or Pt deposited on them [33]. As seen in Fig. (**1**), CQDs efficiently initiate the photocatalytic procedure by harvesting visible photons and converting them into charge transporters on their substratum. Moreover, the catalyst's water solubility allows a photo-reduction process to occur under more ideal homogenous reaction conditions.

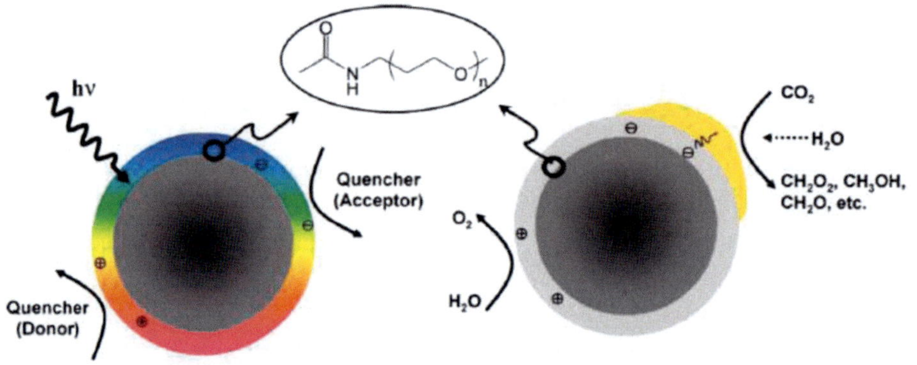

Fig. (1). Prior (left, fluorescence) and afterward (right, photocatalytic for CO_2 reduction in aqueous medium) coated with metal, solubilized PEG-functionalized carbon nanoparticles. For effective Conversion of CO_2 and carbon nanoparticles act as visible-light photocatalysts. Reprinted with permission from ref [55].

Spectral Converter

Various photocatalysts are semiconductors with wide bandgaps and only excite under UV light irradiation, such as TiO_2 and ZnO. As UV radiation accounts for only 4% of total sunlight, this severely limits their applications [56]. To use the complete spectrum of sunlight, wide bandgap semiconductors, including CdTe and CdS, are typically coupled with a traditional TiO_2-based photocatalytic system, but their instability and toxicity limit their use in various practical applications. Therefore more robust technologies are needed to extend the array of absorption from visible light to UV. CQD-based hybrid photocatalysts employ the UCPL property of CQDs as a spectral converter [57 - 62]. Pan *et al.* discussed the photodegradation of Rhodamine B by CQDs/TiO_2 nanocomposite in visible light [59]. The findings imply that the photocatalytic activity of CQDs and TiO_2 nanotubes is significantly greater than that of pure TiO_2. The elevated visible-light photoactivity of CQDs is caused by their UCPL characteristic. The CQDs/TiO_2 nanocomposite emits UCPL emission with excitation wavelengths ranging from 550 to 850nm, while in the case of blank TiO_2, it cannot be excited, which shows that in the composite system, the CQDs worked as a spectral converter.

However, it is clear that TiO_2 does not completely use the UCPL of CQDs, especially when the excitation wavelength of 600nm is used. Under visible light, Fig. (**2**) depicts a photocatalytic process for the CQDs/TiO_2 nanocomposite.

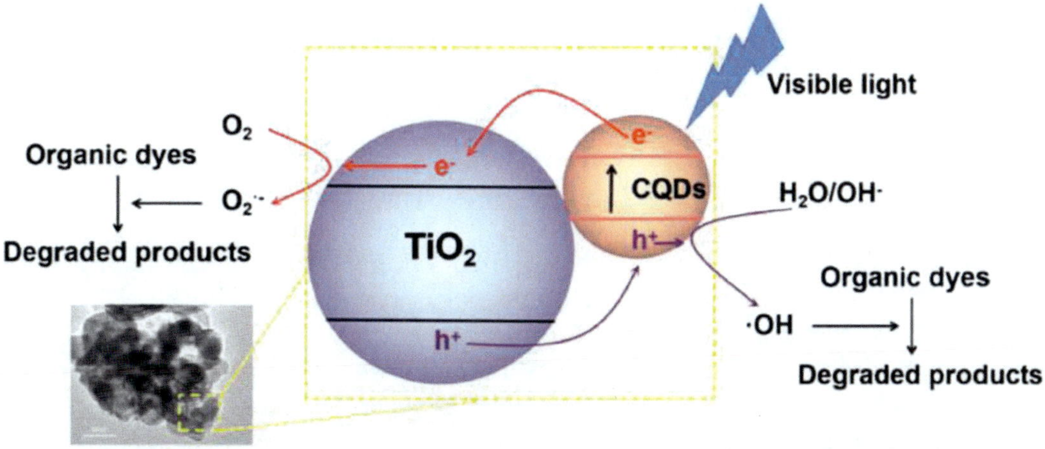

Fig. (2). Graphical representation of the photocatalytic process for the composites of CQDs and TiO_2 nanotubes when exposed to visible light [63].

Sole Photocatalyst

Although the photocatalytic uses of CQDs discussed above include the use of other inorganic materials, it has been found that CQDs could also function as photocatalysts alone [34, 64, 65]. Hu and coworkers have presented a mechanism in which CQDs act as photocatalysts and observe that the oxygenated functional groups present in CQDs enhance the photocatalytic activity [66]. Similarly, CQDs were also synthesized by Kang *et al.* with an electrochemical technique having a size in the range of 1-4nm. These CQDs oxidize benzyl alcohol to form benzaldehyde with the existence of H_2O_2 with 100% selectivity near IR region [34].

Carbonaceous Quantum Dots in Photoelectrochemical Water Splitting

A PEC cell is a particular kind of solar cell that uses solar energy to electrolyze H_2O into oxygen and hydrogen after converting it to electrical energy. Through the PEC water-splitting process, solar energy is transformed into chemical energy, and hydrogen is created. By irradiating a photo anode with solar energy in the form of hydrogen, this cell has been proposed as a way to use solar energy as a fuel.

ZnO, TiO_2 and WO_3 are the most common transition metal oxides utilized in PEC photoanodes. Reduced light absorption and rapid recombination characterize photo anodes constructed from these oxides. However, the use of semiconductor QDs has resulted in light absorption and a lower rate of recombination. Toxicity, poor stability and high production cost are some disadvantages of using semiconductor QDs in photo anodes. CQDs can be utilized in PEC photo anodes instead of semiconductor QDs, resulting in increased stability, lower production costs and reduced toxicity issues [67].

CQDs with a broad spectrum absorption range are useful as photosensitizers. The excited electron in CQDs returns to the initial phase gradually through phosphorescence and fluorescent rather than heat relaxation. As a result, the electrons can be transmitted to the contacting acceptor, which is advantageous for a sensitizer [68]. CQDs act as electro-sinks, transporting electrons from the surface of the semiconductor to the substances. The electron-hole pair will not recombine as a result of this. Additionally, the CQDs function like a semiconductor to produce charge bearers like holes and electrons, which increases the photocatalytic activity in PEC cells [67]. For instance, TiO_2 nanotube arrays in PEC water-splitting devices used CQDs as a sensitizer. At 0V versus Ag/AgCl in the presence of sunlight, the photocurrent density of the cell improved 4 times more than the non-sensitized cell. A 14.01 mol h-1 hydrogen generation rate with almost 100% Faradic efficiency was achieved using CQDs. Fig. (**3**) depicts the working process of DSSC.

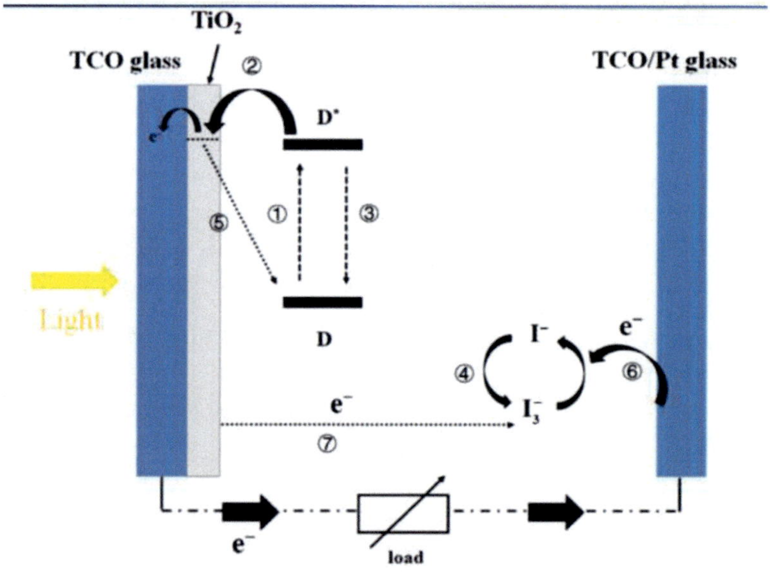

Fig. (3). Working principle of a DSCC device [69].

Similarly, an electron in the HOMO region of the CQD is excited to the LUMO region by a photon with sufficient energy. The conduction band of the nearby TiO_2 nanotube is then entered by the excited electron. The electron is then transported to the counter electrode and employed in the H_2 synthesis process after entering the conduction band of TiO_2. The reagent is oxidized by the defect in the CQD's HOMO state, completing the galvanic cell's circle.

The UV region shows minor CQD sensitization on TiO_2, which could be owing to CQD's reduced light absorption capacity in this region as compared to TiO_2. Only photo anodes sensitized using CQDs and irradiated having 300 W Xenon lamp, with an additional L42 cutoff filter, can detect hydrogen generation. This demonstrates that only the inclusion of CQDs into the photo anode allows hydrogen evolution to occur [68]. For water splitting, CQDs sensitized CQDs were used. When exposed to visible light, the photo anode showed up to five-time greater activity than TiO_2 without CQDs. The higher activity has been linked to the CQD's exceptional absorption of visible light when irradiated. As the concentration improves, the CQDs/TiO_2 photoanodes utilized in PEC absorb more light [67].

α-Fe_2O_3 (Hematite) is a widely studied photo anode with a band gap suitable for visible light absorption. The limited pore diffusion width of hematite can be overcome by using nanotubes, nanorods, and nanowires alongside metal-based co-catalyst. CQDs could be used as a co-catalyst with hematite. N-doped CQDs were used to arrange hematite nanorods on the Ti layer in a PEC cell enabling water oxidation. The photocurrent intensity of N-CDs@ -Fe_2O_3/Ti rose 3.4 times in comparison to the original -Fe_2O_3/Ti photoanode. Greater light wavelengths could be captured by N-doped CQDs due to the up-conversion fluorescence property of CQDs, which leads to reduced wavelength radiation. Additionally, -Fe_2O_3 can be excited by lower wavelength light, which creates electron-hole pair [70]. Additionally, a cocatalytic arrangement with numerous elements may enhance Fe_2O_3 photoanode performance in comparison to a single-component photocatalyst. A 60mV cathodically altered peak voltage for Fe_2O_3 photoanode was produced by a cocatalytic interaction between CQDs and Co_3O_4 in a PEC cell for the oxidation of water [71].

A thin layer of Al2O3 was passivated on the anode of TiO_2 nanorods that were sensitized using CQDs for the PEC water splitting system in order to boost the adherence among CQDS and TiO_2 nanorods. A 5nm thick Al_2O_3 sheet lit with air mass 1.5 worldwide has an IPCE (incident-photon-to-current-efficacy) of 76.12% at the 385nm wavelength. The achieved IPCE value was 6.94 times higher than that of sheer TiO_2 nanorods. Additionally, the IPCE value in the visible light area at 440 nm improved from 3.19 percent to 4.17 percent as a result of the sensitization influence of CQDs. The layer generates a steady photocurrent without deterioration during PEC testing [72].

Carbonaceous Quantum Dots in Photovoltaic Solar Cells

In addition to having the capacity to produce electron-hole pairs, CQDs can act as electron absorbers to stop electron-hole pairs from recombining. Additionally, CQDs have a wide absorbance spectrum, high absorption coefficients, low cost, and toxicity. They could so replace the expensive semiconductors currently employed in solar cells. CQDs could also be utilized in solar cells as a photoanode sensitizer. The dye is connected to TiO_2 by carboxylate ligands in the widely used Ru-based sensitizers. Surface-functionalized CQDs can attach to TiO_2 surfaces as well [73].

Strong photocatalytic efficiency and charge segregation are achieved by CQD/TiO_2 composites, which mix CQD with semiconductors. Due to the perfect band orientations between CQDs and TiO_2, photogenerated electrons from CQDs can be introduced into the adjacent conduction band of TiO_2 [27].

The photoanode of a dye-sensitized solar cell (DSSC) was equipped with CQDs made from Rosemary leaves, which led to a better power transfer efficacy than the photoanode lacking CQDs. The enhancement in nanoparticle connectivity causes electron hopping at the nanoparticle boundaries and charge carrier segregation to improve, increasing the efficacy of power transmission. In reality, electron lifespan improves as a result of electron transfer to the conduction band of TiO_2 and hole accumulation in the valence band of CQDs. CQDs have a broad absorption spectrum from near-IR to UV, which makes them ideal for use in DSSC. This results in an increase in photoelectron injection into TiO_2 nanoparticles, as well as the filling of deep trap sites. When electrons move through the conduction band, the trap site-blocking prevents them from feeling the recombination centers. When 5% CQDs are added to a TiO_2 photo anode, the cell characteristics improve, and the power conversion efficiency increases by 2.2 times [74]. The ZnO NPs functionalized using CQDs were employed in a DSSC in another investigation. The conversion efficacy of the fabricated solar cell was determined using the current-voltage (I-V) characteristics, using the equation $\eta = J_{sc}V_{oc}FF/I_{ph}$, where Jsc is short circuit current density, Voc is open-circuit voltage, FF is filled factor, and Iph is the source of stimulated light at 156 w/m^2. The introduction of 10% CQDs to ZnO NPs boosted conversion efficiency by 7 times. Additionally, the charge transfer resistance of the CQDs hybridized ZnO NPs DSSC was lower than that of the CQDs-free ZnO nanoparticles DSSC. Fig. (**4**) depicts the manufactured anode graphically.

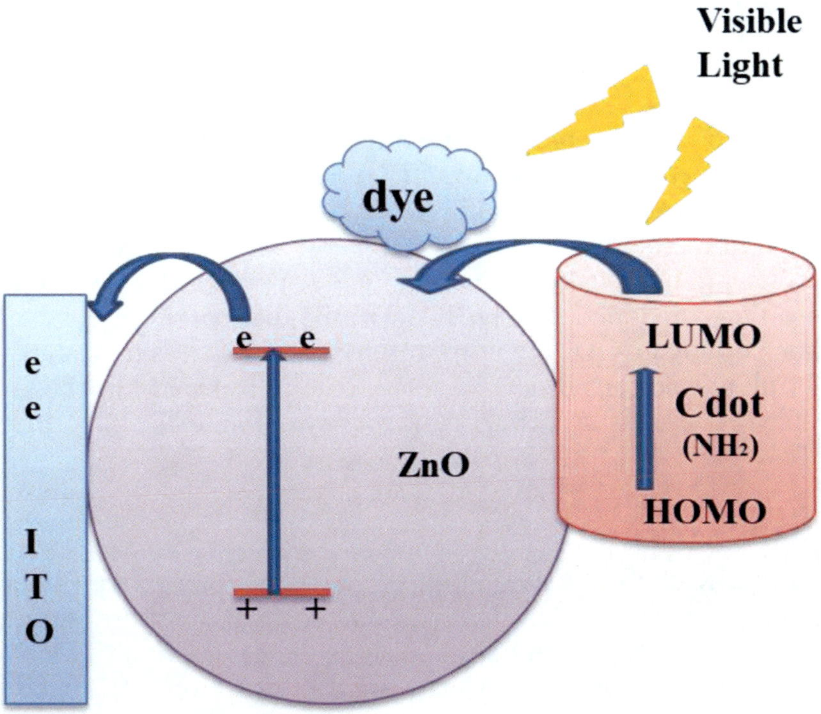

Fig. (4). Depiction of anode in DSSC.

When the dye is triggered by sunshine, the photo energy separates the electron pit pair in ZnO nanoparticles. On the other hand, the CQDs can boost the photovoltaic effect by preventing electron-hole pair recombination in ZnO nanoparticles, transferring electrons from the CQDs donor to the conductive band level of n-type semiconductor ZnO, or donating energy in excited CQDs. The split electrons are then transferred to the ITO. The amine in the CQD also functions as an electron donor.

The sensitizer species, CQDs concentration and ZnO size all have an effect on open-circuit voltage, although these characteristics have little effect on the fill factor. At a concentration of 10w% CQDs, the influencing factors for short-circuit current density and transition efficacy are at their highest levels [75].

Red algae were utilized as light harvesters in mesoscopic solar cell CQDs manufactured from the powder of Gelidium amansii. In order to produce the solar cell, the FTO glass substrate was used for the CQDs/N719/TiO_2 photoanode, FTO glass supported Pt for the counter electrode, and I-/I3- redox electrolyte for the counter electrode. The I-/I3- system's value of redox potential shows that there is a link for hole removal from N719 to the electrolyte, leading to better electron and hole separation, and the CQD's HOMO edge closely matches the HOMO value of the N710 dye [76].

The optical characteristics of CQDs are changed by N-doping. The use of N-doped CQDs in solar cells could improve the efficiency of power transmission. For instance, N-doped CQDs-sensitized TiO_2 nanorods were employed in DSSC because of their increased photocatalytic activity [47]. UV-visible absorption spectra were used to determine the LUMO energy values for synthesized GQDs, which were found to be in the region of 3.5 to 3.7eV [77]. CQDs work function is reduced when they are doped with N. TiO_2 (Rutile) has a 4.45eV conduction band energy. When N-doped CQDs are used to sensitize TiO_2, it is thermodynamically favorable for photoelectrons produced there to be introduced into TiO_2. N-doped CQD-sensitized TiO_2 has a better power conversion efficiency than CQDs-sensitized TiO_2 [47].

CONCLUSION

This chapter studied the implementation of CQDs in solar energy transmission. CQDs were used in various applications due to their various properties. The properties of CQDs, such as generation such as the production of electron voids pairs, recombination, sensitization and light absorption, make them good candidate for photocatalytic and solar energy transmission applications. The optical attributes of CQDs were influenced by methods of doping and functionalization. CQDs absorb light from UV to IR, which provides a wide spectrum range. However, there are some parameters of the absorption of light that require more research. By overcoming such spots, CQDs find economic applications in photocatalysis and solar energy conversion.

REFERENCES

[1] Lu, Y.B.; Li, L.; Su, S.C.; Chen, Y.J.; Song, Y.L.; Jiao, S.J. A novel TiO $_2$ nanostructure as photoanode for highly efficient CdSe quantum dot-sensitized solar cells. *RSC Advances,* **2017**, *7*(16), 9795-9802.
[http://dx.doi.org/10.1039/C6RA26029B]

[2] Li, Z.; Yu, L.; Liu, Y.; Sun, S. Efficient quantum dot-sensitized solar cell based on CdSxSe1-x/M--CdS/TiO2 nanotube array electrode. *Electrochim. Acta,* **2015**, *153*, 200-209.
[http://dx.doi.org/10.1016/j.electacta.2014.11.197]

[3] Peng, Z.; Chen, X.; Liu, Y.; Chen, J.; Chen, J. Balance on the charge generation, separation and transfer performance of different TiO 2 nanostructures in quantum dot sensitized solar cells. *Mater. Res. Bull.,* **2017**, *94*, 463-471.
[http://dx.doi.org/10.1016/j.materresbull.2017.06.041]

[4] Green, M.A.; Ho-Baillie, A.; Snaith, H.J. The emergence of perovskite solar cells. *Nature Photon,* **2014**, *8*, 506-514.
[http://dx.doi.org/10.1038/nphoton.2014.134]

[5] Essner, J.B.; Baker, G.A. The emerging roles of carbon dots in solar photovoltaics: A critical review. *Environ. Sci. Nano,* **2017**, *4*(6), 1216-1263.

[http://dx.doi.org/10.1039/C7EN00179G]

[6] Kouhnavard, M.; Ikeda, S.; Ludin, N.A.; Ahmad Khairudin, N.B.; Ghaffari, B.V.; Mat-Teridi, M.A.; Ibrahim, M.A.; Sepeai, S.; Sopian, K. A review of semiconductor materials as sensitizers for quantum dot-sensitized solar cells. *Renew. Sustain. Energy Rev.,* **2014**, *37*, 397-407. [http://dx.doi.org/10.1016/j.rser.2014.05.023]

[7] Decker, F.; Cattarin, S. Photoelectrochemical cells | Overview, Encycl. In: *Electrochem. Power Sources*; Elsevier, **2009**; pp. 1-9. [http://dx.doi.org/10.1016/B978-044452745-5.00035-6]

[8] Semonin, O.E.; Luther, J.M.; Choi, S.; Chen, H.Y.; Gao, J.; Nozik, A.J.; Beard, M.C. Beard, Peak external photocurrent quantum efficiency exceeding 100% *via* MEG in a quantum dot solar cell. *Science,* **2011**, *334*, 1530-1533.

[9] Emin, S.; Singh, S.P.; Han, L.; Satoh, N.; Islam, A. Colloidal quantum dot solar cells. *Sol. Energy,* **2011**, *85*(6), 1264-1282. [http://dx.doi.org/10.1016/j.solener.2011.02.005]

[10] Kramer, I.J.; Sargent, E.H. The architecture of colloidal quantum dot solar cells: Materials to devices. *Chem. Rev.,* **2014**, *114*(1), 863-882. [http://dx.doi.org/10.1021/cr400299t] [PMID: 24053639]

[11] Kim, M.R.; Ma, D. Quantum-dot-based solar cells: Recent advances, strategies, and challenges. *J. Phys. Chem. Lett.,* **2015**, *6*(1), 85-99. [http://dx.doi.org/10.1021/jz502227h] [PMID: 26263096]

[12] Santra, P.K.; Kamat, P.V. Tandem-layered quantum dot solar cells: Tuning the photovoltaic response with luminescent ternary cadmium chalcogenides. *J. Am. Chem. Soc.,* **2013**, *135*(2), 877-885. [http://dx.doi.org/10.1021/ja310737m] [PMID: 23249280]

[13] Zhu, H.; Song, N.; Lian, T. Charging of quantum dots by sulfide redox electrolytes reduces electron injection efficiency in quantum dot sensitized solar cells. *J. Am. Chem. Soc.,* **2013**, *135*(31), 11461-11464. [http://dx.doi.org/10.1021/ja405026x] [PMID: 23865741]

[14] Jiao, S.; Shen, Q.; Mora-Seró, I.; Wang, J.; Pan, Z.; Zhao, K.; Kuga, Y.; Zhong, X.; Bisquert, J. Band engineering in core/shell ZnTe/CdSe for photovoltage and efficiency enhancement in exciplex quantum dot sensitized solar cells. *ACS Nano,* **2015**, *9*(1), 908-915. [http://dx.doi.org/10.1021/nn506638n] [PMID: 25562411]

[15] Zhang, J.; Gao, J.; Church, C.P.; Miller, E.M.; Luther, J.M.; Klimov, V.I.; Beard, M.C. PbSe quantum dot solar cells with more than 6% efficiency fabricated in ambient atmosphere. *Nano Lett.,* **2014**, *14*(10), 6010-6015. [http://dx.doi.org/10.1021/nl503085v] [PMID: 25203870]

[16] Xu, X.; Ray, R.; Gu, Y.; Ploehn, H.J.; Gearheart, L.; Raker, K.; Scrivens, W.A. Electrophoretic analysis and purification of fluorescent single-walled carbon nanotube fragments. *J. Am. Chem. Soc.,* **2004**, *126*(40), 12736-12737. [http://dx.doi.org/10.1021/ja040082h] [PMID: 15469243]

[17] Sun, Y.P.; Zhou, B.; Lin, Y.; Wang, W.; Fernando, K.A.S.; Pathak, P.; Meziani, M.J.; Harruff, B.A.; Wang, X.; Wang, H.; Luo, P.G.; Yang, H.; Kose, M.E.; Chen, B.; Veca, L.M.; Xie, S.Y. Quantum-sized carbon dots for bright and colorful photoluminescence. *J. Am. Chem. Soc.,* **2006**, *128*(24), 7756-7757. [http://dx.doi.org/10.1021/ja062677d] [PMID: 16771487]

[18] Shen, P.; Xia, Y. Synthesis-modification integration: one-step fabrication of boronic acid functionalized carbon dots for fluorescent blood sugar sensing. *Anal. Chem.,* **2014**, *86*(11), 5323-5329. [http://dx.doi.org/10.1021/ac5001338] [PMID: 24694081]

[19] Yang, S.; Sun, J.; Li, X.; Zhou, W.; Wang, Z.; He, P.; Ding, G.; Xie, X.; Kang, Z.; Jiang, M. Large-

scale fabrication of heavy doped carbon quantum dots with tunable-photoluminescence and sensitive fluorescence detection. *J. Mater. Chem. A Mater. Energy Sustain.,* **2014**, *2*(23), 8660-8667. [http://dx.doi.org/10.1039/c4ta00860j]

[20] Jia, X.; Li, J.; Wang, E. One-pot green synthesis of optically pH-sensitive carbon dots with upconversion luminescence. *Nanoscale,* **2012**, *4*(18), 5572-5575. [http://dx.doi.org/10.1039/c2nr31319g] [PMID: 22786671]

[21] Yu, P.; Lowe, S.E.; Simon, G.P.; Zhong, Y.L. Electrochemical exfoliation of graphite and production of functional graphene. *Curr. Opin. Colloid Interface Sci.,* **2015**, *20*(5-6), 329-338. [http://dx.doi.org/10.1016/j.cocis.2015.10.007]

[22] Sharma, E.; Vashisht, D.; Vashisht, A.; Vats, V.K.; Mehta, S.K.; Singh, K. Facile synthesis of sulfur and nitrogen codoped graphene quantum dots for optical sensing of Hg and Ag ions. *Chem. Phys. Lett.,* **2019**, *730*, 436-444. [http://dx.doi.org/10.1016/j.cplett.2019.06.040]

[23] Sharma, E.; Vashisht, D.; Thakur, V.; Vashisht, A.; Mehta, S.K.; Singh, K. Graphene quantum dots functionalized with bovine serum albumin for sensing of hypochlorite ions. *Mater. Chem. Phys.,* **2021**, *273*, 125088. [http://dx.doi.org/10.1016/j.matchemphys.2021.125088]

[24] Vashisht, D.; Sharma, E.; Kaur, M.; Vashisht, A.; Mehta, S.K.; Singh, K. Solvothermal assisted phosphate functionalized graphitic carbon nitride quantum dots for optical sensing of Fe ions and its thermodynamic aspects. *Spectrochim. Acta A Mol. Biomol. Spectrosc.,* **2020**, *228*, 117773. [http://dx.doi.org/10.1016/j.saa.2019.117773] [PMID: 31740119]

[25] Vashisht, D.; Sangar, S.; Kaur, M.; Sharma, E.; Vashisht, A.; Ibhadon, A.O.; Sharma, S.; Mehta, S.K.; Singh, K. Biosynthesis of silver nanospheres, kinetic profiling and their application in the optical sensing of mercury and chlorite ions in aqueous solutions. *Environ. Res.,* **2021**, *197*, 111142. [http://dx.doi.org/10.1016/j.envres.2021.111142] [PMID: 33865822]

[26] Lim, S.Y.; Shen, W.; Gao, Z. Carbon quantum dots and their applications. *Chem. Soc. Rev.,* **2015**, *44*(1), 362-381. [http://dx.doi.org/10.1039/C4CS00269E] [PMID: 25316556]

[27] Zhang, Z.; Zheng, T.; Li, X.; Xu, J.; Zeng, H. Progress of carbon quantum dots in photocatalysis applications. *Part. Part. Syst. Charact.,* **2016**, *33*(8), 457-472. [http://dx.doi.org/10.1002/ppsc.201500243]

[28] Sharma, E.; Thakur, V.; Sangar, S.; Singh, K. Recent progress on heterostructures of photocatalysts for environmental remediation.*Mater. Today Proc*; Elsevier Ltd, **2020**, pp. 584-593. [http://dx.doi.org/10.1016/j.matpr.2020.02.403]

[29] Fernando, K.A.S.; Sahu, S.; Liu, Y.; Lewis, W.K.; Guliants, E.A.; Jafariyan, A.; Wang, P.; Bunker, C.E.; Sun, Y.P. Carbon quantum dots and applications in photocatalytic energy conversion. *ACS Appl. Mater. Interfaces,* **2015**, *7*(16), 8363-8376. [http://dx.doi.org/10.1021/acsami.5b00448] [PMID: 25845394]

[30] Wang, R.; Lu, K.Q.; Tang, Z.R.; Xu, Y.J. Recent progress in carbon quantum dots: Synthesis, properties and applications in photocatalysis. *J. Mater. Chem. A Mater. Energy Sustain.,* **2017**, *5*(8), 3717-3734. [http://dx.doi.org/10.1039/C6TA08660H]

[31] Martindale, B.C.M.; Hutton, G.A.M.; Caputo, C.A.; Reisner, E. Solar hydrogen production using carbon quantum dots and a molecular nickel catalyst. *J. Am. Chem. Soc.,* **2015**, *137*(18), 6018-6025. [http://dx.doi.org/10.1021/jacs.5b01650] [PMID: 25864839]

[32] Wang, J.; Gao, M.; Ho, G.W. Bidentate-complex-derived TiO2/carbon dot photocatalysts: *in situ* synthesis, versatile heterostructures, and enhanced H2 evolution. *J. Mater. Chem. A Mater. Energy Sustain.,* **2014**, *2*(16), 5703-5709. [http://dx.doi.org/10.1039/c3ta15114j]

[33] Cao, L.; Sahu, S.; Anilkumar, P.; Bunker, C.E.; Xu, J.; Fernando, K.A.S.; Wang, P.; Guliants, E.A.; Tackett, K.N., II; Sun, Y.P. Carbon nanoparticles as visible-light photocatalysts for efficient CO2 conversion and beyond. *J. Am. Chem. Soc.,* **2011**, *133*(13), 4754-4757. [http://dx.doi.org/10.1021/ja200804h] [PMID: 21401091]

[34] Li, H.; Liu, R.; Lian, S.; Liu, Y.; Huang, H.; Kang, Z. Near-infrared light controlled photocatalytic activity of carbon quantum dots for highly selective oxidation reaction. *Nanoscale,* **2013**, *5*(8), 3289-3297. [http://dx.doi.org/10.1039/c3nr00092c] [PMID: 23467384]

[35] Gao, Y.; Hou, F.; Hu, S.; Wu, B.; Wang, Y.; Zhang, H.; Jiang, B.; Fu, H. Graphene quantum-do--modified hexagonal tubular carbon nitride for visible-light photocatalytic hydrogen evolution. *ChemCatChem,* **2018**, *10*(6), 1330-1335. [http://dx.doi.org/10.1002/cctc.201701823]

[36] Himaja, A.L.; Karthik, P.S.; Singh, S.P. Carbon dots: The newest member of the carbon nanomaterials family. *Chem. Rec.,* **2015**, *15*(3), 595-615. [http://dx.doi.org/10.1002/tcr.201402090] [PMID: 25755070]

[37] Yu, H.; Zhao, Y.; Zhou, C.; Shang, L.; Peng, Y.; Cao, Y.; Wu, L.Z.; Tung, C.H.; Zhang, T. Carbon quantum dots/TiO2 composites for efficient photocatalytic hydrogen evolution. *J. Mater. Chem. A Mater. Energy Sustain.,* **2014**, *2*(10), 3344-3351. [http://dx.doi.org/10.1039/c3ta14108j]

[38] Bai, J.; Xiao, N.; Wang, Y.; Li, H.; Liu, C.; Xiao, J.; Wei, Y.; Guo, Z.; Qiu, J. Coal tar pitch derived nitrogen-doped carbon dots with adjustable particle size for photocatalytic hydrogen generation. *Carbon,* **2021**, *174*, 750-756. [http://dx.doi.org/10.1016/j.carbon.2020.10.088]

[39] Yadav, N.; Adolfsson, K.H.; Hakkarainen, M. Carbon dot-triggered photocatalytic degradation of cellulose acetate. *Biomacromolecules,* **2021**, *22*(5), 2211-2223. [http://dx.doi.org/10.1021/acs.biomac.1c00273] [PMID: 33905248]

[40] Cailotto, S.; Negrato, M.; Daniele, S.; Luque, R.; Selva, M.; Amadio, E.; Perosa, A. Carbon dots as photocatalysts for organic synthesis: metal-free methylene–oxygen-bond photocleavage. *Green Chem.,* **2020**, *22*(4), 1145-1149. [http://dx.doi.org/10.1039/C9GC03811F]

[41] Cui, G.; Wang, W.; Ma, M.; Zhang, M.; Xia, X.; Han, F.; Shi, X.; Zhao, Y.; Dong, Y.B.; Tang, B. Rational design of carbon and TiO2 assembly materials: Covered or strewn, which is better for photocatalysis? *Chem. Commun.,* **2013**, *49*(57), 6415-6417. [http://dx.doi.org/10.1039/c3cc42500b] [PMID: 23752661]

[42] Achilleos, D.S.; Kasap, H.; Reisner, E. Photocatalytic hydrogen generation coupled to pollutant utilisation using carbon dots produced from biomass. *Green Chem.,* **2020**, *22*(9), 2831-2839. [http://dx.doi.org/10.1039/D0GC00318B]

[43] Yao, Y.; Li, G.; Ciston, S.; Lueptow, R.M.; Gray, K.A. Photoreactive TiO2/carbon nanotube composites: Synthesis and reactivity. *Environ. Sci. Technol.,* **2008**, *42*(13), 4952-4957. [http://dx.doi.org/10.1021/es800191n] [PMID: 18678032]

[44] Zhang, L.W.; Fu, H.B.; Zhu, Y.F. Efficient TiO_2 photocatalysts from surface hybridization of TiO_2 particles with graphite-like carbon. *Adv. Funct. Mater.,* **2008**, *18*(15), 2180-2189. [http://dx.doi.org/10.1002/adfm.200701478]

[45] Le, N.T.H.; Thanh, T.D.; Pham, V.T.; Phan, T.L.; Lam, V.D.; Manh, D.H.; Anh, T.X.; Le, T.K.C.; Thammajak, N.; Hong, L.V.; Yu, S.C. Structure and high photocatalytic activity of (N, Ta)-doped TiO_2 nanoparticles. *J. Appl. Phys.,* **2016**, *120*(14), 142110. [http://dx.doi.org/10.1063/1.4961718]

[46] Zhang, H.; Huang, H.; Ming, H.; Li, H.; Zhang, L.; Liu, Y.; Kang, Z. Carbon quantum dots/Ag3PO4

complex photocatalysts with enhanced photocatalytic activity and stability under visible light. *J. Mater. Chem.,* **2012**, *22*(21), 10501-10506.
[http://dx.doi.org/10.1039/c2jm30703k]

[47] Zhang, Y.Q.; Ma, D.K.; Zhang, Y.G.; Chen, W.; Huang, S.M. N-doped carbon quantum dots for TiO2-based photocatalysts and dye-sensitized solar cells. *Nano Energy,* **2013**, *2*(5), 545-552.
[http://dx.doi.org/10.1016/j.nanoen.2013.07.010]

[48] Qu, D.; Zheng, M.; Du, P.; Zhou, Y.; Zhang, L.; Li, D.; Tan, H.; Zhao, Z.; Xie, Z.; Sun, Z. Highly luminescent S, N co-doped graphene quantum dots with broad visible absorption bands for visible light photocatalysts. *Nanoscale,* **2013**, *5*(24), 12272-12277.
[http://dx.doi.org/10.1039/c3nr04402e] [PMID: 24150696]

[49] Zeng, W.; Cao, Y.; Bai, Y.; Wang, Y.; Shi, Y.; Zhang, M.; Wang, F.; Pan, C.; Wang, P. Efficient dye-sensitized solar cells with an organic photosensitizer featuring orderly conjugated ethylenedioxythiophene and dithienosilole blocks. *Chem. Mater.,* **2010**, *22*(5), 1915-1925.
[http://dx.doi.org/10.1021/cm9036988]

[50] Hara, K.; Sato, T.; Katoh, R.; Furube, A.; Yoshihara, T.; Murai, M.; Kurashige, M.; Ito, S.; Shinpo, A.; Suga, S.; Arakawa, H. Novel conjugated organic dyes for efficient dye-sensitized solar cells. *Adv. Funct. Mater.,* **2005**, *15*(2), 246-252.
[http://dx.doi.org/10.1002/adfm.200400272]

[51] Hua, Y.; Lin Lee, L.T.; Zhang, C.; Zhao, J.; Chen, T.; Wong, W.Y.; Wong, W.K.; Zhu, X. Co-sensitization of 3D bulky phenothiazine-cored photosensitizers with planar squaraine dyes for efficient dye-sensitized solar cells. *J. Mater. Chem. A Mater. Energy Sustain.,* **2015**, *3*(26), 13848-13855.
[http://dx.doi.org/10.1039/C5TA01665G]

[52] Yeh, T.F.; Syu, J.M.; Cheng, C.; Chang, T.H.; Teng, H. Graphite oxide as a photocatalyst for hydrogen production from water. *Adv. Funct. Mater.,* **2010**, *20*(14), 2255-2262.
[http://dx.doi.org/10.1002/adfm.201000274]

[53] Yeh, T.F.; Teng, C.Y.; Chen, S.J.; Teng, H. Nitrogen-doped graphene oxide quantum dots as photocatalysts for overall water-splitting under visible light illumination. *Adv. Mater.,* **2014**, *26*(20), 3297-3303.
[http://dx.doi.org/10.1002/adma.201305299] [PMID: 24677453]

[54] Pan, Z.; Zhang, H.; Cheng, K.; Hou, Y.; Hua, J.; Zhong, X. Highly efficient inverted type-I CdS/CdSe core/shell structure QD-sensitized solar cells. *ACS Nano,* **2012**, *6*(5), 3982-3991.
[http://dx.doi.org/10.1021/nn300278z] [PMID: 22509717]

[55] Cao, L.; Sahu, S.; Anilkumar, P.; Bunker, C.E.; Xu, J.; Fernando, K.A.S.; Wang, P.; Guliants, E.A.; Tackett, K.N., II; Sun, Y.P. Carbon nanoparticles as visible-light photocatalysts for efficient CO2 conversion and beyond. *J. Am. Chem. Soc.,* **2011**, *133*(13), 4754-4757.
[http://dx.doi.org/10.1021/ja200804h] [PMID: 21401091]

[56] Zhang, Y.; Tang, Z.R.; Fu, X.; Xu, Y.J. TiO2-graphene nanocomposites for gas-phase photocatalytic degradation of volatile aromatic pollutant: Is TiO2-graphene truly different from other TiO2-carbon composite materials? *ACS Nano,* **2010**, *4*(12), 7303-7314.
[http://dx.doi.org/10.1021/nn1024219] [PMID: 21117654]

[57] Li, H.; He, X.; Liu, Y.; Yu, H.; Kang, Z.; Lee, S.T. Synthesis of fluorescent carbon nanoparticles directly from active carbon *via* a one-step ultrasonic treatment. *Mater. Res. Bull.,* **2011**, *46*(1), 147-151.
[http://dx.doi.org/10.1016/j.materresbull.2010.10.013]

[58] Ming, H.; Ma, Z.; Liu, Y.; Pan, K.; Yu, H.; Wang, F.; Kang, Z. Large scale electrochemical synthesis of high quality carbon nanodots and their photocatalytic property. *Dalton Trans.,* **2012**, *41*(31), 9526-9531.
[http://dx.doi.org/10.1039/c2dt30985h] [PMID: 22751568]

[59] Pan, J.; Sheng, Y.; Zhang, J.; Wei, J.; Huang, P.; Zhang, X.; Feng, B. Preparation of carbon quantum

dots/ TiO_2 nanotubes composites and their visible light catalytic applications. *J. Mater. Chem. A Mater. Energy Sustain.,* **2014**, *2*(42), 18082-18086.
[http://dx.doi.org/10.1039/C4TA03528C]

[60] Yu, X.; Liu, J.; Yu, Y.; Zuo, S.; Li, B. Preparation and visible light photocatalytic activity of carbon quantum dots/TiO2 nanosheet composites. *Carbon,* **2014**, *68*, 718-724.
[http://dx.doi.org/10.1016/j.carbon.2013.11.053]

[61] Wang, H.; Wei, Z.; Matsui, H.; Zhou, S. Fe_3O_4 /carbon quantum dots hybrid nanoflowers for highly active and recyclable visible-light driven photocatalyst. *J. Mater. Chem. A Mater. Energy Sustain.,* **2014**, *2*(38), 15740-15745.
[http://dx.doi.org/10.1039/C4TA03130J]

[62] Guo, Y.; Yao, P.; Zhu, D.; Gu, C. A novel method for the development of a carbon quantum dot/carbon nitride hybrid photocatalyst that responds to infrared light irradiation. *J. Mater. Chem. A Mater. Energy Sustain.,* **2015**, *3*(25), 13189-13192.
[http://dx.doi.org/10.1039/C5TA02262B]

[63] Jin, X.; Che, R.; Yang, J.; Liu, Y.; Chen, X.; Jiang, Y.; Liang, J.; Chen, S.; Su, H. Activated carbon and carbon quantum dots/titanium dioxide composite based on waste rice noodles: Simultaneous synthesis and application in water pollution control. *Nanomaterials,* **2022**, *12*(3), 472.
[http://dx.doi.org/10.3390/nano12030472] [PMID: 35159817]

[64] Ma, Z.; Ming, H.; Huang, H.; Liu, Y.; Kang, Z. One-step ultrasonic synthesis of fluorescent N-doped carbon dots from glucose and their visible-light sensitive photocatalytic ability. *New J. Chem.,* **2012**, *36*(4), 861-864.
[http://dx.doi.org/10.1039/c2nj20942j]

[65] Yang, P.; Zhao, J.; Wang, J.; Cui, H.; Li, L.; Zhu, Z. Pure carbon nanodots for excellent photocatalytic hydrogen generation. *RSC Advances,* **2015**, *5*(27), 21332-21335.
[http://dx.doi.org/10.1039/C5RA01924A]

[66] Hu, S.; Tian, R.; Wu, L.; Zhao, Q.; Yang, J.; Liu, J.; Cao, S. Chemical regulation of carbon quantum dots from synthesis to photocatalytic activity. *Chem. Asian J.,* **2013**, *8*(5), 1035-1041.
[http://dx.doi.org/10.1002/asia.201300076] [PMID: 23441085]

[67] Xie, S.; Su, H.; Wei, W.; Li, M.; Tong, Y.; Mao, Z. Remarkable photoelectrochemical performance of carbon dots sensitized TiO_2 under visible light irradiation. *J. Mater. Chem. A Mater. Energy Sustain.,* **2014**, *2*(39), 16365-16368.
[http://dx.doi.org/10.1039/C4TA03203A]

[68] Zhang, X.; Wang, F.; Huang, H.; Li, H.; Han, X.; Liu, Y.; Kang, Z. Carbon quantum dot sensitized TiO2 nanotube arrays for photoelectrochemical hydrogen generation under visible light. *Nanoscale,* **2013**, *5*(6), 2274-2278.
[http://dx.doi.org/10.1039/c3nr34142a] [PMID: 23417112]

[69] Yuan, Y.; Wan, C. Dual application of waste grape skin for photosensitizers and counter electrodes of dye-sensitized solar cells. *Nanomaterials,* **2022**, *12*(3), 563.
[http://dx.doi.org/10.3390/nano12030563]

[70] Huang, X.; Yang, L.; Hao, S.; Zheng, B.; Yan, L.; Qu, F.; Asiri, A.M.; Sun, X. N-Doped carbon dots: A metal-free co-catalyst on hematite nanorod arrays toward efficient photoelectrochemical water oxidation. *Inorg. Chem. Front.,* **2017**, *4*(3), 537-540.
[http://dx.doi.org/10.1039/C6QI00517A]

[71] Zhang, P.; Wang, T.; Chang, X.; Zhang, L.; Gong, J. Synergistic cocatalytic effect of carbon nanodots and co_3o_4 nanoclusters for the photoelectrochemical water oxidation on hematite. *Angew. Chem. Int. Ed.,* **2016**, *55*(19), 5851-5855.
[http://dx.doi.org/10.1002/anie.201600918] [PMID: 27011264]

[72] Feng, M.; Liu, Y.; Wei, N.; Ma, S.; Li, Z.; Li, H.; Chen, S.; Liu, J.; Wang, D. Alumina anchored CQDs/TiO_2 nanorods by atomic layer deposition for efficient photoelectrochemical water splitting

under solar light. *J. Mater. Chem. A Mater. Energy Sustain.,* **2018**, *6*(37), 18293-18303. [http://dx.doi.org/10.1039/C8TA05092A]

[73] Mirtchev, P.; Henderson, E.J.; Soheilnia, N.; Yip, C.M.; Ozin, G.A. Solution phase synthesis of carbon quantum dots as sensitizers for nanocrystalline TiO_2 solar cells. *J. Mater. Chem.,* **2012**, *22*(4), 1265-1269.
[http://dx.doi.org/10.1039/C1JM14112K]

[74] Rezaei, B.; Irannejad, N.; Ensafi, A.A.; Kazemifard, N. The impressive effect of eco-friendly carbon dots on improving the performance of dye-sensitized solar cells. *Sol. Energy,* **2019**, *182*, 412-419. [http://dx.doi.org/10.1016/j.solener.2019.02.072]

[75] Efa, M.T.; Imae, T. Effects of carbon dots on ZnO nanoparticle-based dye-sensitized solar cells. *Electrochim. Acta,* **2019**, *303*, 204-210.
[http://dx.doi.org/10.1016/j.electacta.2019.02.012]

[76] Dou, D.; Duan, J.; Zhao, Y.; He, B.; Tang, Q. Cubic carbon quantum dots for light-harvesters in mesoscopic solar cells. *Electrochim. Acta,* **2018**, *275*, 275-280.
[http://dx.doi.org/10.1016/j.electacta.2018.04.124]

[77] Yan, X.; Li, B.; Cui, X.; Wei, Q.; Tajima, K.; Li, L. Independent tuning of the band gap and redox potential of graphene quantum dots. *J. Phys. Chem. Lett.,* **2011**, *2*(10), 1119-1124.
[http://dx.doi.org/10.1021/jz200450r] [PMID: 26295312]

CHAPTER 7

Carbonaceous Quantum Dots as Efficient Zero-dimensional Nanomaterials for Sensing Applications

Alma Mejri[1,*], **Abdelmoneim Mars**[1] and **Hamza Elfil**[1]

[1] *Laboratory of Desalination and Natural Water Valorisation (LADVEN), Water Researches and Technologies Center (CERTE), Techno-park BorjCedria, BP 273, 8020, Soliman, Tunisia*

Abstract: Carbonaceous quantum dots (QDs), including carbon and graphene quantum dots, have been widely used recently in various sensing fields, such as healthcare sensing, environmental monitoring, and food safety. Owing to their exceptional electronic, fluorescent, photoluminescent, chemiluminescent, and electrochemiluminescent properties, carbonaceous QDs are essential tools for designing an ultra-sensitive sensing platform. In this chapter, we summarized the applications of carbonaceous QDs in the detection of various target analysts, citing heavy metals, toxic compounds, pesticides, and proteins (DNA, aptamer, and RNA). In this regard, the authors described the effects of synthetic methods and surface functionalization on the properties of carbonaceous QDs and the analytical performance of sensors. We believe that understanding these parameters gives us better sensors that could not be obtained by other means. To give the reader a clear vision of the implementation of these zero-dimensional nanomaterials in sensor architectures, a comparative study has been developed.

Keywords: Analytic performances, Carbonaceous quantum dots, Electrochemical proprieties, Optical proprieties, Sensing applications.

INTRODUCTION

Carbonaceous nanomaterials have been a focal point of interest in the construction of various analytical tools, especially sensors and biosensors, because of their exceptional properties that are of great interest in the sensing process. Scientists have paid a great deal of attention to graphene and carbon quantum dots as viable sub-class because they have special features like high conductivity, nontoxicity, excellent chemical photoluminescence durability, and tuneable fluorescence emissions. Moreover, the existence of a variety of functional groups on the carbon

* **Corresponding author Alma Mejri:** Laboratory of Desalination and Natural Water Valorisation (LADVEN), Water Researches and Technologies Center (CERTE), Techno-park BorjCedria, BP 273, 8020, Soliman, Tunisia; Email: abdelmoneim.mars@gmail.com

Kulvinder Singh & Vineet Kumar (Eds.)

core, including epoxy, hydroxyl, carboxyl, and carbonyl moieties, facilitate the functionalization of the carbonaceous QDs surface with various species, which can further open the opportunity for the fabrication of a huge variety of (bio)sensors.

Structurally, carbonaceous QDs have dissimilar structures, sizes, and surface functional groups. In regards to graphene QDs, they are made up of a thin layer (less than 10 layers) of graphene having a diameter <10 nm and are decorated with functional groups on the edges. These groups originated from carbon sources or subsequent reactions. However, carbon QDs owned spherical shapes with a crystalline graphite-like lattice. On the other hand, the methods for synthesizing carbonaceous QDs have been well reviewed in the literature, and extreme potential has been invested in pursuing the production of simple, low-cost, environmentally friendly, and controlled properties which can be monitored with QDs morphology, size, and doping type and amount [1 - 3].

To date, the literature on the applications of carbonaceous QDs in sensing fields has grown rapidly. Indeed, they have been applied in the detection of various target analytes, citing heavy metals, toxic compounds, pesticides, and proteins (DNA, aptamer, and RNA). The number of reports, described in the literature that mentioned sensing and carbonaceous quantum dots terms has increased exponentially in these ten recent years (Fig. **1**). Thus, the summary of their use in detection is of considerable interest to the reading community. This chapter is thematically focused on the exploitation of the fluorescence, chemiluminescence, electroluminescence, and electrochemical proprieties of the carbonaceous QDs in the sensing and biosensing applications reaching the picomolar, nanomolar, or femtomolar region. Additionally, sensing techniques such as photo-induced electron transfer (PET), fluorescence resonance energy transfer (FRET), dynamic quench, stable quenching, energy transmission, and quenching of fluorescence, were examined.

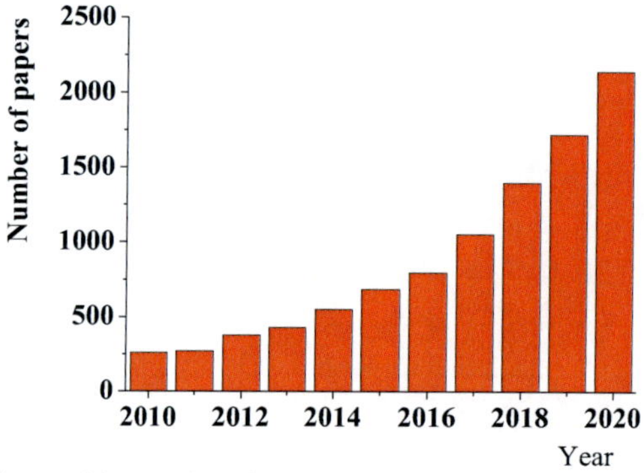

Fig. (1). Number of publications with mentioned "sensing and carbonaceous quantum dots" terms.

Photoluminescence (Bio) Sensors

Carbonaceous QDs, with their different structures, have identical optical features based on their absorption and fluorescence behaviors (Fig. **2**). Indeed, carbonaceous QDs' UV-Vis spectrum displays high optic absorption in the UV range (between 230-320 nm), having a tail ascending into the visible region. The π-π* transmission of aromatic C-C bonds present in the core carbon is responsible for the maximal spike at 230 nm, while the n-π* transmission of C=O bonds attached to the edges or other linked groups is responsible for the shoulder at 300 nm. We emphasize that discovered discrepancies in absorption spectrum data suggest variations in the compositions or structures of various derivatives of hybridized carbonaceous QDs. Concerning the emission spectra of carbon QDs, the position of the emission peaks is related to the excitation wavelengths [4]. This property can be obtained from a broad range of QD sizes, surface chemistry, and emissive traps. Interestingly, the photoluminescence color of QDs is relative to the surface groups and their molecular environment rather than the size. Fortunately, this behavior can be applied in (bio)sensing applications.

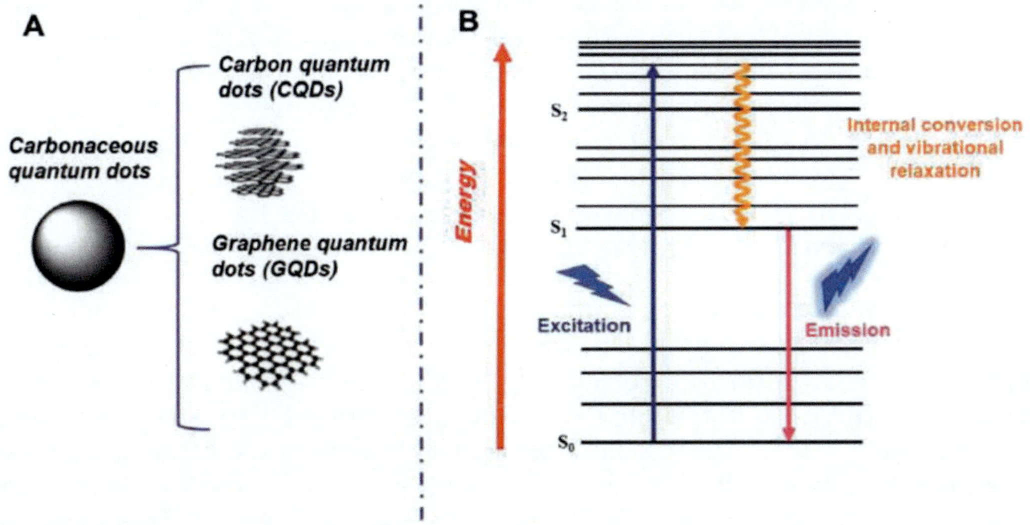

Fig. (2). (**A**) Structure of carbon quantum dot (CQDs) and graphene quantum dot (GQDs) (**B**) Schematic representation of photoluminescence principle.

Heavy Metal Ions Sensing

Currently, heavy metal pollution has become an emergent concern for global sustainability because they are highly toxic and carcinogenic, even at a trace level. These metals are nonbiodegradable and can stack up in the food chain, posing a significant environmental and human health risk [5]. Therefore, it is crucial to monitor heavy metal concentrations in drinking water, foods, and biological fluids. In literature, carbonaceous QDs-based sensors are widely used to sense several heavy metal ions citing Fe^{3+} [6 - 8], Pb^{2+} [9, 10], Hg^{2+} [11 - 13], Cd^{2+} [14], Co^{2+} [15] and Ni^{2+} [16]. This tremendous interest could be illustrated by the ability of these zero-dimensional nanomaterials to detect target ions in a highly sensitive and selective manner. In this context, Zhang and collaborators [17] reported the use of graphene QDs as an efficient fluorescent label-free sensing platform for Hg^{2+} ions in water. The GQDs were prepared *via* an ultrasonic route starting from graphene. Mercury detection induced a proportional decrease of the GQDs fluorescence concentration with the enhancement of Hg^{2+} ion amount. The sensor exhibited a broad region of concentrations across 8.10^{-7} to 9.10^{-6} M. As regards of sensing process, the authors described that it relies on the quench of the QDs fluorescent signal by target ions owing to a formation of an aggregation of GQDs-Hg^{2+} complexes seen that Hg^{2+} have a high affinity to carboxylate groups present on GQDs surface. As a result, this might encourage an effective electron transmission mechanism that leads to non-radiative electron/hole interaction destruction [18]. Moreover, Nandi's group [19] fabricated a carbon QDs paper-based strips detector for the sensing of lead ions (Pb^{2+}) in human cells. The carbon QDs were prepared to utilize basic microwave thermal treatment of a potato-dextrose agar as a biocompatible biological media and were then immobilized on the cellulose-based filter paper. XPS study revealed that prepared CQDs contain Nitrogen atoms with a % atom ratio of C:O: N = 58.64:34.89:6.47. Furthermore, fluorescence results revealed that the quenching of the CQDs fluorescence signal was related to the production of a molecule across Pb^{2+} and carbon QDs. The applicability of the CQD paper-based sensor considered the Pb^{2+} assay to be rapid and simple in human cells having a LOD of 110 pM.

Several reports stated that the quantum yield of carbonaceous QDs is relatively low, limiting their sensitivity to detection. As a promising solution, the scientists propose functionalization and/or doping surface of carbonaceous QDs with various heteroatoms citing sulfur (S), nitrogen (N), phosphorous (P) and boron (B), which enhance the fluorescence features by changing the bandgap of these zero-dimension nanomaterials. Based on this, a large range of preparation methods for doped carbonaceous QDs has been studied to apply them in sensing applications [20]. Ren *et al.* [21] used nitrogen-doped CQDs as a fluorescence sensor to measure sensitively and selectively Ag^{+} and Hg^{2+} ions. According to reports, strengthening and increasing the quantum yields of CQDs by doping the surface with nitrogen atoms results in a 56% quantum yield. As shown in Fig. (**3**), the sensing of Hg^{2+} and Ag^{+} ions induced a significant quenching of fluorescence response under the UV lamp. Since the fluorescence quenching mechanism of carbonaceous QDs with heavy metal ions is still unclear at present, the authors attributed the quenching to the nonradiative electron-transmission using the CQDs to the metal ions, and this is based on the previous reports. Indeed, Hg^{2+} and Ag^{+}

have excellent affinities and rapid chelating dynamics, including the carboxyl, hydroxyl, and amino group present on N-doped CQDs surface, changing the electronic structure of QDs and affecting the distribution of excitons.

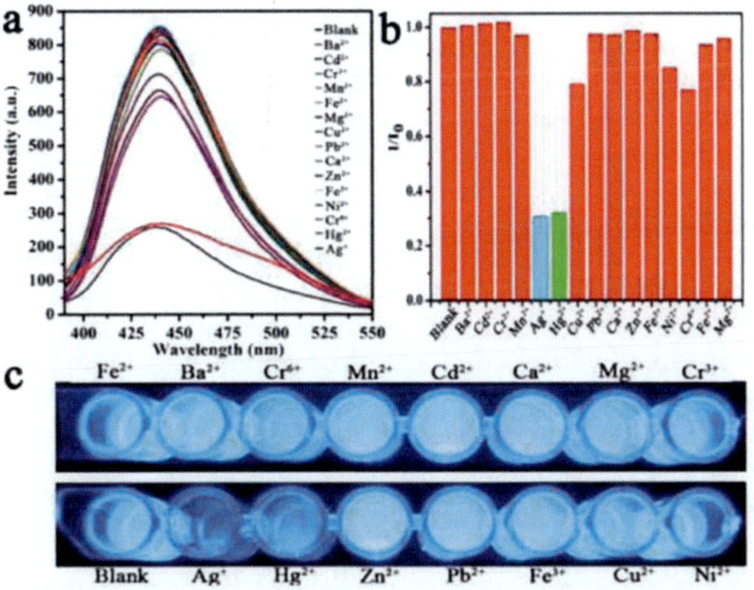

Fig. (3). **(A)** Photoluminescence spectrum of N-CQDs solution followed the introduction of 100 μM Ag^+ ion and 100 μM other ions **(B)** Relative fluorescence intensity of N-CQDs medium in the context of 100 μM of several metal ions (λ_{ex}=370 nm) **(C)** The corresponding photo images of detection under UV lamp [21].

Interestingly, other studies have promoted the advancement of the development of sulfur-doped carbonaceous QDs-based sensors. In this area, Bian *et al.* [22] describe the application of sulfur-doped graphene QDs for the sensing of Ag^+ ions. They report that S-GQDs were obtained by a one-pot hydrothermal procedure. Similar to graphene, the 1,3,6-trinitropyrene molecule was used as a carbon source, and 3-mercaptopropionic acid served as a supplier of sulphur and a carboxylic group. It has been demonstrated that sulphur doping significantly enhances the electrical and chemical characteristics of GQDs. There have been reports of a quick reaction, a broad linear spectrum of 0.1-130 mM, and a minimal LOD of 0.03 mM. Moreover, Xu *et al.* [23] synthesized sulfur-doped carbon QDs using a hydrothermal approach from sodium citrate and sodium thiosulfate as sulfur and carbon sources, correspondingly. Fluorescence studies indicate that a significant fluorescence quantum yield of 67% was determined. As an application, S-CQDs were applied for Fe^{3+} detection. Similarly, a quenching behavior was observed after the formation of the S-CQDs- Fe^{3+} complex. As for doping of carbonaceous QDs with boron, the simplest precursor is the exfoliation of graphite in a borax buffer ($Na_2B_4O_7 \cdot 10H_2O$). In this concern, Fan *et al.* [24]

prepared boron-doping of graphene QDs using an electrochemical exfoliation of graphite. In borax buffer, the quantum yield was increased from 3.6% to 13%. They reported that obtained B-GQDs possess a high fluorescence sensitivity to aluminum ion (Al^{3+}).

Furthermore, co-doping routes of carbonaceous QDs by two or more heteroatoms have been extensively described because of their exceptionally sensitive interaction with the surrounding media. For short, two applications of co-doped carbonaceous QDs in the sensing of heavy metal ions will be given [25, 26]. Omer and collaborators [27] have reported the use of a hydrothermal precursor to synthesize N, P-CQDs by mixing urea, citric acid and phosphoric acid. The generated QDs were used in a highly sensitive and selective manner to detect Fe3+ ions. The test has a limit of detection of 50 nM and responds linearly in the concentration region of 0.1 M to 0.9 M. The authors also said that absorption spectroscopy and energy state computations were used to explore the quenching mechanism. The outcomes demonstrated that the selective coordination of Fe3+ by the functional groups on the surface of the CQDs is what causes the quenching action. The photo-induced electron transmission from the CQDs to the d orbitals of Fe3+ is made easier as a result. Regarding nitrogen and sulfur co-doped CQDs, Mohapatra *et al.* [28] proposed an efficient sensing method of mercury (II) ion using N, S-CQDs. Co-doping CQDs with nitrogen and sulphur, according to their findings, enhances the electrical characteristics and external chemical properties of QDs with a quantum yield of 69%. Through a soft-soft interface between the nitrogen and sulphur co-doped carbon dot and the mercury core, the excited state's nonradiative electron transmission to the metal is responsible for the quenching process.

While the functionalization of CQDs has been elaborated to be an efficient way to improve the analytical performances of carbonaceous QDs-based sensors. In this context, Gao *et al.* [29] proposed a simple method to prepare amino-functionalized CQDs. The fluorescence sensitivity has been increased by 36% due to the functionalization process. The synthesized QDs have been exploited to quantify mercury (II) in water specimens with a LOD of 20 nmol.L^{-1}. Moreover, Dong *et al.* [30] discussed the utilization of the poly(ethylenimine) functionalized CQDs as a rapid fluorescent detecting system for the sensing of copper (II). They demonstrated that a cupric amine complex was formed after the sensing of Cu(II) ions. The authors proposed the internal filter effect (IFE) as a quenching mechanism that was monitored by an amine-copper complex. To prove the IFE mechanism, the authors compared the fluorescent response of the amine-based CQDs toward the presence of copper (II) ions with the fluorescent responses of bare CQDs and hydroxyl-functionalized CQDs. It was discovered that neither of the CQDs without amines responded substantially to Cu2+, demonstrating that cupric amine production is crucial for the CQDs containing poly(ethylenimine) caps to respond to Cu2+. As a second variant of the functionalization of carbonaceous QDs, many scientists have used chelating moieties to decorate the surface of these QDs. For instance, an extremely selective and sensitive fluorescent platform based on 1,4,8,11-tetraazacyclotetradecane (Cyclam)-functionalized carbon QDs for the copper (II) detection was discussed by Chen *et al.* [31], the Cyclam was served as a potential acceptor to chelate Cu^{2+} ions. The results revealed an acceptable linear concentration region of 0.7 µM to 4.0 µM with a LOD of 130 nM. The developed CQDs based-sensor could be employed for monitoring copper ion concentration inside live cells.

Recently, another carbonaceous QDs-based-sensing strategy was reported by Yuan *et al.* [32]. Indeed, they have used copper complex-based-carbon quantum dots ($CuDTC_2$-CQDs) to sense mercury (II). The sensing mechanism was monitoring by a complex displacement reaction (exchange ion between copper-based complex and mercury-based complex) profiting from the highest affinity of the chelate, bis-dithiocarbamate, to Hg^{2+}. This later could immediately activate the fluorescence signal of the CQDs-based-sensor by taking the place of Cu^{2+} in the complex, and this is owing to both electron transmission and energy transmission pathways. Interestingly, the authors have explored this method to develop a paper-based detector which was prepared by fabricating the $CuDTC_2$-CQDs solution ink over the cellulose acetate papers for screening of Hg^{2+} (Fig. **4**).

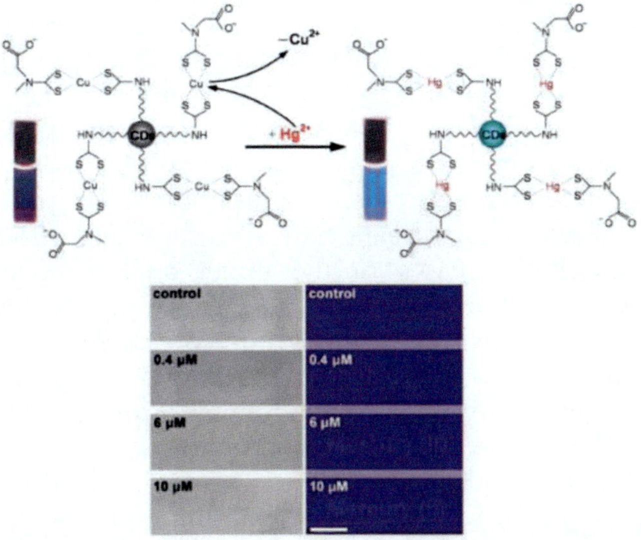

Fig. (4). Schematic presentation of the detection process by the Hg^{2+}- Causing Relocation of Cu^{2+} in the Interface Bound CuDTC2 molecule Paper-based visual detection of Hg^{2+} ion at three concentrations of 0.4, 6, and 10 µM and the corresponding fluorescence images under the 365 nm UV lamp [32].

Organic Compound Sensing

Nowadays, a wide number of references describe the application of carbonaceous QDs-based fluorescent sensors for the sensing of several organic complexes having vitamins, pesticides, and toxic compounds. Yao and collaborators [33] have recently reported the exploitation of the quenching impact of copper (II) upon the fluorescence signal of GQDs to manufacture a turn-on ascorbic acid detector, known as vitamin C. Indeed, the quenched fluorescence signal of GQDs is efficiently recovered after the addition of vitamin C and the realization of Cu^{2+}.

The findings showed a rectilinear relation among relative fluorescence intensity and vitamin C quantity in the region of 0.3-10 μM. The LOD was founded to be 94 nM. As an alternative method, Liu *et al.* [34] proposed the detection of vitamin C without the use of copper (II) complex in the quenching of the fluorescence signal of GQDs. Indeed, catechol quenches the fluorescence indicator of GQDs in the context of hydrogen peroxide (H_2O_2) and horse radish peroxidase, according to the stated turn-on fluorescence process (HRP). They explained that catechol could be oxidized using hydroxyl radicals formed by the reaction between HRP and H_2O_2, producing o-benzoquinone. The above could considerably reduce GQD fluorescence intensity. When AA is added to the solution, it can absorb hydroxyl and H_2O_2 radicals, which inhibits the synthesis of o-benzoquinone and restores luminosity. Furthermore, riboflavin (RF), referred to be vitamin B2, was detected using the carbon QDs-based assay method developed by Kundu *et al.* [35]. The fluorescence resonance energy transfer (FRET) method may be used as a detection method because the authors noted that the absorption spectra of RF and the emission spectra of the manufactured CQDs exhibit an inherent overlap. This occurrence is dependent on the transmission of energy *via* a non-radiative association between carbon QDs (as the donor) to RF (as the acceptor). The results demonstrated that the fluorescence emission intensity of carbon QDs was gradually decreased with a significant blue shift, and that of RF was increased proportionally with the increases of the target analyte concentration. They explained that the combination formed among the hydroxyl and carboxyl moieties of the CQDs and the carbonyl and amine groups of the RF was what caused the quenching action. In aqueous solutions, the LOD was established to be 1.2 nM. While, a series of aniline derivatives comprising N,N′-dimethylaniline, aniline, N-methylaniline, N-ethylaniline, N,N′-diphenylaniline and N,N′-diethylaniline was detected using a GQDs-based sensor. Indeed, Tian *et al.* [36] have reported that GQDs, prepared using GO, reveal an excellent green photoluminescence signal observed at 510 nm upon photoexcitation at 440 nm. The fluorescence signals decreased when aniline derivatives were present, and this was caused by the photoinduced electron transmission quenching process. According to the authors, an electron transmission to an acceptor molecule quenches the fluorescence signal of stimulated GQDs acting as chromophores. As a result, the aniline derivatives functioned as electron donors and interacted with stimulated graphene quantum dots to cause a clear dampening of luminescence sensitivity. Moreover, nitro-based derivatives were widely sensed by carbonaceous quantum dots. For instance, Wang's group [37] has manufactured a specific paper sensor that relied on amino-functionalized CQDs for the sensing of 2,4,6-trinitrotoluene (TNT) explosives residues in groundwater. The results revealed that the luminescence intensity of CQDs decreases with the increase of the concentration of TNT compound. The authors claim that a photoinduced electron transfer mechanism involving the amino groups of CQDs and the nitro groups of TNT was what caused the sensor's fluorescence response to be quenched. The paper sensor, prepared by immobilization of the NH_2-CQDs onto cellulose paper, was applied for visual and on-site detection of TNT traces having a LOD of 0.213 μM (Fig. **5**).

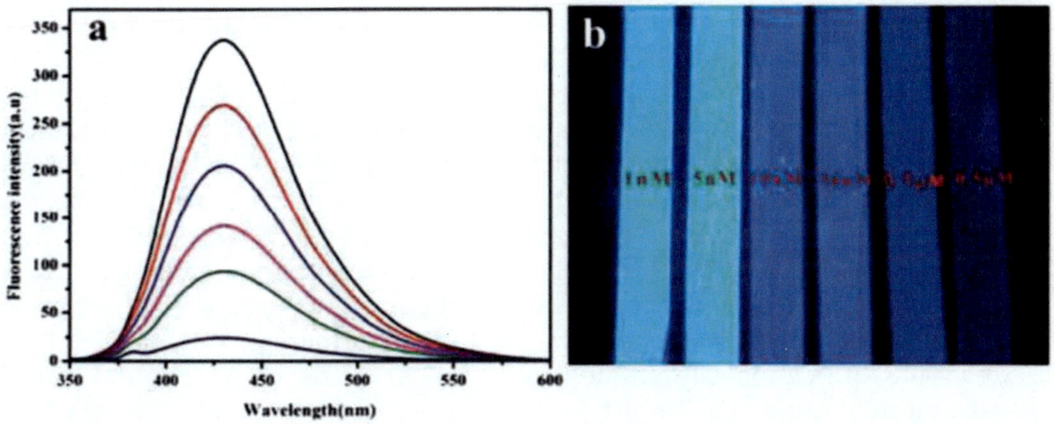

Fig. (5). (A) Fluorescence spectra of practical water samples for paper nanosensor **(B)** and the corresponding images of the paper nanosensor in the 365 nm UV lamp and upon the addition of several concentrations of TNT (1.10^{-9}, 5.10^{-9}, 1.10^{-8}, 5.10^{-8}, 1.10^{-7}, 5.10^{-7} M) [37].

Concerning the detection of pesticides, Merkoçi's group [38] has designed GQDs-modified silica beads for the sensing of tributyltin pesticides. They adopted the molecularly imprinted polymer as a sensing method due to its high selectivity toward the target analyte. The fluorescence studies demonstrated that the fluorescence strength of GQDs decreases proportionally to the increase of the pesticide concentrations from 0.0 to 10 ppm. The quenching phenomenon is essentially linked to the filling of the imprinted voids presented upon the interface of the GQDs, which prevents the excitation of QDs. The limit of detection was found to be 13.54 and 41.24 ppb, in seawater and water, respectively. In the same context, Amjadi *et al.* [39] developed imprinted mesoporous silica decorated having carbon QDs and semiconductor QDs for diniconazole sensing in specimens, including water, wastewater and soil specimen. The authors revealed the designed sensor associated with the high selectivity of the molecular imprinting technique and the sensitive fluorescence response of the modified CQDs. This method is able to sense diniconazole in the site of 20 to 160 $\mu g.L^{-1}$ having a LOD of 6.4 $\mu g.L^{-1}$. Additionally, organophosphate pesticides which are widely applied in agriculture, have potentially been detected by carbon QD-based sensors. For instance, Lin's group [40] used CQDs-modified AuNPs for the recognition of paraoxon. Since Au NPs can quench the fluorescence emission of CQDs using the FRET process, the researcher utilized gold nanoparticle aggregation to release the adsorbed CQDs on the substrate of AuNPs, culminating in a turn-on fluorescence strength of CQDs. The reported method allowed the paraoxon quantification to have a broad width of 0.05 to 50 μgL^{-1} and a LOD of 0.05 μgL^{-1}. Furthermore, a paraoxon sensor based on MnO_2 nanosheet-CQDs was developed by the same group [41]. Indeed, they used the FRET quenching mechanism to monitor the context of the pesticide. Furthermore, the enzymatic hydrolysate thiocholine (ATCh), which is generated by the reaction of butyrylcholinesterase (BChE) with acetylthiocholine, could effectively cause the MnO_2 nanofilms degradation, culminating in the retraction of CDs fluorescence. Meanwhile, pesticides containing organophosphorus inhibit the degradation of MnO_2 and the development of ATCh that follow the "turn-off" of fluorescence. The linear range of the FRET-based sensing platform was 0.04 to 4.9 ng mL^{-1}, having a LOD of 0.015 ng mL^{-1}. Additionally, the suggested method exhibited a visual colour depending on the concentration of paraoxon pesticide, allowing the naked eye to monitor the paraoxon sensing event. Furthermore, Han *et al.* [42] described the utilization of N,S-codoped CQDs to detect a potential water pollutant, nitrobenzene. The sensing process was based on an electron-transfer quenching mechanism between the thiol groups of N,S-CQDs, as electron takers, and nitrobenzene analyte, as electron receiver. The results showed a large linear region of 5.10-4 to 1 mg. L^{-1} had a significant association between the quenched luminescence strength and the target analyte content.

Biological Molecules Sensing

Nowadays, carbonaceous QDs are extensively used in the development of biosensor platforms due to their highly sensitive photoluminescence behaviours. A general detection technique for enzymatic activities whenever the materials that may produce p-nitrophenol during the enzymatic processes were designed in this field using phenylboronic acid functionalized CQDs (PBA-CQDs). The authors picked β-glucosidase and α-galactosidase as two examples to track the enzymatic function. It has been proven that the presence of p-nitrophenol causes CQDs to exhibit less photoluminescence and forms a conjugate *via* B-O bonds that are not luminous. It was discovered that the detection limit was less than 0.33 U/L, which is sufficiently minimal to enable the identification of α-glucosidase concentrations in human specimens [43]. More interestingly, Wu' group [44] developed an α-glucosidase detection method that relied on the internal filtration effect (IFE) quenching of N-doped CQDs to monitor the anti-diabetic drug. Using α-glucosidase enzyme activity, 4-nitrophenyl-α-D-glucopyranoside ejected 4-nitrophenol, which is used as α-glucosidase substrate. Exploring the overlap between the 4-nitrophenol absorption and the N-doped CQDs excitation, 4-nitrophenol was used as an important absorber of IFE to deactivate the fluorescence signals. The authors reported that the sensing strategy was considered a rapid, sensitive and eco-friendly method for enzyme activity quantification with a LOD of 10^{-8} M. On the other hand, Sun and collaborators [45] have reported the application of β-cyclodextrin functionalized CQDs to the sensing of cholesterol. In this work, the host-guest evaluation between cholesterol and β-cyclodextrin and the photoluminescence of CQDs were explored as a cholesterol quantification strategy and a transduction method, respectively. Prior to the recognition of cholesterol, p-nitrophenol was used to fill the cavities of β-cyclodextrin, which exhibits deactivation of the CQD fluorescence signal. Subsequently, the addition of cholesterol induces the replacement of p-nitrophenol, exhibiting the recovery of the fluorescence signal because of the high affinity of cholesterol for β-cyclodextrin compared to that of p-nitrophenol. Under the optimized sensing conditions, a good analytical performance in cholesterol detection was obtained, having an LOD of 0.7 ± 0.1 μM. Similarly, Luo *et al.* [46] examined the utilization of β-cyclodextrin functionalized N,S-codoped CQDs as a photoinduced electron transfer (PET) fluorescent detector for testosterone evaluation in environmental H_2O. The authors used (ferrocenylmethyl) trimethylammonium iodide (Fc^+) as a β-cyclodextrin (β-CD) guest. During the fluorescence excitation, the photo-induced excited electron would effectively transfer to the empty molecular orbits of Fc^+. As a result, the fluorescent intensity of β-CD-CQDs decreased with an increase in the amount of Fc^+. Nonetheless, the Fc^+ guest can be removed from the β-CD cavity by a competing encounter across hosts and guests β-CD and testosterone exhibiting a cut off of the PET process and turning on the fluorescence signal of β-CD-CQDs. The results exhibit a good encounter across the fluorescent strength of the sensor and the testosterone content in a linear concentration spectrum within 0 μM to 280 μM. The LOD was estimated to be 0.51 μM. Recently, Sony and collaborators developed a bilirubin sensing platform based on S,N-codoped CQDs. These later were synthesized utilizing carbon for the citric acid source and sulfur and nitrogen source to be L-cysteine and *via* a microwave-assisted method. The authors have explored the quenching effect of iron (III) ion on CQDs and the ability of bilirubin in the selectively restoring of the fluorescence signal of CQDs. The sensor showed considerable sensitivity and selectivity for bilirubin, having a LOD of 0.12 nM.

As a result, the procedure was used to assess bilirubin levels in the urine specimen and human serum. Interestingly, the method was exploited to design a paper-based test stripe for visual sensing of bilirubin.

Currently, many reported works interestingly exploited the competitive binding affinity of thiolate biomolecules, which play an essential role in human physiology, in turn off/on fluorescent mechanism. In this context, Luo and collaborators [47] described the use of CQDs as a turn-off/on fluorescent sensor. The authors selected the mercury (II) ion to quench the fluorescence response of CQDs. As reported, the recovery of the signal was assisted by the introduction of homocysteine and cysteine as biothiols. A broad width of concentration of 0.5-10 μM has been found to have a LOD of 168 nM. As an alternative to metal ion quenchers, quinone was also applied in the biosensing strategies. For example, Wang *et al.* [48] developed a metal ions-free method for the measurement of thiols using L-tyrosine methyl ester decorated CQDs (Tyr-CDs). Indeed, they demonstrated that Tyr-CQDs revealed extreme quenching efficacy owing to the effective energy transmission across the quinone products and Tyr-CDs. Tyrosinase's catalytic activity in the synthesis of quinone was delayed throughout the biothiol molecules' measurement period, allowing the fluorescence indication to restore. The sensor-based CQDs were used in the field of detecting biothiols in specimens of human serum. Furthermore, Dong's group [49] has exploited the quenching impact of the gold nanoparticles over the luminescence signal of the CQDs. Indeed, they developed an effective turn-on luminescence detector for the glutathione quantification that relied on FRET quenching mechanism among N,S-codoped carbon QDs and AuNPs. After the addition of glutathione into the solution of N,S-codoped carbon QDs/AuNPs, the fluorescence response of N,S-codoped carbon QDs was significantly recovered. To explain the results, the authors speculated that glutathione can coordinate AuNPs by a strong Au-S bond, thus causing the release of N,S-codoped carbon QDs from the AuNPs surface. The developed detector was effectively introduced to measure glutathione in serum specimens with a LOD of 3.8 to 415.1 μM with a limit of detection of 0.21 μM. Likewise, Cai *et al.* [50] created a rapid fluorescence "switch-on" platform based on CQDs-MnO_2 nanocomposites to detect glutathione. The fluorescence response of CQDs was quenched using MnO_2 nanosheets by employing FRET quenching process. The presence of glutathione in the environment reduced the MnO_2 nanosheets to Mn^{2+} ions, thus resulting in the sufficient recovery of the fluorescence response of CQDs. The method enabled an ultra-sensitive recognition of glutathione having a LOD of 300 nM.

While, Carbonaceous QDs-based sensors were also used to detect proteins, including DNA, aptamer, and RNA. As an example of DNA sensing, Liang *et al.* [51] discussed the design of a fluorescence DNA detector that relied on CQDs and

3-mercapropionic acid (MTX) coated cadmium telluride (CdTe) quantum dots. The fluorescence signal of CdTe-carbon QDs was quenched by MTX *via* an electron transfer quenching mechanism. The presence of the target DNA strand allows the release of the MTX from the surface of QDs and its intercalation into the dual helix configuration of the double-strand DNA. As a result, the monitor of the DNA strand targets was provided by the restoration of the fluorescent signal. With a LOD of 1 nM, the reported approach demonstrated a linear response in the concentration range of 0 to 50 nM. Moreover, a simple CQDs-based detection method was developed by Khakbaz and collaborators [52] for the sensing of mir9-1 (a marker for breast cancer) micro-RNA as a model. The reported method was based on a FRET quenching mechanism. The CQDs were functionalized by exploiting the π-π interaction with fluorescein amidites (FAM) labeled single-stranded DNA (ssDNA). The energy is transmitted from the CQDs as a contributor to the FAM as a receiver *via* a non-radiative resonance mechanism. The luminescence signal of CQDs was restored in the context of mir 9-1 as a result of the creation of a duplex helix between mir 9-1 and ssDNA, eliminating the π-π interaction and stopping the luminescence resonance energy transmission. Last but not least, Yang *et al.* [53] have developed a fluorescence detector relying on molybdenum disulfide and GQDs nanosheets for the sensing of epithelial cell adhesion molecule (EpCAM) protein. The "turn-on" fluorescence biosensor was developed by the decoration of the MoS_2 nanosheets with GQDs labelled EpCAM aptamer. Prior to the sensing phase, the fluorescent output of GQD was suppressed by MoS2 nanosheets by the FRET method. However, the addition of EpCAM protein caused the GQD-labeled EpCAM aptamer to separate from MoS2 nanosheets. This was achieved by a greater selective propensity contact among the EpCAM protein and aptamer, which restored the system's ability to fluoresce. Using a linear concentration spectrum of 3 nM to 54 nM and LOD of 0.45 nM, the target EpCAM protein could be located. The EpCAM-expressed breast cancer design was used to test the applicability of the sensing concept.

CHEMILUMINESCENCE (BIO) SENSORS

Chemiluminescence (CL), which is dependent on the electromagnetic radiation that an exothermic chemical reaction emits, is created when electrically excited entities of byproducts revert to the initial state. A chemical reaction that produces enough energy (> 41 kcal/mol) to cause an electron to move from its initial state to its active state is often required for CL. When the electronically stimulated species is deactivated to its ground state, it either emits light or transmits its energy to some other species that produce the discharge. The CL reaction has been divided into two main groups based on whether the light radiation mechanism is implicit or explicit, as defined below:

$$A+B \rightarrow P^* + Q \text{ (R1)}$$

$$P^* \rightarrow P + h\nu \text{ (direct CL) (R2)}$$

$$\text{Or} \quad P^* + F \rightarrow P+F^* \text{ (R3)}$$

$$F^* \rightarrow F + h\nu \text{ (indirect CL) (R4)}$$

In direct chemiluminescence, the chemical reaction across the reagents A and B, where A is normally B is a effective oxidant and CL a agent, produces an excited state product by exhibiting sufficiently excitation energy to motivate the transfer of an electron from the initial phase to the excited phase. (P*). The energy loss of the excited molecule through a light discharge was termed as direct CL [54, 55]. The excited state product P* that forms in the indirect CL scenario releases energy when it is transferred to a fluorophore (F), creating an excited state fluorophore (F*). Eventually, the latter releases radiation that characterises the transfer of chemiluminescence resonance energy (CRET). It is important to note that excited molecules may lose energy *via* radiation-free processes such as internal conversion, collisional deactivation, intersystem crossover, or engaging in other reactions. Unfortunately, limitations, including poor selectivity and low CL intensity plagued typical CL reagents. Therefore, novel CL systems have previously been explored to increase the CL system's intensity. Due to their outstanding electrical characteristics, such as their superb electron donor and acceptor character producing luminous qualities with high quantum yields, CQDs have become CL-based sensors in this context. They could be used in chemiluminescence resonance energy transfer (CRET) procedures as potent oxidants or fluorescence acceptors. It has been demonstrated that CRET could only occur between a chemiluminescent supplier and a fluorophore receiver at close ranges (less than 10 nm) and under the condition that the donor CL emission spectrum and the acceptor CL emission spectrum overlap the fluorescent acceptor's absorption [56]. In this section, the authors discuss various CL carbonaceous QDs based-sensors and their potential mechanisms in order to better understand how carbonaceous QDs amplified CL.

Recently, Shan and colleagues [57] have used the system peroxalate, H_2O_2, bis(2,4,6-trichlorophenyl) oxalate to enhance the CL response of CQDs. The detector design was used to sense hydrogen peroxide and glucose. They have reported that the CQDs sensor induced a bright and persistent deep-red (DR) chemiluminescence with a quantum yield of $(8.22 \pm 0.30) \times 10^{-3}$. In this process, the authors attributed the CL behavior to the chemically induced electron transfer luminescence process. The process between the peroxalate and hydrogen peroxide yields a free-energy intermediate, 1,2-dioxetanedione. The latter excites CQDs *via* an intermolecular electron transfer between them and CQDs. In a final step, the return of CQDs to the ground state releases photons producing a CL emission. Similarly, a powerful CL formaldehyde HCHO sensor was reported by Joo and collaborators [58]. In this investigation, the catalytic impact of GQDs on sensitizing effect of $HSO3^-$ on the RB-H_2O_2-Sodium Dodecyl Sulfonate system. It was demonstrated that GQDs in the presence of bisulfite promoted the weak CL discharge of the SDS-RB-H_2O_2 process. However, the presence of HCOH quenched the CL system, and the decrease in CL intensity was proportional to HCHO content in the linear region of 0.02-11 $\mu g.L^{-1}$ having a LOD of 6 $ng.L^{-1}$. In another work, the chemiluminescence (CL) emission from peroxynitrous acid (ONOOH)-carbonate framework to sense nitrite ions in tap H_2O. They reported that the presence of sodium carbonate and CQDs together in ONOOH solution which was generated from $NaNO_2$-H_2O_2, was accompanied by a very highest strong CL. This is not the case in the absence of nitrite, where a weak CL signal was observed. The authors revealed that CQDs functioned as an energy receiver in enhancing the intensity of CL response of the $NaNO_2$-H_2O_2-carbonate system. The CL discharge and the concentration of the nitrite analyte used for ONOOH formation were found to have a great linear relationship. Further, the sensing approach was reported to be sensitive, having a low LOD of 5 nM [59]. Furthermore, Lin and collaborators [60] used N-doped CQDs for the evaluation of iron (II) ions in environmental specimens. Following its interaction with the inducing hydroxyl radicals ($HO^{\bullet}$), N-CQDs exhibited a strong chemiluminescence emission guaranteed by the formation of an electron-hole transition. Indeed, Fe^{2+} ions can generate $HO^{\bullet}$ radical from hydrogen peroxide (H_2O_2). Therefore, the authors have used a pre-mixed N-CQDs/H_2O_2 solution to selectively monitor the Fe^{2+} concentration *via* CL emission. The N-CQDs/H_2O_2 sensor is enabled to measure Fe (II) ions at a lower concentration of 0.2 nM, having a linear dynamic state between 1 nM to 1 μM.

Furthermore, Azizi and colleagues [61] have investigated the impact of potassium permanganate on the CL response of S,N-codoped CQDs. The reported studies revealed that $KMnO_4$ could synthesis the highest CL emission than other oxidants such as hydrogen peroxide, potassium ferricyanide, and sodium periodate. They reported that the chemiluminescence response of S,N-CQDs is monitored by a direct CL mechanism described below:

$$\text{S,N-CQD} + KMnO_4 \rightarrow \text{S,N-CQDs}^* + \text{Mn(II)}$$

$$\text{S,N-CQDs}^* \rightarrow \text{S,N-CQDs} + h\nu\ (\sim 500\ \text{nm})$$

Mechanically, the CL process of CQDs is characterized by the creation of an excited-state CQDs* *via* electron-transmission annihilation, leading to the energy transmitted in the sort of CL emission. Similarly, potassium permanganate was also used as an oxidant in the $KMnO_4$/Ce(IV) CL system. Indeed, S. Han *et al.* [62] reported that the aforementioned CL system directly oxidizes nitrogen-doped CQDs producing CL emission. They concluded that the excited state of N-CQDs (N-CQDs*) was the primary luminophore in the N-CQDs-$KMnO_4$/Ce(IV) phase. The CL mechanism for the investigated gadget was thought to be the optical reintegration of the oxidant-injected voids and electrons of the N-CQDs. In order to create N-CQDs in the oxidised phase (N-CQDs•+), the scientists reasoned that Ce(IV) or KMnO4 may introduce voids into the N-CQDs. When N-CQDs•+ and the dynamically stimulated electrons in the high-energy band annihilate to create N-CQDs*, energy was released in the yield of CL emissions as the system reverted to its initial state (Fig. **6**). The CL strength was raised in the addition of thiourea (TU) or tannic acid (TA), and the findings show that the emission remained N-CQDs* in the N-CQDs-KMnO4-TU/TA complex. Additionally, the outcomes demonstrated that the oxidant KMnO4 may interact with TU and TA to produce the stable intermediate isothiocarbamido radicals. As a consequence, the N-CQDs might receive gaps from these intermediates charges, producing more N-CQDs•+ and a stronger CL signal. However, they proposed that the $KMnO_4$-TU/TA redox reaction produced energy that moved to Mn(II), generating additional Mn(II)*, which resulted in an improvement in the reaction's overall quantum efficiency. The results demonstrated a linear relationship between CL intensity and TU and TA concentrations in the spectrum of 0.1 to 10.0 M and 0.07 to 7.0 M, correspondingly, having LOD of 0.038 and 0.026 M. The CL approach has been used to find traces of TU and TA in a variety of actual matrix, such as oranges, peels, and Chinese gall.

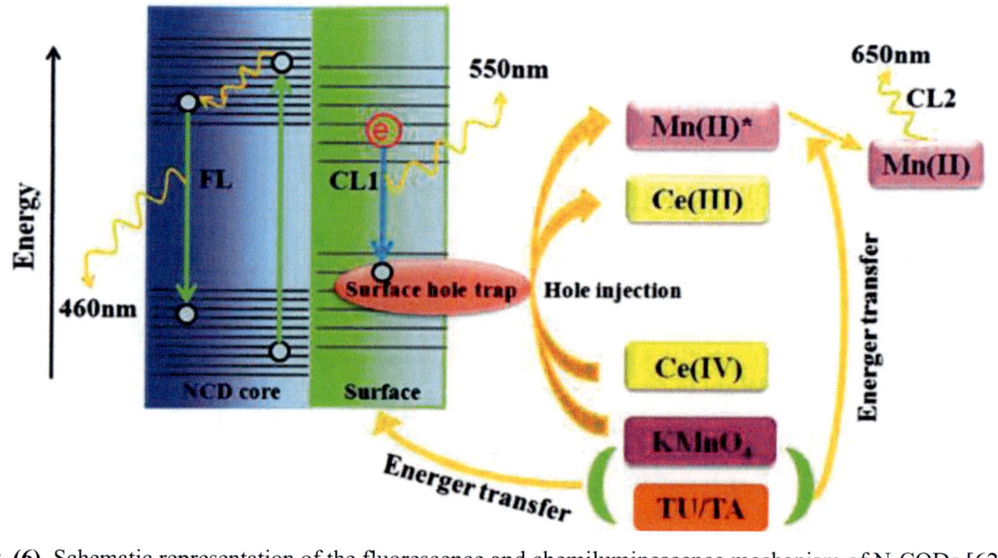

Fig. (6). Schematic representation of the fluorescence and chemiluminescence mechanism of N-CQDs [62].

On the other hand, a novel dual-mode "off-on" chemiluminescence technique was constructed to determine iodide and mercuric ions in kelp, tea, and water samples. The method relied on the employment of sulfur and nitrogen co-doped GQDs. They reported that under acidic conditions, a stable CL intensity issued from the reaction of N,S-GQDs, and Cerium (IV) was observed. It was demonstrated that the presence of iodine ions quenched the CL response of the system N,S-GQD--Ce(IV). Nevertheless, the quenched CL signal was recovered by mercury (II) ions. In this way, the bimodal sensor was used to determine the iodine and mercury (II) amounts in the linear spectrum of 0.04-3 and 0.3-7 μM, having a LOD of 4.23 and 24.7 nM, respectively [63]. As an application of the cerium (IV)-sodium bisulfite CL system, Zhang *et al.* [64] have discussed the influence of the GQDs in the enhancement of the CL process of Ce (IV) using $NaHSO_3$ *via* electron and energy transmit process in the sensing of tyrosinase. Indeed, they showed that the chemiluminescence resonance energy transfer (CRET) involving SO2*/1O2 and GQDs is where CL originates. The merging of the hole-injected GQDs (GQDs•+) and electron-injected GQDs (GQDs•-) results in radiative electron-hole destruction. They claimed that tyrosinase is readily oxidized by the OH• radical and Ce (IV), causing the levels of these two substances to drop. As a result, the rise in tyrosinase quantity causes a corresponding reduction in the CL strength of the GQDs-Ce(IV)-NaHSO3 combination. In order to identify tyrosinase in the spectrum of 0.02-8 M having a LOD of 4 nM, the researchers have adopted the GQDs-based CL as a selective and sensitive approach. Mirbirang's group [65] explored the luminol-Mn(IV) CL system for the sensing of uric acid in biological fluids. S,N-impact CQDs on the luminol-Mn(IV) CL process were investigated. According to the findings, S,N-CQDs cause a 13-fold enhancement in the CL reaction's strength. Additionally, it was discovered that uric acid considerably reduced the CL amplitude of the S, N-CQDs-lumin-l-Mn(IV) system. According to scientists, the struggle between uric acid and luminol to react to Mn may be the cause of the reported behaviour. A novel uric acid monitoring sensor was created using the magnification phenomena. In fact, it was discovered that there was a logarithmic connection between the CL amplitude and the uric acid level in the intensity region of 0.05 to 1.5 M having a LOD of 17 nM. More interestingly, Cheng *et al.* [66] have recently described the usage of CQDs dots in the catalysis of the chemiluminescence process across the sulfite and bromate. The outcomes did, in fact, show that bromate and sulfite interact quickly to produce an ultraweak CL release that lasts for less than 5 seconds. However, following the addition of CQDs, the CL amplitude of the BrO3- -SO32- system improved 50-fold. The acquired high light discharge also achieved its peak within 7.5 seconds before progressively reverting to its baseline following almost two minutes. They proposed that the production of CQDs with electron and hole injection leads to the CL process. The formation of CL is due to the radiative recombination of holes and electrons within CQDs. Considering a LOD of 0.1 mol/L, a dynamic bromate level spectrum of 0.3 to 10 mol/L was found. The technique constituted a novel tool for measuring bromate, a suspected malignancy.

ELECTROCHEMICAL (BIO) SENSORS

One of the most recent breakthroughs in research on carbon-based nanomaterials is carbonaceous quantum dots, namely GQDs, which have drawn a lot of interest in the creation of electrochemical sensors for environmental, food, and healthcare monitoring. These 0D nanomaterials are defined as good electron transporters increasing the electrochemically active electrode area. Typically, the edges and basal planes of GQDs have moieties like hydroxyl, carboxyl, carbonyl, and epoxide, which enable them to operate as reaction and coordination sites. With regard to CQDs, these quasi-spherical nanoparticles have high solubility in water, good biocompatibility, and non-toxicity, which make them an excellent candidates in the functionalization of electrode surfaces [67]. The discussion of cutting-edge electrode modification techniques and pertinent applications for the monitoring of heavy metal ions, organic contaminants, and biological molecules will be discussed in this section.

Heavy Metal Ions Sensing

Frequently, carbon QDs have been applied in photoluminescence applications. Nevertheless, these nanomaterials have recently been utilized in the fabrication of electrochemical sensors for heavy metals due to their unique properties citing easy functionalization and a larger active surface.

Recently, a simple GQDs based-sensor was developed by Via's group [68] for the quantification of lead (II) ion concentration in liquid phases. In this research, GQDs were manufactured by cutting GO powder and reducing it using L-ascorbic acid. The glassy carbon electrode functionalized with prepared GQDs was used then in the sensing of Pb^{2+} ions. The ability to accumulate metal ions was enhanced by their demonstration that the increased specific surface region of GQDs may significantly increase the contact opportunity of active regions and target ions. In another work, Cui and colleagues [69] developed a sensitive electrochemical sensor that relied upon GQDs decorated graphene-GCE for the sensing of copper (II) ions in real water specimens. The electrochemical platform was obtained by drop-coating modification strategy and GQDs were created through the pyrolysis of the citric acid. The authors described that functionalized GQDs-graphene enhanced the electrochemical responses of the GCE, which is due to the rapid electron transmit of graphene and the effective Cu(II) attraction of GQD. The sensing process was evaluated by differential pulse anodic stripping voltammetry. Indeed, the results showed a broad linear level spectrum of 0.015-8.775 μM having a LOD of 1.34 nM.

Moreover, a sensitive design based on polyaniline/GQDs-modified SPCE was reported by Chailapakul and colleagues [70] for a rapid determination of chromium (VI) ions. In this work, polyaniline was electropolymerized using anode potentials, and GQDs were prepared using the bottom-up method *via* citric acid pyrolysis. The electrochemical studies revealed that the determination of Cr(VI) was obtained with a broad linear spectrum of 0.1-10 $mg.L^{-1}$, and a LOD of 0.097 $mg.L^{-1}$.

Furthermore, the ability of GQDs to simultaneously detect various heavy metals is also discussed in the literature. For instance, the reported work by Chen *et al.* [71] described the usage of the conjugates of AuNPs and GQDs in the electrochemical sensing of Hg^{2+} and Cu^{2+}. The authors used the 1-ethyl-3-[3-dimethylaminopropyl] carbodiimide/ N-hydroxysulfosuccinimide (EDC/NHS) covalent dual procedure to functionalize cysteamine-gold nanoparticles with acid carboxylic-GQDs. The final nanoconjugate was used to decorate the GCE. An ultralow limit of detection (0.02 nM for Hg^{2+} and 0.05 nM for Cu^{2+}) and elevated sensitivity (2.47 μA/nM for Hg^{2+} and 3.69 μA/nM for Cu^{2+}) were attained.

More interestingly, Kong *et al.* [72] have described the application of the nanocomposites of GQDs and chitosan (CS) in the simultaneous detection of Zn^{2+}, Cd^{2+}, and Pb^{2+}. The design was prepared with a simple and reliable approach of forming the nanocomposite *via* the establishment of hydrogen bonds between GQDs and CS. The CS-GQDs modified glassy carbon electrode (GCE) presented highly sensitive responses opening a new avenue for monitoring heavy metal ion traces.

Organic Toxic Compounds Sensing

The determination of toxic organic compounds is of considerable interest because of their dangerous effects on the environment and health. Currently, carbonaceous QDs are widely used in the sensing of trace levels of various analytes citing organophosphorus pesticides, hydroquinone and catechol derivatives, L-tyrosine, bisphenol A, acetamidophenol, *etc.*

In this context, Gervasio and collaborators [73] developed a simple detection framework centred on nitrogen-doped GQDs for the electrochemical determination of the trace level of trinitrotoluene (TNT). The oxidative ultrasonication of GO is used to create N-GQDs, which are then chemically reduced with hydrazine. The N-GQDs enhanced GCE was incubated in TNT fluid as part of the detection technique. The linear scan voltammetry (LSV) was chosen as a sensitivity monitor approach following TNT accumulated on the surface of the electrode. The results showed that TNT content in the aqueous solution varied linearly from 1 to 400 ppb having a LOD of 0.2 ppb. Using fenthion as a paradigm, Ai *et al.* [74] recently announced a novel method for the electrochemical sensing of non-electroactive organophosphorus insecticides. In this research, the electrochemical platform was developed in two steps. Firstly, the GCE was decorated by GQDs prepared from graphene oxide. To immobilize pralidoxime iodide moieties on the surface of GQDs-GCE, the authors have chosen to use the electrochemical route using the chronoamperometry technique. They demonstrated that functionalization of the surface increases the electrode reaction site, enhances the adsorption capacity, amplifies the current response of PAM therefore and improves the determination sensitivity. The electrochemical platform exhibited significant specificity, better reproducibility, and excellent performance in real sample sensing.

As a combination of GQDs and metallic nanoparticles, Chen's group [75] deliberated the use of GQDs-modified silver nanoparticles to decorate the surface of GCE. The altered GCE was introduced in the simultaneous determination of guanine and adenine. The electrochemical results displayed excellent electrochemical performances in the oxidation of selected analytes. Linear interaction across oxidation current and the quantity were attained in the spectrum of 0.015 to 430 μM and 0.015 to 390 μM for guanine and adenine, having LOD of10 nM and 12 nM, respectively.

More interestingly, Chen and collaborators [76] deliberated the fabrication of a novel electrochemical framework that relied upon molecularly imprinted polypyrrole/graphene QDs composite for sensing bisphenol A pollutant in H_2O specimens. The authors chose the molecularly imprinted polymer sensing strategy due to its high selectivity. Additionally, they claimed that the electrical conductance and electron transmission efficiency were greatly improved by the inclusion of GQDs. The oxidation activity and BPA content for the given sensor were linearly correlated with a LOD of 0.04 M and ranged from 0.1 M to 50 M. With recoveries of 94.5 percent and 93.7 percent, correspondingly, the sensor's usability was examined in specimens of tap and sea water.

Biological Compounds Sensing

Nowadays, Carbonaceous QDs have been frequently used in the domain of electrochemical biodetector due to their interesting properties presented for both electrocatalysis of many compounds used in biological processes (H_2O_2, O_2, dopamine, *etc.*) and immobilization of bioreceptors. In the literature, these QDs are more frequently applied in immunosensing, DNA and microRNAs sensing but rarely in aptasensing. In 2018, Mars and collaborators [77] commended the usage of GQDs to perform an electrochemical immunosensor for the screening of toxins of scorpion venin in human urine and serum. The sensing strategy was based on the development of an immunosandwich based on the use of nanobodies and immobilized upon the substrate of GQDs functionalized SPCE. The formation of the immunosandwich was monitored by the electrocatalytic reaction between horseradish peroxidase (HRP), hydrogen peroxide, and hydroquinone. In this work, the authors described that GQDs amplified dramatically the electrochemical response of the platform. The findings displayed a sensitive electrochemical response toward the presence of the toxins with a picomolar limit of detection. Additionally, extremely positive metrological results were obtained in terms of consistency, repeatability, sensitivity, and long-term stability. In another work, the same author described the use of GQDs to modify ITO electrodes to simultaneously detect the apolipoprotein E4 DNA by electrochemical and fluorescence techniques.

Furthermore, a sensitive electrochemical biosensor for the sensing of miRNA 155 was created by Hu *et al.* [78]. They reported that GQDs were used as a carrier for the HRP enzyme, which was immobilized by a noncovalent assembly. The HRP-altered RNA detector could efficiency electrocatalyze the oxidation of the substrate 3,3′,5,5′-tetramethylbenzidine (TMB). The proposed biosensor may screen miRNA-155 in a linear spectrum of 1 fM to 100 pM, having a LOD of 0.14 fM. Moreover, Li and colleagues [79] reported the use of GQDs altered pyrolytic graphite electrode bioconjuguated having sequence ssDNA for the sensing of DNA as a model. The authors described that the platform induced an electrochemical response as a result of the excellent conductivity of GQDs. Before the sensing of DNA, the electrochemical response was inhibited by a significant contact across the free ssDNA probe and electrode graphitic surface. Indeed, the electron transfer rate between the electrochemical active moiety $[Fe(CN)_6]^{3-/4-}$ and the electrode was considerably decreased. While, the introduction of the DNA molecules induced the release of the GQDs and the capturing with the ssDNA probe. As a consequence, the relative current spike of $[Fe(CN)_6]^{3-/4-}$ was recovered and increased with the target molecules. More recently, Roushani's group [80] described the application of AgNPs and thiol graphene QDs nanocomposite decorated GCE for the determination of the hepatitis C virus. In this investigation, thiolated GQDs were produced from the oxidation of the multiwalled carbon nanotubes sequenced by a coupling process across acid carboxylic groups of GQDs and amine group of cysteamine. They reported that the electrochemical response of the riboflavin mediator was notably decreased with the increase of the concentration of HCV core antigen. Further, they attributed the observed behavior to the formation of an important number of antigen/antibody complexes hindering the electron transmit process of riboflavin on the detecting surface. The electrochemical platform showed an acceptable linear spectrum from 0.05 pg.mL^{-1} to 60 ng.mL^{-1} having a LOD of 3 fg.mL^{-1}. Interestingly, Dong and collaborators [81] reported the application of label-free electrochemical immunosensors for the sensing of carcinoembryonic antigen (CEA). They used N-GQDs@AuNPs-supported platin-palladium Bimetallic Nanoparticles for the decoration of the GCE surface. In this work, the authors exploited the electrocatalytic activities of the nanocomposites on the reduction of hydrogen peroxide to monitor the formation of anti-CEA-CEA (antibody-antigen) complex. Meanwhile, the sensing of CEA was accompanied by the inhibition of the electron transfer caused by the formation of an antigen-antibody complex on the PtPd/N-GQDs@Au modified GCE. A logarithmic relationship between the electrochemical response and the CEA content from the spectrum of 5 fg/mL to 50 ng/mL was observed.

Several carbonaceous QDs-based sensors and biosensors for other biological compounds have been discussed in the literature utilizing different electrochemical techniques and regrouped in Table **1**.

Table. 1. Selective applications of carbonaceous QDs-based sensors and biosensors for the detection of biological compounds

Platform	*Analyte*	*Electrochemical method*	*Detection limit*	*Linear range*	*References*
N-carbon QDs/ferrocene@ β-cyclodextrin/GCE	Uric acid	DPV	0.08 μM	5.0-120.0 μM	[82]
AuNPs- carbon QDs/chitosan /GCE	Dopamine	DPV	1.0 nM	0.01-100.0 μM	[83]
Carbon QDs- chitosan /GCE	Dopamine	DPV	11.2 nM	0.1-30.0 μM	[84]
Reduced graphene oxide - Carbon QDs /GCE	Dopamine	DPV	1.5 nM	0.01-450.0 μM	[85]
B-cyclodextrin-Graphene QDs/GCE	L-cysteine	DPV	-	0.01-2.0 mM	[86]
Graphene QDs-AgNPs/GCE	Guanine Adenine	DPV	10 nM 12.0 nM	15 nM-430μM	[75]
Carbon QDs-SPCE	Dopamine	CV	0.099 μM	1-7 μM	[87]
rAuNPs/Graphene QDs/GCE	Dopamine	SWV	3 10^{-10} M	10^{-9} -10^{-7} M	[88]
GCE/Graphene QDs/AuNPs	Norepinephrine	SWSV	0.15 μM	0.5-7.5 μM	[89]
Graphene QDs/IL/CPE	Norepinephrine	DPV	0.06 μM	0.2-400.0 μM	[90]
Graphene QDs-$RuCl_3$/CCE	L-Tyrosine	Amperometry	0.23 μM	1-937 μM	[91]

(Table 1) cont.....

Poly(β-cyclodextrin)/CQDs/GCE	Dopamine Uric acid Tryptophan	DPV	0.14 μM 0.01 μM 0.16 μM	4- 220 μM 0.3 -200 μM 5 - 270 μM	[92]

SPCE : Screen-parinted carbon electrode. Il : Ionic liquid. CPE : Carbon paste electrode. GCE : Glassy carbon electrode. CCE: Carbon paste electrode. NPs : Nanoparticles.

ELECTROCHEMILUMINESCENCE (BIO) SENSORS

Due to its numerous benefits, including the lack of a source of light, precise regulation over the potential, location, and timing of the light-emitting procedure, flexible luminophores, responsiveness, a straightforward operating procedure, and affordable devices, electrochemiluminescence (ECL) has recently attracted a lot of focus in chemical sensing, biodetection, image analysis, and electron transfer (ET) exploring. The story begins in 1960, when Hercules and Bard published groundbreaking research outlining the processes of ECL. They described ECL as a type of chemiluminescence in which an electrochemical process comes before the chemiluminescent process that emits light. Redox processes at electrodes, homogenous chemical reactions, excited-state species creation, and light emission are typically the four steps that regulate the ECL operations (Fig. **7**).

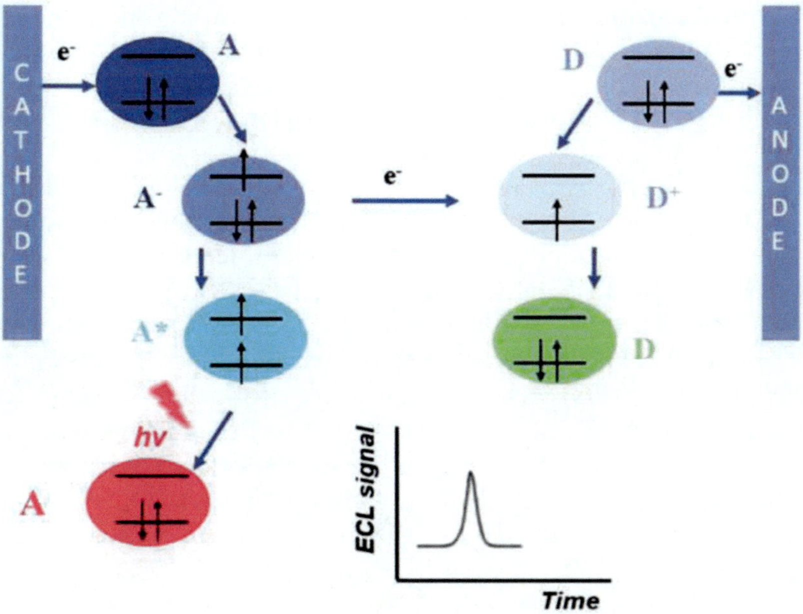

Fig. (7). Schematic presentation of general principles of electrochemiluminescence (A and D could be the same product)

In ECL (bio)sensing, any molecules that can inhibit or promote one of the aforementioned steps will change the final ECL emission and, therefore can be quantitatively detected. Based on this, ECL detection strategies could be classified into five main types, including i/ direct emission, caused by an ECL emitter as a signal label, ii/ steric hindrance inhibition, or impedance effect caused by biorecognition reactions, iii/ emission caused by interaction between analytes and ECL species such as ECL luminophores, ECL luminophore radicals or excited ECL species, iv/ emission caused by interaction between analytes and coreactive species, which is suitable for coreactive ECL systems and v/ inhibition by ECL resonance energy transfer (ECL-RET) where RET is a non-radiative process by which energy transfers *via* long-range dipole-dipole interactions from an excited state donor to a proximal ground state acceptor. Nonetheless, various ECL luminophores were described in the literature, including metal-organic complexes and organics, typically $Ru(bpy)_3^{2+}$ and its derivatives, and luminol. Despite the excellent ECL activities, the above luminophores have difficulty labeling analytes [93]. Recently, carbonaceous QDs have emerged as a new alternative group of ECL luminophores exploiting both annihilation and coreactive ECL mechanisms and have been widely applied in (bio) detection applications. In this context, J. Zhu and colleagues [94] explained that the mechanism of electron transfer annihilation is applied when active $CQD^{\bullet+}$ and $CQDs^{\bullet-}$ are generated by a potential sweep from positive and negative voltages, respectively. Meanwhile, the co-reactant ECL mechanism relies on the potential of scanning in the negative range using an electroactive co-reactant citing peroxydisulfate $S_2O_8^{2-}$, sulfite SO_3^{2-}, tris(2,2'-bipyridine)ruthenium(II) ion $Ru(bpy)_3^{2+}$, oxalate, nicotinamide adenine dinucleotide (NADH), 2-(dibutylamino)ethanol (DBAE), and benzoyl peroxide (BPO) [93].

The authors choose to present in this section various examples of ECL (bio)sensors based on carbonaceous QDs using some of the ECL mechanisms described above. As an ECL label, Yang *et al.* [95] developed an ultrasensitive immunosensor based on Pt nanoparticles dotted graphene-carbon nanotubes composites (Pt/GrCNTs) as a platform and carbon dots functionalized Pt/Fe nanoparticles (Pt/Fe@CDs) as bionanolabels for the sensing of human carcinoembryonic antigen (CEA). In this work, the ECL mechanism involves the formation of excited-state CQDs (CQDs*) *via* electron-transfer annihilation of negatively charged ($CQDs^{\bullet-}$) and positively charged CDs ($CQDs^{\bullet+}$) and the use of $S_2O_8^{2-}$ as a co-reactant was proposed as follows:

$$CQDs + e^- \rightarrow CQDs^{\bullet-}$$

$$S_2O_8^{2-} + e^- \rightarrow SO_4^{2-} + SO_4^{\bullet-}$$

$$CQDs^{\bullet-} + SO_4^{\bullet-} \rightarrow CQD^* + SO_4^{2-}$$

$$CQDs^* \rightarrow CQDs + h\upsilon$$

The described method provided a good linear relationship between the intensity of the ECL emission and the concentration range from 0.003 to 600 ng mL^{-1} with a low limit of detection of 0.8 pg.mL^{-1}. The ECL platform showed good specificity, acceptable stability and reproducibility. Han *et al.* [100] reported the use N-doped CQDs-$K_2S_2O_8$ system as an electrochemiluminescence platform for sensitive detection of human IgG (HIgG) in the concentration range of 0.1-4 ng. mL^{-1} with a low detection limit of the order of 0.05 ng.mL^{-1}. The developed study showed that the N-doped CQDs with abundant carboxy groups reveal more important and stable electrochemiluminescence response than CQDs because of the easier electrochemical reduction of N-CQDs to generate more N-$CQDs^{\bullet-}$ radicals. This nanomaterial can be considered as an excellent ECL luminophore for protein detection.

Recently, sulfite (SO_3^{2-}) has been used as a new type of co-reactant. Indeed, it has been demonstrated that GQDs/SO_3^{2-} system exhibited a sensitive ECL emission in the presence of hydrogen peroxide, which was used as ECL signal quencher. As an example, Y. Chi *et al.* [96] reported the development of an ECL biosensor platform based on the interaction between GQDs and sulfite as co-reactant for the sensitive sensing of hydrogen peroxide. It has been demonstrated that GQDs induced much higher ECL activity than CQDs when SO_3^{2-} was used as a co-reactant. The authors reported that the observed phenomenon is due to the more abundant surface states of GQDs than CQDs. Additionally, the investigation showed that ECL emission of GQDs/SO_3^{2-} the system was sensitively quenched by H_2O_2 at physiological pH. As an interesting application, the ECL platform was applied for the detection of H_2O_2 and biologically interesting molecules such as glucose.

Furthermore, nitrogen-doped graphene QDs (N-GQDs) were used to develop a sensitive monitoring approach for zearalenone (ZEN) aptasensing by You *et al.* [97]. The reported method was based on the combination of amine-functionalized $Ru(bpy)_3^{2+}$ doped silica nanoparticles (NH_2-Ru@ SiO_2 NPs) and N-GQDs. Since the emitter (N-GQDs) and co-reactant ($Ru(bpy)_3^{2+}$) simultaneously existed in the same nanocomposites, a shortened electron-transfer distance and decreased energy loss were obtained. The self-enhanced ECL strategy exhibited a wide linear range of 10 fg.mL^{-1} - 10 ng.mL^{-1} and a low detection limit of 1 fg mL^{-1} for ZEN detection. More interestingly, the same collaborators have reported the use of $Ru(bpy)_3^{2+}$ nanosheets (RuNSs) as a novel form of co-reactant. The study described the development of a multifunctional solid-state electrochemiluminescence (ECL) platform based on poly(ethylenimine) (PEI) capped N-doped carbon QDs (NCQDs@PEI). In this strategy, RuNSs was used as a luminophor, and NCQDs@PEI, was served as a co-reactant. The authors proposed as ECL mechanism the following steps:

$$RuNSs(II) - e^- \rightarrow RuNSs(III)$$

$$N\text{-}CQDs@PEI - 2e^- \rightarrow 2[N\text{-}CQDs@PEI]^{\cdot +} \rightarrow 2[N\text{-}CQDs@PEI]^{\cdot} + 2H^+$$

$$[N\text{-}CQDs@PEI]^{\cdot +} + RuNSs(III) \rightarrow RuNSs(II)^* + products$$

$$RuNSs(II)^* \rightarrow RuNSs(II) + h\upsilon$$

The developed ECL sensor (NCDs@PEI-rGO/RuNSs/GCE) exhibited enhanced ECL emission, with about 69-fold the ECL intensity of RuNSs/ GCE. The prepared sensor showed a high sensitivity for the detection of dopamine in a wide linear range of 0.01-50 µM with a detection limit of 10 nM. Moreover, the sensing platform was also applied to detect other compounds citing catechol, hydroquinone, and bisphenol A [98]. In the author's study, Li and collaborators [99] described the use of the ECL-RET mechanism to develop an ECL aptasensor based on graphene oxide modified by gold nanoparticles and carbon QDs for the detection of the lincomycin analyte. Following the recognition of lincomycin, CQDs exhibited an enhanced ECL signal monitored by the ECL-RET mechanism between AuNPs-GO and CQDs. Indeed, they explained that the binding of lincomycin to the modified surface induces a blockage of energy transfer resulting in a decrease in the intensity of the ECL [100]. This approach was applied to monitor lincomycin residuals in real meat samples.

CONCLUSION

Carbonaceous Quantum Dots (CQDs) have emerged as a promising class of zero-dimensional nanomaterials with unique optical, electrical, and physicochemical properties that make them highly attractive for sensing applications. CQDs are a type of carbon-based nanoparticles that typically range in size from 1 to 10 nm and are characterized by their high surface area, excellent photoluminescence, and biocompatibility. One of the key advantages of CQDs is their tunable optical properties. They exhibit strong and stable fluorescence, making them ideal for use in fluorescence-based sensing applications. Additionally, their unique surface chemistry enables functionalization with different chemical groups, allowing for the selective detection of specific targets. CQDs have shown promising results in a variety of sensing applications, including the detection of heavy metal ions, environmental pollutants, and biological molecules such as glucose, proteins and DNA. They have also been explored for use in sensing applications in the fields of healthcare, food safety, and environmental monitoring.

However, there are still challenges that need to be addressed before CQDs can be widely adopted for sensing applications. One of the biggest challenges is the development of efficient and cost-effective methods for their production and purification. Additionally, the optimization of their sensing performance for specific applications and the development of new sensing platforms that can exploit their unique properties are also critical areas of research. Despite these challenges, the future of CQDs in sensing applications looks promising. Their unique properties make them highly attractive for a range of sensing applications, and their potential for further optimization and functionalization opens up new possibilities for sensing in a variety of fields.

REFERENCES

[1] Xu, Y.; Liu, J.; Gao, C.; Wang, E. Applications of carbon quantum dots in electrochemiluminescence: A mini review. *Electrochem. Commun.,* **2014**, *48*, 151-154. [http://dx.doi.org/10.1016/j.elecom.2014.08.032]

[2] Yuan, F.; Li, S.; Fan, Z.; Meng, X.; Fan, L.; Yang, S. Shining carbon dots: Synthesis and biomedical and optoelectronic applications. *Nano Today,* **2016**, *11*(5), 565-586. [http://dx.doi.org/10.1016/j.nantod.2016.08.006]

[3] Molaei, M.J. Carbon quantum dots and their biomedical and therapeutic applications: A review. *RSC Advances,* **2019**, *9*(12), 6460-6481. [http://dx.doi.org/10.1039/C8RA08088G] [PMID: 35518468]

[4] Hu, S. Tuning optical properties and photocatalytic activities of carbon-based *quantum dots* through their surface groups. *Chem. Rec.,* **2016**, *16*(1), 219-230. [http://dx.doi.org/10.1111/tcr.201500225] [PMID: 26663656]

[5] Pan, L.; Wang, Y.; Ma, J.; Hu, Y.; Su, B.; Fang, G.; Wang, L.; Xiang, B. A review of heavy metal pollution levels and health risk assessment of urban soils in Chinese cities. *Environ. Sci. Pollut. Res. Int.,* **2018**, *25*(2), 1055-1069. [http://dx.doi.org/10.1007/s11356-017-0513-1] [PMID: 29134530]

[6] Liu, M.; Xu, Y.; Niu, F.; Gooding, J.J.; Liu, J. Carbon quantum dots directly generated from electrochemical oxidation of graphite electrodes in alkaline alcohols and the applications for specific ferric ion detection and cell imaging. *Analyst,* **2016**, *141*(9), 2657-2664. [http://dx.doi.org/10.1039/C5AN02231B] [PMID: 26878217]

[7] Aslandaş, A.M.; Balcı, N.; Arık, M.; Şakiroğlu, H.; Onganer, Y.; Meral, K. Liquid nitrogen-assisted synthesis of fluorescent carbon dots from Blueberry and their performance in Fe^{3+} detection. *Appl. Surf. Sci.,* **2015**, *356*, 747-752. [http://dx.doi.org/10.1016/j.apsusc.2015.08.147]

[8] Yang, Q.; Wei, L.; Zheng, X.; Xiao, L. Single particle dynamic imaging and Fe3+ sensing with bright carbon dots derived from bovine serum albumin proteins. *Sci. Rep.,* **2015**, *5*(1), 17727. [http://dx.doi.org/10.1038/srep17727] [PMID: 26634992]

[9] Liu, Y.; Zhou, Q.; Yuan, Y.; Wu, Y. Hydrothermal synthesis of fluorescent carbon dots from sodium citrate and polyacrylamide and their highly selective detection of lead and pyrophosphate. *Carbon,* **2017**, *115*, 550-560. [http://dx.doi.org/10.1016/j.carbon.2017.01.035]

[10] Liu, Y.; Zhou, Q.; Li, J.; Lei, M.; Yan, X. Selective and sensitive chemosensor for lead ions using fluorescent carbon dots prepared from chocolate by one-step hydrothermal method. *Sens. Actuators B Chem.,* **2016**, *237*, 597-604. [http://dx.doi.org/10.1016/j.snb.2016.06.092]

[11] Xu, Y.; Tang, C.J.; Huang, H.; Sun, C.Q.; Zhang, Y.K.; Ye, Q.F.; Wang, A.J. Green synthesis of fluorescent carbon quantum dots for detection of Hg^{2+}. *Chin. J. Anal. Chem.,* **2014**, *42*(9), 1252-1258. [http://dx.doi.org/10.1016/S1872-2040(14)60765-9]

[12] Guo, Y.; Wang, Z.; Shao, H.; Jiang, X. Hydrothermal synthesis of highly fluorescent carbon nanoparticles from sodium citrate and their use for the detection of mercury ions. *Carbon,* **2013**, *52*, 583-589.
[http://dx.doi.org/10.1016/j.carbon.2012.10.028]

[13] Guo, Y.; Zhang, L.; Cao, F.; Leng, Y. Thermal treatment of hair for the synthesis of sustainable carbon quantum dots and the applications for sensing Hg2+. *Sci. Rep.,* **2016**, *6*(1), 35795.
[http://dx.doi.org/10.1038/srep35795] [PMID: 27762342]

[14] Gaddam, R.R.; Vasudevan, D.; Narayan, R.; Raju, K.V.S.N. Controllable synthesis of biosourced blue-green fluorescent carbon dots from camphor for the detection of heavy metal ions in water. *RSC Advances,* **2014**, *4*(100), 57137-57143.
[http://dx.doi.org/10.1039/C4RA10471D]

[15] Shi, J.; Lu, C.; Yan, D.; Ma, L. High selectivity sensing of cobalt in HepG2 cells based on necklace model microenvironment-modulated carbon dot-improved chemiluminescence in Fenton-like system. *Biosens. Bioelectron.,* **2013**, *45*, 58-64.
[http://dx.doi.org/10.1016/j.bios.2013.01.056] [PMID: 23454738]

[16] Tan, X.W.; Romainor, A.N.B.; Chin, S.F.; Ng, S.M. Carbon dots production via pyrolysis of sago waste as potential probe for metal ions sensing. *J. Anal. Appl. Pyrolysis,* **2014**, *105*, 157-165.
[http://dx.doi.org/10.1016/j.jaap.2013.11.001]

[17] Wang, B.; Zhuo, S.; Chen, L.; Zhang, Y. Fluorescent graphene quantum dot nanoprobes for the sensitive and selective detection of mercury ions. *Spectrochim. Acta A Mol. Biomol. Spectrosc.,* **2014**, *131*, 384-387.
[http://dx.doi.org/10.1016/j.saa.2014.04.129] [PMID: 24840485]

[18] Liu, R.; Li, H.; Kong, W.; Liu, J.; Liu, Y.; Tong, C.; Zhang, X.; Kang, Z. Ultra-sensitive and selective Hg2+ detection based on fluorescent carbon dots. *Mater. Res. Bull.,* **2013**, *48*(7), 2529-2534.
[http://dx.doi.org/10.1016/j.materresbull.2013.03.015]

[19] Gupta, A.; Verma, N.C.; Khan, S.; Tiwari, S.; Chaudhary, A.; Nandi, C.K. Paper strip based and live cell ultrasensitive lead sensor using carbon dots synthesized from biological media. Sensor. Actuat. *Biol. Chem.,* **2016**, *232*, 107-114.

[20] Shtepliuk, I.; Caffrey, N.M.; Iakimov, T.; Khranovskyy, V.; Abrikosov, I.A.; Yakimova, R. On the interaction of toxic Heavy Metals (Cd, Hg, Pb) with graphene quantum dots and infinite graphene. *Sci. Rep.,* **2017**, *7*(1), 3934.
[http://dx.doi.org/10.1038/s41598-017-04339-8] [PMID: 28638144]

[21] Ren, G.; Zhang, Q.; Li, S.; Fu, S.; Chai, F.; Wang, C.; Qu, F. One pot synthesis of highly fluorescent N doped C-dots and used as fluorescent probe detection for Hg^{2+} and Ag^{+} in aqueous solution. *Sens. Actuators B Chem.,* **2017**, *243*, 244-253.
[http://dx.doi.org/10.1016/j.snb.2016.11.149]

[22] Bian, S.; Shen, C.; Qian, Y.; Liu, J.; Xi, F.; Dong, X. Facile synthesis of sulfur-doped graphene quantum dots as fluorescent sensing probes for Ag+ ions detection. *Sens. Actuators B Chem.,* **2017**, *242*, 231-237.
[http://dx.doi.org/10.1016/j.snb.2016.11.044]

[23] Xu, Q.; Pu, P.; Zhao, J.; Dong, C.; Gao, C.; Chen, Y.; Chen, J.; Liu, Y.; Zhou, H. Preparation of highly photoluminescent sulfur-doped carbon dots for Fe(III) detection. *J. Mater. Chem. A Mater. Energy Sustain.,* **2015**, *3*(2), 542-546.
[http://dx.doi.org/10.1039/C4TA05483K]

[24] Fan, Z.; Li, Y.; Li, X.; Fan, L.; Zhou, S.; Fang, D.; Yang, S. Surrounding media sensitive photoluminescence of boron-doped graphene quantum dots for highly fluorescent dyed crystals, chemical sensing and bioimaging. *Carbon,* **2014**, *70*, 149-156.
[http://dx.doi.org/10.1016/j.carbon.2013.12.085]

[25] Gu, D.; Zhang, P.; Zhang, L.; Liu, H.; Pu, Z.; Shang, S. Nitrogen and phosphorus co-doped carbon dots derived from lily bulbs for copper ion sensing and cell imaging. *Opt. Mater.,* **2018**, *83*, 272-278. [http://dx.doi.org/10.1016/j.optmat.2018.06.012]

[26] Omer, K.M.; Mohammad, N.N.; Baban, S.O.; Hassan, A.Q. Carbon nanodots as efficient photosensitizers to enhance visible-light driven photocatalytic activity. *J. Photochem. Photobiol. Chem.,* **2018**, *364*, 53-58. [http://dx.doi.org/10.1016/j.jphotochem.2018.05.041]

[27] Omer, K.M.; Tofiq, D.I.; Hassan, A.Q. Solvothermal synthesis of phosphorus and nitrogen doped carbon quantum dots as a fluorescent probe for iron(III). *Mikrochim. Acta,* **2018**, *185*(10), 466. [http://dx.doi.org/10.1007/s00604-018-3002-4] [PMID: 30229316]

[28] Mohapatra, S.; Sahu, S.; Sinha, N.; Bhutia, S.K. Synthesis of a carbon-dot-based photoluminescent probe for selective and ultrasensitive detection of Hg^{2+} in water and living cells. *Analyst,* **2015**, *140*(4), 1221-1228. [http://dx.doi.org/10.1039/C4AN01386G] [PMID: 25537483]

[29] Gao, Z.; Lin, Z.; Chen, X.; Lai, Z.; Huang, Z. Carbon dots-based fluorescent probe for trace Hg^{2+} detection in water sample. *Sens. Actuators B Chem.,* **2016**, *222*, 965-971. [http://dx.doi.org/10.1016/j.snb.2015.09.032]

[30] Dong, Y.; Wang, R.; Li, G.; Chen, C.; Chi, Y.; Chen, G. Polyamine-functionalized carbon quantum dots as fluorescent probes for selective and sensitive detection of copper ions. *Anal. Chem.,* **2012**, *84*(14), 6220-6224. [http://dx.doi.org/10.1021/ac3012126] [PMID: 22686413]

[31] Chen, J.; Li, Y.; Lv, K.; Zhong, W.; Wang, H.; Wu, Z.; Yi, P.; Jiang, J. Sensor. Cyclam-functionalized carbon dots sensor for sensitive and selective detection of copper(II) ion and sulfide anion in aqueous media and its imaging in live cells. Actuat. *Biol. Chem.,* **2016**, *224*, 298-306.

[32] Yuan, C.; Liu, B.; Liu, F.; Han, M.Y.; Zhang, Z. Fluorescence *turn on* detection of mercuric ion based on bis(dithiocarbamato)copper(II) complex functionalized carbon nanodots. *Anal. Chem.,* **2014**, *86*(2), 1123-1130. [http://dx.doi.org/10.1021/ac402894z] [PMID: 24377316]

[33] Liu, J.J.; Chen, Z.T.; Tang, D.S.; Wang, Y.B.; Kang, L.T.; Yao, J.N. Graphene quantum dots-based fluorescent probe for turn-on sensing of ascorbic acid. *Sens. Actuators B Chem.,* **2015**, *212*, 214-219. [http://dx.doi.org/10.1016/j.snb.2015.02.019]

[34] Liu, H.; Na, W.; Liu, Z.; Chen, X.; Su, X. A novel turn-on fluorescent strategy for sensing ascorbic acid using graphene quantum dots as fluorescent probe. *Biosens. Bioelectron.,* **2017**, *92*, 229-233. [http://dx.doi.org/10.1016/j.bios.2017.02.005] [PMID: 28222367]

[35] Kundu, A.; Nandi, S.; Das, P.; Nandi, A.K. Facile and green approach to prepare fluorescent carbon dots: Emergent nanomaterial for cell imaging and detection of vitamin B2. *J. Colloid Interface Sci.,* **2016**, *468*, 276-283. [http://dx.doi.org/10.1016/j.jcis.2016.01.070] [PMID: 26852351]

[36] Ghosh, T.; Chatterjee, S.; Prasad, E. Photoinduced electron transfer from various aniline derivatives to graphene quantum dots. *J. Phys. Chem. A,* **2015**, *119*(49), 11783-11790. [http://dx.doi.org/10.1021/acs.jpca.5b08522] [PMID: 26580460]

[37] Tian, X.; Peng, H.; Li, Y.; Yang, C.; Zhou, Z.; Wang, Y. Highly sensitive and selective paper sensor based on carbon quantum dots for visual detection of TNT residues in groundwater. *Sens. Actuators B Chem.,* **2017**, *243*, 1002-1009. [http://dx.doi.org/10.1016/j.snb.2016.12.079]

[38] Zor, E.; Morales-Narváez, E.; Zamora-Gálvez, A.; Bingol, H.; Ersoz, M.; Merkoçi, A. Graphene quantum dots-based photoluminescent sensor: A multifunctional composite for pesticide detection. *ACS Appl. Mater. Interfaces,* **2015**, *7*(36), 20272-20279.

[http://dx.doi.org/10.1021/acsami.5b05838] [PMID: 26313947]

[39] Amjadi, M.; Jalili, R. Molecularly imprinted mesoporous silica embedded with carbon dots and semiconductor quantum dots as a ratiometric fluorescent sensor for diniconazole. *Biosens. Bioelectron.,* **2017**, *96*, 121-126. [http://dx.doi.org/10.1016/j.bios.2017.04.045] [PMID: 28477568]

[40] Wu, X.; Song, Y.; Yan, X.; Zhu, C.; Ma, Y.; Du, D.; Lin, Y. Carbon quantum dots as fluorescence resonance energy transfer sensors for organophosphate pesticides determination. *Biosens. Bioelectron.,* **2017**, *94*, 292-297. [http://dx.doi.org/10.1016/j.bios.2017.03.010] [PMID: 28315592]

[41] Yan, X.; Song, Y.; Zhu, C.; Li, H.; Du, D.; Su, X.; Lin, Y. MnO_2 nanosheet-carbon dots sensing platform for sensitive detection of organophosphorus pesticides. *Anal. Chem.,* **2018**, *90*(4), 2618-2624. [http://dx.doi.org/10.1021/acs.analchem.7b04193] [PMID: 29237266]

[42] Han, M.; Wang, L.; Li, S.; Bai, L.; Zhou, Y.; Sun, Y.; Huang, H.; Li, H.; Liu, Y.; Kang, Z. High-bright fluorescent carbon dot as versatile sensing platform. *Talanta,* **2017**, *174*, 265-273. [http://dx.doi.org/10.1016/j.talanta.2017.05.067] [PMID: 28738577]

[43] Ao, H.; Feng, H.; Huang, X.; Zhao, M.; Qian, Z. A reversible fluorescence nanoswitch based on dynamic covalent B-O bonds using functional carbon quantum dots and its application for α-glucosidase activity monitoring. *J. Mater. Chem. C Mater. Opt. Electron. Devices,* **2017**, *5*(11), 2826-2832. [http://dx.doi.org/10.1039/C7TC00223H]

[44] Kong, W.; Wu, D.; Xia, L.; Chen, X.; Li, G.; Qiu, N.; Chen, G.; Sun, Z.; You, J.; Wu, Y. Carbon dots for fluorescent detection of α-glucosidase activity using enzyme activated inner filter effect and its application to anti-diabetic drug discovery. *Anal. Chim. Acta,* **2017**, *973*, 91-99. [http://dx.doi.org/10.1016/j.aca.2017.03.050] [PMID: 28502432]

[45] Sun, Q.; Fang, S.; Fang, Y.; Qian, Z.; Feng, H. Fluorometric detection of cholesterol based on β-cyclodextrin functionalized carbon quantum dots via competitive host-guest recognition. *Talanta,* **2017**, *167*, 513-519. [http://dx.doi.org/10.1016/j.talanta.2017.02.060] [PMID: 28340753]

[46] Luo, M.; Hua, Y.; Liang, Y.; Han, J.; Liu, D.; Zhao, W.; Wang, P. Synthesis of novel β-cyclodextrin functionalized S, N codoped carbon dots for selective detection of testosterone. *Biosens. Bioelectron.,* **2017**, *98*, 195-201. [http://dx.doi.org/10.1016/j.bios.2017.06.056] [PMID: 28683411]

[47] Liang, J.Y.; Han, L.; Liu, S.G.; Ju, Y.J.; Li, N.B.; Luo, H.Q. Carbon dots-based fluorescent turn off/on sensor for highly selective and sensitive detection of Hg^{2+} and biothiols. *Spectrochim. Acta A Mol. Biomol. Spectrosc.,* **2019**, *222*, 117260. [http://dx.doi.org/10.1016/j.saa.2019.117260] [PMID: 31226617]

[48] Wang, Y.; Wang, S.; Ge, S.; Wang, S.; Yan, M.; Zang, D.; Yu, J. Facile and sensitive paper-based chemiluminescence DNA biosensor using carbon dots dotted nanoporous gold signal amplification label. *Anal. Methods,* **2013**, *5*(5), 1328-1336. [http://dx.doi.org/10.1039/c2ay26485d]

[49] Duan, Y.; Huang, Y.; Chen, S.; Zuo, W.; Shi, B. Cu-doped carbon dots as catalysts for the chemiluminescence detection of glucose. *ACS Omega,* **2019**, *4*(6), 9911-9917. [http://dx.doi.org/10.1021/acsomega.9b00738] [PMID: 31460081]

[50] Cai, Q.Y.; Li, J.; Ge, J.; Zhang, L.; Hu, Y.L.; Li, Z.H.; Qu, L.B. A rapid fluorescence *switch-on* assay for glutathione detection by using carbon dots-MnO_2 nanocomposites. *Biosens. Bioelectron.,* **2015**, *72*, 31-36. [http://dx.doi.org/10.1016/j.bios.2015.04.077] [PMID: 25957074]

[51] Liang, S.S.; Qi, L.; Zhang, R.L.; Jin, M.; Zhang, Z.Q. Ratiometric fluorescence biosensor based on CdTe quantum and carbon dots for double strand DNA detection. *Sens. Actuators B Chem.,* **2017**, *244*,

585-590.
[http://dx.doi.org/10.1016/j.snb.2017.01.032]

[52] Khakbaz, F.; Mahani, M. Micro-RNA detection based on fluorescence resonance energy transfer of DNA-carbon quantum dots probes. *Anal. Biochem.,* **2017**, *523*, 32-38.
[http://dx.doi.org/10.1016/j.ab.2017.01.025] [PMID: 28159568]

[53] Xu, S.; Su, Z.; Zhang, Z.; Nie, Y.; Wang, J.; Ge, G.; Luo, X. Rapid synthesis of nitrogen doped carbon dots and their application as a label free sensor array for simultaneous discrimination of multiple proteins. *J. Mater. Chem. B Mater. Biol. Med.,* **2017**, *5*(44), 8748-8753.
[http://dx.doi.org/10.1039/C7TB02129A] [PMID: 32264268]

[54] Haile Fereja, T.; Hymete, A.; Gunasekaran, T. A Recent review on chemiluminescence reaction, principle and application on pharmaceutical analysis. ISRN Spectroscopy, **2013**; pp. 1-12.

[55] Baeyens, W.R.G.; Schulman, S.G.; Calokerinos, A.C.; Zhao, Y.; García Campaña, A.M.; Nakashima, K.; De Keukeleire, D. Chemiluminescence-based detection: Principles and analytical applications in flowing streams and in immunoassays. *J. Pharm. Biomed. Anal.,* **1998**, *17*(6-7), 941-953.
[http://dx.doi.org/10.1016/S0731-7085(98)00062-4] [PMID: 9884186]

[56] Huang, X.; Li, L.; Qian, H.; Dong, C.; Ren, J. A resonance energy transfer between chemiluminescent donors and luminescent quantum-dots as acceptors (CRET). *Angew. Chem. Int. Ed.,* **2006**, *45*(31), 5140-5143.
[http://dx.doi.org/10.1002/anie.200601196] [PMID: 16826612]

[57] Shen, C.L.; Zheng, G.S.; Wu, M.Y.; Wei, J.Y.; Lou, Q.; Ye, Y.L.; Liu, Z.Y.; Zang, J.H.; Dong, L.; Shan, C.X. Chemiluminescent carbon nanodots as sensors for hydrogen peroxide and glucose. *Nanophotonics,* **2020**, *9*(11), 3597-3604.
[http://dx.doi.org/10.1515/nanoph-2020-0233]

[58] Khataee, A.; Hassanzadeh, J.; Lotfi, R.; Joo, S.W. Graphene quantum dots/bisulfite assisted chemiluminescence of rhodamine B-H_2O_2 system for sensitive recognition of HCHO. *Sens. Actuators B Chem.,* **2018**, *254*, 402-410.
[http://dx.doi.org/10.1016/j.snb.2017.07.098]

[59] Lin, Z.; Dou, X.; Li, H.; Ma, Y.; Lin, J.M. Nitrite sensing based on the carbon dots-enhanced chemiluminescence from peroxynitrous acid and carbonate. *Talanta,* **2015**, *132*, 457-462.
[http://dx.doi.org/10.1016/j.talanta.2014.09.046] [PMID: 25476331]

[60] Shah, S.N.A.; Dou, X.; Khan, M.; Uchiyama, K.; Lin, J.M. N-doped carbon dots/H_2O_2 chemiluminescence system for selective detection of Fe^{2+} ion in environmental samples. *Talanta,* **2019**, *196*, 370-375.
[http://dx.doi.org/10.1016/j.talanta.2018.12.091] [PMID: 30683378]

[61] Hallaj, T.; Amjadi, M.; Manzoori, J.L.; Azizi, N. A novel chemiluminescence sensor for the determination of indomethacin based on sulfur and nitrogen co□doped carbon quantum dot-KMnO $_4$ reaction. *Luminescence,* **2017**, *32*(7), 1174-1179.
[http://dx.doi.org/10.1002/bio.3306] [PMID: 28524362]

[62] Liu, Y.; Han, S. Chemiluminescence of nitrogen-doped carbon quantum dots for the determination of thiourea and tannic acid. *Food Anal. Methods,* **2017**, *10*(10), 3398-3406.
[http://dx.doi.org/10.1007/s12161-017-0911-5]

[63] Zhang, J.; Li, Y.; Han, S. Simultaneous detection of iodide and mercuric ions by nitrogen-sulfur co-doped graphene quantum dots based on flow injection *turn off-on* chemiluminescence analysis system. *Microchem. J.,* **2019**, *147*, 1141-1146.
[http://dx.doi.org/10.1016/j.microc.2019.04.039]

[64] Pang, C.; Han, S.; Li, Y.; Zhang, J. Graphene quantum dot□enhanced chemiluminescence through energy and electron transfer for the sensitive detection of tyrosine. *J. Chin. Chem. Soc.,* **2018**, *65*(12), 1504-1509.
[http://dx.doi.org/10.1002/jccs.201800141]

[65] Hallaj, T.; Amjadi, M.; Mirbirang, F. S, N-doped carbon quantum dots enhanced Luminol-Mn(IV) chemiluminescence reaction for detection of uric acid in biological fluids. *Microchem. J.,* **2020**, *156*, 104841.
[http://dx.doi.org/10.1016/j.microc.2020.104841]

[66] Li, L.; Lai, X.; Xu, X.; Li, J.; Yuan, P.; Feng, J.; Wei, L.; Cheng, X. Determination of bromate via the chemiluminescence generated in the sulfite and carbon quantum dot system. *Mikrochim. Acta,* **2018**, *185*(2), 136.
[http://dx.doi.org/10.1007/s00604-017-2653-x] [PMID: 29594442]

[67] Sinha, A.; Dhanjai, ; Jain, R.; Zhao, H.; Karolia, P.; Jadon, N. Voltammetric sensing based on the use of advanced carbonaceous nanomaterials: A review. *Mikrochim. Acta,* **2018**, *185*(2), 89.
[http://dx.doi.org/10.1007/s00604-017-2626-0] [PMID: 29594390]

[68] Zhou, C.; Jiang, W.; Via, B.K. Facile synthesis of soluble graphene quantum dots and its improved property in detecting heavy metal ions. *Colloids Surf. B Biointerfaces,* **2014**, *118*, 72-76.
[http://dx.doi.org/10.1016/j.colsurfb.2014.03.038] [PMID: 24732395]

[69] Wang, Y.; Zhao, S.; Li, M.; Li, W.; Zhao, Y.; Qi, J.; Cui, X. Graphene quantum dots decorated graphene as an enhanced sensing platform for sensitive and selective detection of copper(II). *J. Electroanal. Chem.,* **2017**, *797*, 113-120.
[http://dx.doi.org/10.1016/j.jelechem.2017.05.031]

[70] Punrat, E.; Maksuk, C.; Chuanuwatanakul, S.; Wonsawat, W.; Chailapakul, O. Polyaniline/graphene quantum dot-modified screen-printed carbon electrode for the rapid determination of Cr(VI) using stopped-flow analysis coupled with voltammetric technique. *Talanta,* **2016**, *150*, 198-205.
[http://dx.doi.org/10.1016/j.talanta.2015.12.016] [PMID: 26838400]

[71] Ting, S.L.; Ee, S.J.; Ananthanarayanan, A.; Leong, K.C.; Chen, P. Graphene quantum dots functionalized gold nanoparticles for sensitive electrochemical detection of heavy metal ions. *Electrochim. Acta,* **2015**, *172*, 7-11.
[http://dx.doi.org/10.1016/j.electacta.2015.01.026]

[72] Ou, J.; Tao, Y.; Ma, J.; Kong, Y. Well-dispersed chitosan-graphene quantum dots nanocomposites for electrochemical sensing platform. *J. Electrochem. Soc.,* **2015**, *162*(12), H884-H889.
[http://dx.doi.org/10.1149/2.0421512jes]

[73] Cai, Z.; Li, F.; Wu, P.; Ji, L.; Zhang, H.; Cai, C.; Gervasio, D.F. Synthesis of nitrogen-doped graphene quantum dots at low temperature for electrochemical sensing trinitrotoluene. *Anal. Chem.,* **2015**, *87*(23), 11803-11811.
[http://dx.doi.org/10.1021/acs.analchem.5b03201] [PMID: 26545150]

[74] Dong, J.; Hou, J.; Jiang, J.; Ai, S. Innovative approach for the electrochemical detection of non-electroactive organophosphorus pesticides using oxime as electroactive probe. *Anal. Chim. Acta,* **2015**, *885*, 92-97.
[http://dx.doi.org/10.1016/j.aca.2015.05.033] [PMID: 26231893]

[75] Wang, G.; Shi, G.; Chen, X.; Yao, R.; Chen, F. A glassy carbon electrode modified with graphene quantum dots and silver nanoparticles for simultaneous determination of guanine and adenine. *Mikrochim. Acta,* **2015**, *182*, 31-322.
[http://dx.doi.org/10.1007/s00604-015-1568-7]

[76] Tan, F.; Cong, L.; Li, X.; Zhao, Q.; Zhao, H.; Quan, X.; Chen, J. An electrochemical sensor based on molecularly imprinted polypyrrole/graphene quantum dots composite for detection of bisphenol A in water samples. *Sens. Actuators B Chem.,* **2016**, *233*, 599-606.
[http://dx.doi.org/10.1016/j.snb.2016.04.146]

[77] Mars, A.; Bouhaouala-Zahar, B.; Raouafi, N. Ultrasensitive sensing of androctonus australis hector scorpion venom toxins in biological fluids using an electrochemical graphene quantum dots/nanobody-based platform. *Talanta,* **2018**, *190*, 182-187.
[http://dx.doi.org/10.1016/j.talanta.2018.07.087] [PMID: 30172496]

[78] Hu, T.; Zhang, L.; Wen, W.; Zhang, X.; Wang, S. Enzyme catalytic amplification of miRNA-155 detection with graphene quantum dot-based electrochemical biosensor. *Biosens. Bioelectron.,* **2016**, *77*, 451-456.
[http://dx.doi.org/10.1016/j.bios.2015.09.068] [PMID: 26453906]

[79] Zhao, J.; Chen, G.; Zhu, L.; Li, G. Graphene quantum dots-based platform for the fabrication of electrochemical biosensors. *Electrochem. Commun.,* **2011**, *13*(1), 31-33.
[http://dx.doi.org/10.1016/j.elecom.2010.11.005]

[80] Valipour, A.; Roushani, M. Using silver nanoparticle and thiol graphene quantum dots nanocomposite as a substratum to load antibody for detection of hepatitis C virus core antigen: Electrochemical oxidation of riboflavin was used as redox probe. *Biosens. Bioelectron.,* **2017**, *89*(Pt 2), 946-951.
[http://dx.doi.org/10.1016/j.bios.2016.09.086] [PMID: 27818057]

[81] Yang, Y.; Liu, Q.; Liu, Y.; Cui, J.; Liu, H.; Wang, P.; Li, Y.; Chen, L.; Zhao, Z.; Dong, Y. A novel label-free electrochemical immunosensor based on functionalized nitrogen-doped graphene quantum dots for carcinoembryonic antigen detection. *Biosens. Bioelectron.,* **2017**, *90*, 31-38.
[http://dx.doi.org/10.1016/j.bios.2016.11.029] [PMID: 27871047]

[82] Zhang, H.; Dai, P.; Huang, L.; Huang, Y.; Huang, Q.; Zhang, W.; Wei, C.; Hu, S. A nitrogen-doped carbon dot/ferrocene@β-cyclodextrin composite as an enhanced material for sensitive and selective determination of uric acid. *Anal. Methods,* **2014**, *6*(8), 2687-2691.
[http://dx.doi.org/10.1039/C4AY00140K]

[83] Huang, Q.; Zhang, H.; Hu, S.; Li, F.; Weng, W.; Chen, J.; Wang, Q.; He, Y.; Zhang, W.; Bao, X. A sensitive and reliable dopamine biosensor was developed based on the Au@carbon dots-chitosan composite film. *Biosens. Bioelectron.,* **2014**, *52*, 277-280.
[http://dx.doi.org/10.1016/j.bios.2013.09.003] [PMID: 24064477]

[84] Huang, Q.; Hu, S.; Zhang, H.; Chen, J.; He, Y.; Li, F.; Weng, W.; Ni, J.; Bao, X.; Lin, Y. Carbon dots and chitosan composite film based biosensor for the sensitive and selective determination of dopamine. *Analyst,* **2013**, *138*(18), 5417-5423.
[http://dx.doi.org/10.1039/c3an00510k] [PMID: 23833763]

[85] Hu, S.; Huang, Q.; Lin, Y.; Wei, C.; Zhang, H.; Zhang, W.; Guo, Z.; Bao, X.; Shi, J.; Hao, A. Reduced graphene oxide-carbon dots composite as an enhanced material for electrochemical determination of dopamine. *Electrochim. Acta,* **2014**, *130*, 805-809.
[http://dx.doi.org/10.1016/j.electacta.2014.02.150]

[86] Shadjou, N.; Hasanzadeh, M.; Talebi, F.; Marjani, A.P. Graphene quantum dot functionalized by beta-cyclodextrin: a novel nanocomposite toward amplification of L -cysteine electro-oxidation signals. *Nanocomposites,* **2016**, *2*(1), 18-28.
[http://dx.doi.org/10.1080/20550324.2016.1187428]

[87] Devi, N.R.; Kumar, T.H.V.; Sundramoorthy, A.K. Electrochemically exfoliated carbon quantum dots modified electrodes for detection of dopamine neurotransmitter. *J. Electrochem. Soc.,* **2018**, *165*(12), G3112-G3119.
[http://dx.doi.org/10.1149/2.0191812jes]

[88] Yola, M.L.; Atar, N. Functionalized graphene quantum dots with bi-metallic nanoparticles composite: sensor application for simultaneous determination of ascorbic acid, dopamine, uric acid and tryptophan. *J. Electrochem. Soc.,* **2016**, *163*(14), B718-B725.
[http://dx.doi.org/10.1149/2.1191614jes]

[89] Fajardo, A.; Tapia, D.; Pizarro, J.; Segura, R.; Jara, P. Determination of norepinephrine using a glassy carbon electrode modified with graphene quantum dots and gold nanoparticles by square wave stripping voltammetry. *J. Appl. Electrochem.,* **2019**, *49*(4), 423-432.
[http://dx.doi.org/10.1007/s10800-019-01288-0]

[90] Mohammadzadeh Jahani, P.; Jafari, M.; Gupta, V.K.; Agarwal, S. Graphene quantum dots/ionic liquid-Modified Carbon Paste Electrode-Based Sensor for Simultaneous voltammetric determination

of norepinephrine and acetylcholine. *Int. J. Electrochem. Sci.,* **2020**, *15*, 947-958. [http://dx.doi.org/10.20964/2020.01.45]

[91] Habibi, E.; Heidari, H. Renewable surface carbon-composite electrode bulk modified with GQD-RuCl3 nano-composite for high sensitive detection of l-tyrosine. *Electroanalysis,* **2016**, *28*(10), 2559-2564.
[http://dx.doi.org/10.1002/elan.201600010]

[92] Chen, J.; He, P.; Bai, H.; He, S.; Zhang, T.; Zhang, X.; Dong, F. Poly(β-cyclodextrin)/carbon quantum dots modified glassy carbon electrode: Preparation, characterization and simultaneous electrochemical determination of dopamine, uric acid and tryptophan. *Sens. Actuators B Chem.,* **2017**, *252*, 9-16.
[http://dx.doi.org/10.1016/j.snb.2017.05.096]

[93] Xu, Y.; Liu, J.; Gao, C.; Wang, E. Applications of carbon quantum dots in electrochemiluminescence: A mini review. *Electrochem. Commun.,* **2014**, *48*, 151-154.
[http://dx.doi.org/10.1016/j.elecom.2014.08.032]

[94] Chen, Y.; Cao, Y.; Ma, C.; Zhu, J.J. Carbon-based dots for electrochemiluminescence sensing. *Mater. Chem. Front.,* **2020**, *4*(2), 369-385.
[http://dx.doi.org/10.1039/C9QM00572B]

[95] Deng, W.; Liu, F.; Ge, S.; Yu, J.; Yan, M.; Song, X. A dual amplification strategy for ultrasensitive electrochemiluminescence immunoassay based on a Pt nanoparticles dotted graphene-carbon nanotubes composite and carbon dots functionalized mesoporous Pt/Fe. *Analyst,* **2014**, *139*(7), 1713-1720.
[http://dx.doi.org/10.1039/C3AN02084C] [PMID: 24519411]

[96] Zhou, C.; Chen, Y.; You, X.; Dong, Y.; Chi, Y. An electrochemiluminescent biosensor based on interactions between a graphene quantum dot–sulfite co□reactant system and hydrogen peroxide. *Chem. Electro. Chem.,* **2017**, *4*, 1783-1789. Available from:https://www.sciencedirect.com/science/article/abs/pii/S0308814619307058

[98] Li, L.; Liu, D.; Mao, H.; You, T. Multifunctional solid-state electrochemiluminescence sensing platform based on poly(ethylenimine) capped N-doped carbon dots as novel co-reactant. *Biosens. Bioelectron.,* **2017**, *89*(Pt 1), 489-495.
[http://dx.doi.org/10.1016/j.bios.2016.03.069] [PMID: 27048890]

[99] Li, S.; Liu, C.; Yin, G.; Zhang, Q.; Luo, J.; Wu, N. Aptamer-molecularly imprinted sensor base on electrogenerated chemiluminescence energy transfer for detection of lincomycin. *Biosens. Bioelectron.,* **2017**, *91*, 687-691.
[http://dx.doi.org/10.1016/j.bios.2017.01.038] [PMID: 28119249]

[100] Han, T.; Yan, T.; Li, Y.; Cao, W.; Pang, X.; Huang, Q.; Wei, Q. Eco-friendly synthesis of electrochemiluminescent nitrogen-doped carbon quantum dots from diethylene triamine pentacetate and their application for protein detection. *Carbon,* **2015**, *91*, 144-152.
[http://dx.doi.org/10.1016/j.carbon.2015.04.053]

CHAPTER 8

Application of Quantum Dots in Wastewaters Treatment

Garima Kumari[1], Eder Lima[2], Mika Sillanpaa[3], Nitesh Kumar[4], Anupam Guleria[5], Dinesh Kumar[5], Atma Rai[6] and Ashish Guleria[6,*]

[1] *Department of Biotechnology, Eternal University, Baru Sahib, Sirmaur, Himachal Pradesh, India*

[2] *Institute of Chemistry, Federal University of Rio Grande, Rio Grande, Brazil*

[3] *Department of Green Chemistry, Aarhus University, Nordre Ringgade, Denmark*

[4] *Department of Chemistry, Govt. College Jukhala, Himachal Pradesh, India*

[5] *Centre of Biomedical Research, SGPGIMS Campus, Lucknow, Uttar Pradesh, India*

[6] *Department of Applied Sciences, WIT Dehradun, Uttarakhand, India*

Abstract: Water stress and scarcity are among the primary issue that the world is witnessing in this era, as pollution of freshwater resources are increasing due to various emerging pollutants, such as pharmaceuticals, personal care products, pesticides, and household and industrial chemicals. Efficient treatment of wastewater is an important aspect of fresh water supply, and such water can also be used for different household activities and other purposes. So for wastewater treatment, different eco-friendly as well as economic approaches have been analyzed, and the use of carbon quantum dots (CQDs) for the treatment of wastewater is efficient and effective technique extensively studied in the last few years. CQDs are promising nanomaterials for water pollution treatment due to their small particle sizes, tuneable fluorescent properties and containing oxygen-based functional groups. In this chapter, the chemical and physical attributes of CQDs, raw substances and methodologies being utilized in the synthesis, and stability of CQDs, along with their effective employment in wastewater remediation and treatment, has discussed in detail.

Keywords: Heavy metals, Organic pollutants, Photo-catalysis, Quantum dots, Water pollution.

INTRODUCTION

In this current era, water contamination is an emerging issue as waste water from manufacturing and chemical processes in industries contributes to it. Heavy

* **Corresponding author Ashish Guleria:** Department of Applied Sciences WIT Dehradun Uttarakhand, India; Tel: +91-9805848660, Email: ashishchem786@gmail.com

Kulvinder Singh & Vineet Kumar (Eds.)

metals, organic dyes, phenols, pesticides, and detergents have all been linked to water contamination in past years, according to reports from all around the world. Heavy metal pollution is a global issue, and living things have been exposed to toxic metals either directly by drinking water contamination or indirectly *via* eating foods and drinking agricultural irrigation water polluted by toxic metals. These contaminants have a strong propensity to bioaccumulate in the food supply chain, possess significant ecological motility, and exhibit excellent resilience to chemical or biochemical destruction [1, 2]. Owing to their toxicity, perhaps at minute concentrations, the release of these contaminants into water bodies poses a serious hazard to the ecosystem and to people's health. Exposure of humans to toxic metals can cause a number of serious diseases or even death. Exposure to cadmium causes acute and chronic intoxications, and it can replace zinc ions in some metallo-enzymes, there by affecting enzyme activity [3]. Cr(VI) compounds exposure to humans is associated with a higher incidence of respiratory cancers [4]. Humans who consume excessive amounts of copper suffer lung disease, mucous inflammation, extensive vascular injury, melancholy, and necrosis alterations in the kidney and liver [5]. Additionally, microbiological degradation of water sources driven by human operations, including the release of organic waste and fecal matter contaminants, is a prevalent issue. In contrast to other forms of cleaning, chlorination is a frequently used approach for microbial rehabilitation [6, 7]. Sodium chlorite is disinfection frequently employed in drinking water treatment, fruits, vegetables and livestock cleaning, and for filtration reasons [8 - 10]. A crucial factor that is frequently overlooked is the quantity of bleach to be used. Ultrahigh disinfectant doses may result either in inefficient decontamination or major health problems such as erythrocytes and endocrine disruption [11]. Additionally, relatively tiny amounts of these toxins may result in a number of health problems, such as mortality, cancers, renal failure, eye malfunction with visual problems, neurological damage, and decreased muscle synchronization [12 - 14].

In recent years, water pollution due to inorganic and organic pollutants has appeared as global trouble as it goes through terrestrial and aquatic ecosystems imposing a relentless threat to the survival of both biotic and abiotic components. So, in order to reduce the concentrations of pollutants from wastewater, various techniques have been used, which include filtration, adsorption, ion exchange, chemical precipitation, coagulation, electrolytic treatment, electron precipitation, evaporation, electrolysis, liquid extraction, magnetic separation, membrane separation, crystallization, activated sludge, electrodialysis, ultrafiltration, chemical oxidation, aerobic and anaerobic treatment [15 - 17]. However, due to their limitation of practical utility in subsequent metal removal as a result of exacting procedures, low binding abilities, and cost-effectiveness, there remains a void for suitable nanomaterial with high metal loading capacity and facile synthesis, in an effort to extract heavy metal and organic pollutants through the ecosystem. Among the various nanomaterials, carbon quantum dots have been recognized as an important nanomaterial that finds extensive applications in wastewater treatment [18]. Small carbon quantum dots have an atom configuration comparable to that of bulk counterparts, only because of 3-dimensional truncation; they contain more atoms on surfaces [19]. In particular, CQDs are semiconductor NPs that display a number of uncommon qualities, including size-relied emission spectra, wide stimulation spectrum, and the ability to emit brilliant light whenever excited using UV light [20]. In recent years, the development of high-quality quantum dots seems to be an effective and more efficient approach for the treatment of wastewater and removing various pollutants from aquatic systems [21, 22]. In this chapter, we have complied the literature on the fabrication of CQDs, characterization techniques for CQDs, along with their specific applications in wastewater treatment.

APPLICATIONS OF CQDS FOR WASTEWATERS REMEDIATION

CQDs find extensive applications in the detection of metal ions in solutions, extraction of heavy metals from effluents and photocatalytic degradation of organic pollutants [23 - 31]. Due to their hydrophilicity, CQDs are a great choice for ecological water assessment. The usage of CQDs in wastewater cleanup is covered in the sections below. Carbon quantum dots (CQDs) are a new class of carbon-based nanomaterials that have gained attention due to their unique properties, such as high quantum yield, good biocompatibility, and low toxicity. The applications of CQDs in various fields, including wastewater remediation, have been extensively researched in recent years. Wastewater remediation is the process of treating contaminated water to remove pollutants and make it safe for discharge into the environment [32 - 36]. CQDs offer a promising alternative for wastewater remediation due to their unique properties, including high surface area, strong adsorption capacity, and photocatalytic activity. The applications of CQDs for wastewater remediation are based on their unique properties and mechanisms. The following section discusses the mechanism of CQDs for wastewater remediation. Photocatalysis is a process in which a catalyst is used to accelerate a chemical reaction using light. CQDs can act as photocatalysts due to their high quantum yield, which means they can absorb light and emit electrons efficiently [37 - 45]. In wastewater treatment, CQDs can be used to degrade organic pollutants such as dyes, pharmaceuticals, and pesticides. The mechanism of CQDs for photocatalysis involves the following steps:

Absorption of Light: CQDs can absorb light and convert it into excited electrons.

Generation of Reactive Oxygen Species (ROS): The excited electrons react with oxygen molecules to produce ROS, such as hydroxyl radicals (OH•), which are strong oxidants [46 - 51].

Degradation of Pollutants: The ROS generated by CQDs can react with organic pollutants, leading to the degradation of the pollutants.

The photocatalytic activity of CQDs can be influenced by various factors, including their size, shape, surface functionalization, and the properties of the target pollutants. Adsorption is a process in which a material adsorbs or attracts molecules or particles from a fluid onto its surface. CQDs have a high surface area, which means they can adsorb pollutants efficiently. The adsorption mechanism of CQDs involves the following steps:

Surface Functionalization: CQDs can be functionalized with various groups, including amine, carboxylic acid, and hydroxyl groups, which can enhance their adsorption properties.

Adsorption of Pollutants: The functionalized CQDs can adsorb pollutants, including organic compounds, heavy metals, and other contaminants, onto their surface [52 - 54].

Desorption of Pollutants: The adsorbed pollutants can be desorbed from the surface of the CQDs using various methods, such as changing the pH, temperature, or concentration of the desorbing agent.

The adsorption capacity of CQDs can be influenced by various factors, including their surface functionalization, surface area, and the properties of the target pollutants. Fluorescence sensing is a process in which a fluorescent material is used to detect pollutants. CQDs can be used as fluorescence sensors due to their high fluorescence intensity, and their fluorescence properties can be affected by the presence of pollutants [55 - 61]. The fluorescence sensing mechanism of CQDs involves the following steps:

Adsorption of Pollutants: The functionalized CQDs can adsorb pollutants onto their surface.

Quenching of Fluorescence: The presence of pollutants on the surface of the CQDs can quench their fluorescence, leading to a decrease in the fluorescence intensity.

Detection of Pollutants: The decrease in the fluorescence intensity can be used to detect the presence of pollutants.

In this section, we will discuss the several usage of CQDs in wastewater treatment.

Carbon Quantum Dots for Removal of Heavy Metals from Wastewater

Heavy metals are one of the most concerning pollutants in wastewater, posing severe health and environmental risks. These heavy metals can enter the human body through various sources like contaminated water, food, and air, and can cause several adverse health effects like neurological disorders, liver damage, and cancer [62 - 64]. Heavy metals, like lead, cadmium, arsenic, and mercury, are known to be toxic, and their toxicity can accumulate in living organisms through the food chain. Hence, it is crucial to remove these heavy metals from wastewater before discharging them into the environment. In this regard, several methods like chemical precipitation, coagulation-flocculation, ion-exchange, and adsorption have been developed to remove these pollutants from wastewater. However, the adsorption method is considered the most effective method for removing heavy metals from wastewater as it is simple, efficient, and cost-effective [65 - 67]. In recent years, carbon quantum dots (CQDs) have been gaining attention as effective adsorbent material for the removal of heavy metals from wastewater. Surface functional groups of CQDs, like carboxyl, amino, and hydroxyl groups, can adsorb heavy metal ions by electrostatic interaction and coordination. The adsorption capacity of CQDs is mainly determined by the functional groups present on their surfaces. The functional groups on the surface of CQDs can be controlled by various methods like oxidation, hydrothermal treatment, and chemical functionalization. These functional groups can be tailored to increase the adsorption capacity of CQDs towards heavy metals. For example, the presence of carboxyl groups on the surface of CQDs enhances the adsorption of heavy metal ions through the formation of metal-carboxylate complexes [68 - 70].

Moreover, CQDs can also chelate heavy metal ions by forming chemical bonds with them. Chelation is a process where a ligand (CQDs) forms a complex with a metal ion by donating electrons to the metal ion. The chelation process enhances the stability of the heavy metal-CQD complex, preventing the release of heavy metals back into the environment [71]. The chelation process is mainly governed by the size, shape, and surface chemistry of CQDs. The size and shape of CQDs can be controlled by various methods like solvothermal, microwave-assisted, and hydrothermal synthesis methods. These methods can control the morphology and size of CQDs, which can affect the chelation process. Furthermore, CQDs can also be used as a photocatalyst in the presence of ultraviolet (UV) light to enhance the removal of heavy metals from wastewater. Under UV light, CQDs can produce reactive oxygen species (ROS) that can oxidize and degrade organic and inorganic pollutants, including heavy metals [72]. The photocatalytic activity of

CQDs is mainly governed by the surface functional groups, size, and bandgap of CQDs. The bandgap of CQDs can be controlled by various methods like doping, which can enhance the photocatalytic activity of CQDs.

The surface performance of nitrogen-doped CQDs was assessed by Sabet and Madhavi to remove Cd(II) and Pb(II) ions from wastewater [18]. Nitrogen-doped quantum dots had huge surface regions, modified surfaces, and non-corrosive features that effectively extracted toxic metal ions in the water. N-CQDs adsorbed 75% of Pb(II) and 37% of Cd(II) through the aquatic media owing to the huge specific region [73]. Fig. (**1**) illustrates the extraction of Cd(II) and Pb(II) using nitrogen-doped-CQDs.

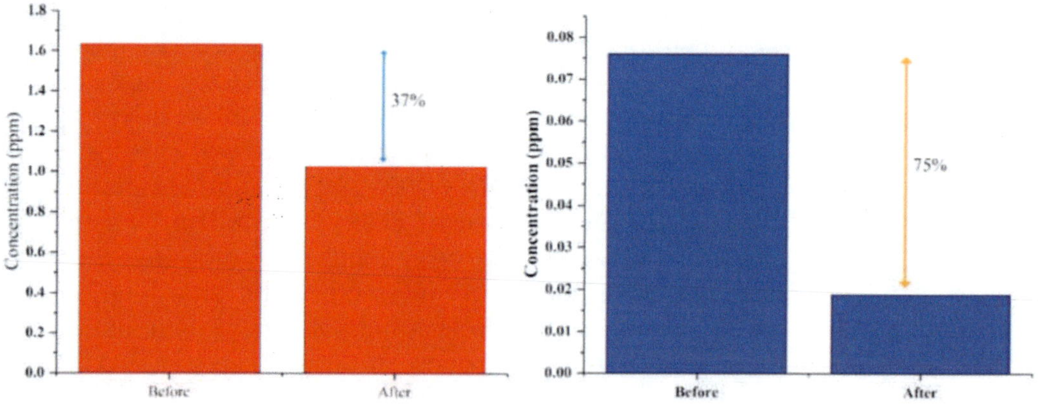

Fig. (1). Extraction of Cd(II) and Pb(II) by utilizing nitrogen-doped-CQDs. (Reproduced with permission from [18], copyright 2019, Elsevier).

In addition to increasing physical and chemical linkages between particles, CQD surfaces and edges could be tuned to narrow the energy band gap of QDs, which improves the capacity of metal ions to adsorb in wastewater [74]. Additionally, the interfaces of N-CQDs may operate as active spots for the sorption process due to the inclusion of nitrogen components and oxygenated moieties [75].

However, it has been found that if several functional groups are induced on the surface of CQDs, they can operate as effective adsorbents in the elimination of harmful compounds from wastewater. Numerous functional sites, including amine and carboxylate, could serve as active centers to attach the metal ions *via* electrostatic interaction and the linkages between atoms in the "pie-pie" configuration [76, 77]. In order to create a nanocomposite (PECQDs/$MnFe_2O_4$) adsorbent, Hu *et al.* synthesized polyethyleneimine-functionalized CQDs upon the magnetic elements ($MnFe_2O_4$) [78]. The nanocomposite fabricated was used for the removal of uranium from wastewater. Strong magnetic field nanocomposite with carbon quantum dots showed a high adsorption capacity of 91%, for uranium. However, nanocomposite $MnFe_2O_4$ without carbon quantum dots as an adsorbent showed an adsorption capacity of 25% for uranium. Uranium and the positively charged surface of the adsorbent were strongly electrostatically attracted to one another in the manufactured PECQDs/$MnFe_2O_4$. However, linkages among surface moieties and uranium ions on the nanocomposite, like OH, NH_2, and COOH, improve uranium sorption [79]. Functional groups mentioned above increase the surface hydrophilicity and hence increase the adsorption capacity of the adsorbent.

Abidin *et al.* synthesized CQDs using rice husk through a hydrothermal process and functionalized CQDs by employing ethylenediamine and ascorbic acid compounds [80]. Ethylenediamine and ascorbic acid incorporated amino and carboxyl functional groups in CQDs, and were verified by TEM analysis that functionalized CQDs offer sizes <10 nm. The presence of oxygen moieties in functionalized CQDs and their enhanced propensity for cationic attachment was revealed by FTIR findings. Fig. (**2**) displays the FTIR and UV-visible spectrum of functionalized CQDs. Functionalized carbon quantum dots showed more than 70% adsorption capability for the extraction of cadmium metal ions. In the case of carboxyl carbon quantum dots, the adsorption capability was 573 mg/g, while in amino carbon quantum dots, the adsorption capacity was 575mg/g for cadmium metal ions.

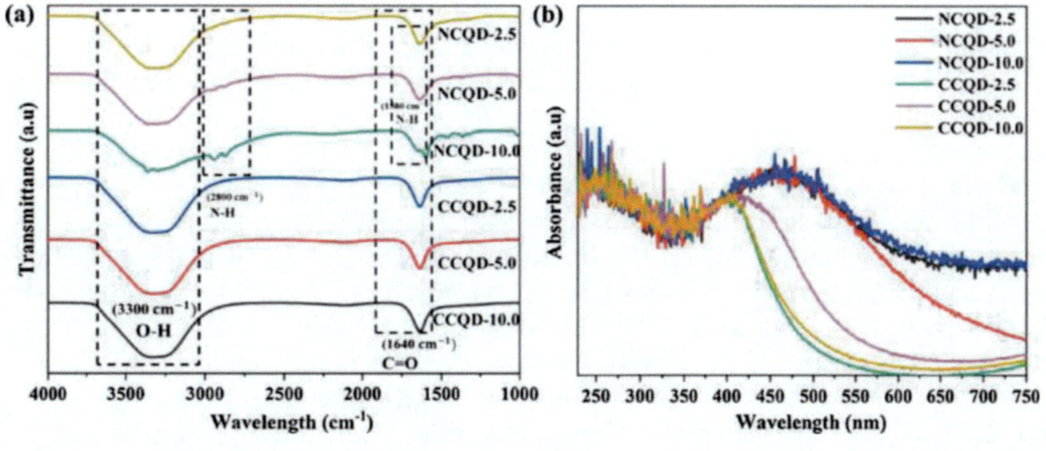

Fig. (2). (**a**) FTIR and (**b**) UV–vis spectrum of carboxyl CQDs functionalized and amino-functionalized CQDs. (Reproduced with permission from [80], copyright 2019, Elsevier).

Adsorption parameters, including time, pH, content of pollutants and temperature, *etc.*, influence adsorption capacity. Adsorption of pollutants highly relied upon the pH of the solution. It revealed that at low solution pH, the adsorption capacity was also low [81]. This may be due to the fact that in an acidic solution, H^+ and metallic ions compete for the active adsorption sites. Since hydrogen ions and metallic ions possess an identical positive charge, a greater amount of hydrogen ions in the solvent at low pH could prevent metallic ions from adhering to the surface of the adsorbent [82]. However, in elevated pH solutions, deprotonation by hydroxyl ions causes the adsorbent surface to turn negatively charged, which strengthens the attractive interactions between the adsorbent and metal ions. At higher pH values, both carboxyl and amino-functionalized CQDs showed higher adsorption capacity to remove Cd^{2+} ions. The percentage removal of cadmium increases slowly from 75 to 76% for carboxyl functionalized CQDs, while amino-functionalized CQDs enhances the extraction across 75.8 to 76.8% when using pH from 2 to 8 [80]. As a result, electrostatic forces might encourage the mixing of adsorbent and sorbent, leading to efficient adsorption.

Toxic metal ions were also removed from aqueous solutions using inorganic metal oxides coupled with quantum dots. Iron oxide (Fe_3O_4) was coupled using nitrogen-doped CQDs made using dicyandiamide and citric acid to create adsorption for lead elimination. The interactions involving lone pair electrons on nitrogen atoms in nitrogen-doped QDs and the vacant d-orbitals of lead ions could be utilized to describe the significant extraction efficiency of lead ions and adsorption [83]. It is anticipated that the amino groups in nitrogen-doped QDs will act as adsorbent surfaces to increase the ability of adsorbents to remove contaminants [84].

In another study, the calcination technique was used to synthesize double-layer oxides/CD nanocomposites (LDO-C) for efficient removal of U(VI) from an aqueous solution [85]. Magnesium chloride, aluminium chloride and chitosan were used as starting materials for the preparation of LDO-C. The adsorption of U(VI) on LDO-C was investigated, and it was found that LDO-C displayed an adsorption capability of 354.2 mg/g greater then pristine LDO 237.6 mg/g. Increased adsorption owing to the inclusion of CDs, and improved surface area and oxygen functional groups in LDO-C as proved by BET, XPS and EXAFS characterization techniques. In addition, LDO-C also exhibited a greater adsorption capability of 241Am(III) than LDO. The impulsive and endothermic adsorption mechanisms of U(VI) on LDO and LDO-C may be accurately predicted by a pseudo-second-order framework. Radioactive contaminants are highly adsorbed due to their large surface oxygen moieties and increased effective surface area.

Heavy metal ion extraction in wastewater has been accomplished using composites built on CQDs. The adsorptive elimination of Cd(II) metal ions in water was achieved using a composite made of CQDs and zinc aluminium layer double hydroxide (CQD/ZnAl-LDH) [86]. By precipitating metal nitrates in the context of CQDs, a CQD/ZnAl-LDH composite was created. The chemical content and architecture of the composite CQD/ZnAl-LDH were then examined using TEM, FTIR, EDX, and XRD methods. To assess the impacts of the Cd(II) metal ion absorption by the CQDs/ZnAl-LDH composite material, a number of variables, including pH, initial concentration, and baseline content, were investigated. At pH 6.0, the CQDs/ZnAl-LDH had a maximal capability of 12.60 mg/g at 20 minutes for Cd(II) metallic ions. The Freundlich isotherm framework and the pseudo-second-order dynamic framework both provided good fits for the adsorption dynamics and isotherm models of Cd(II) on CQDs and ZnAl-LDH, accordingly.

Starch-water system was also used for the synthesis of CQDs utilizing rapid microwave-assisted heating [86, 87]. Microwave heating was carried out for 10 minutes, and CQDs prepared were used on the surface of poly(anthranilic acid-formaldehyde-phthalic acid) (PAFP) polymeric material. CQDs@PAFP nanobiosorbent was utilized to adsorptive remove U(VI) from aqueous solution and had an exterior coverage of 28.79 m^2 g^{-1} BET. For 30 mg L^{-1} of nanobiosorbent, 95.5 to 80.0 percent were achieved. For 30–90 mg L^{-1}, the CQDs@PAFP nanobiosorbent absorbed 95.5–98.0% U(VI), and the adsorption process is based on a pseudo-second-order model. CQDs@PAFP nanobiosorbent had an excellent removal efficiency for U(VI) as it removed 97.3% from wastewater and 96.0% from sea water.

Song *et al.* prepared a new CQDs-based nanofibrillated cellulose (NFC) composite aerogel using green chemical synthesis and used it for the detection and adsorption of Cr^{3+} heavy metal ions in water [88]. Poplar cellulose was isolated, and it was used as a carbon source to fabricate CQDs. Ethylenediamine was used as a modifier for carbon quantum dots and the hydrothermal method was used for CQDs preparation. CQDs/NFC composite aerogel was prepared by adding 5 mL of carbon dots solution into NFC suspension in a freeze-dryer under a vacuum atmosphere. The mixture was magnetically stirred for 3 h for uniform mixing, and the mixed suspension was then placed in a refrigerator at -15 °C for complete freezing. Fig. (**3**). shows the adsorption and detection of Cr^{3+} metal ions by CQDs/NFC composite aerogel and optical pictures of composite aerogel under UV light before and after adsorption.

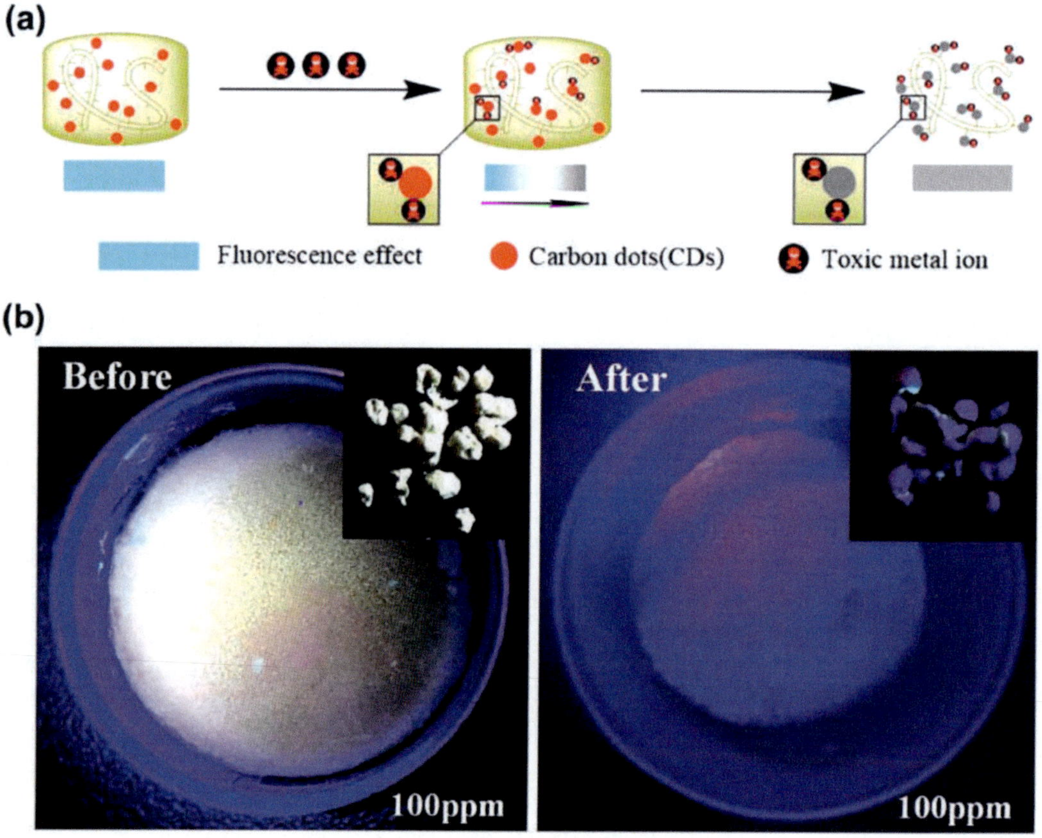

Fig. (3). (**A**) Adsorption and detection of Cr^{3+} metal ions by CQDs/NFC composite aerogel. (**B**) Optical pictures of composite aerogel under UV light (365 nm) before and after adsorption of Cr^{3+} metal ions (Reproduced with permission from [88], copyright 2019, Elsevier).

CQDs for Sensing of Heavy Metals in Water

Carbon quantum dots are small carbonaceous and fluorescent nanostructures that have gained significant interest in designing sensing devices for heavy metals owing to their excellent attributes, including optical tunability, photostability, facile surface modification, and environmental friendliness [89]. With regard to their luminescent properties, CQDs possess surface characteristics that include oxygen-based molecules like hydroxy, carboxylic, and carboxylic groups. O, N and S heteroatoms might also engage anions and cations by coordinating and electrical interaction science in such interface functions. The fluorescence strength of pure CQDs might be increased, additional adsorption sites might be added, and the surface functions might be altered, in contrast to interface modification/ passivation and loading before or after synthesis. The sensitivities and specificity of sensing probes for inorganic metallic ions can be improved by changing CQDs.

A number of different heavy metals/metalloids, such as arsenic, cadmium, copper, aluminum, chromite, zinc, and mercury, have also been effectively detected using CQDs. A luminous chemical sensor's sensitivities are heavily influenced by its level of illumination, quantum efficiency, and functional properties. Due to their superior refractive and luminous characteristics, CQDs could be employed as a fluorescence probe for a number of metallic ions.

For label-free detection of Fe (III) in H_2O having a LOD of 0.32 mM, CQDs made from dopamine were utilized. Fe (III) ions from water specimens are preferentially bound by catechol surface functional groups of CDs. On the basis of a competitive process and a "turn-on" read-out method, the Fe (III)/CDs substrate has been additionally proven for dopamine sensing in biomolecules [90]. Citric acid was utilized to make blue luminous pure CQDs, which were utilized to identify Fe (III) ions in H_2O to a concentration of 1.3 mM and exhibited a quantum efficiency of 52%. The fluorescent quench of CQDs *via* chelating or coordination reactions among Fe (III) ions and the hydroxyl group on CQDs was identified as the sensing element.

Zhou and colleagues showed how to use CQDs made using pyrolyzed EDTA to identify Hg (II). The mixing of non-radiative electron/hole groups was anticipated to result from the attachment of Hg (II) to the interface of CQDs [91]. The created substance effectively detected Hg(II) in river water, filtered water, and across a collection of interference ions down to a level of 4.2 nM. Additionally, the sensitivity limits of CQDs made using sodium citrate, folic acid, polyethylene glycol, and casein precursor for Hg (II) in freshwater were 10 nM, 1 fM, 0.23 mM, and 6.5 nM, correspondingly [92 - 95]. When Hg(II) was sensed by CQDs, a luminous quench took place. This mechanism resulted in an electrostatic attraction, non-radiative recombine, and excitations *via* electron transport among the oxygen moieties [96].

Nitric acid was used to create CQDs using activated carbon, and they were able to recognize Cu (II) ions with an LOD of 6.7 nM in an aqueous phase [97]. The sensor content effectively quenched luminescence after each introduction of Cu (II), which was then reversed with the introduction of oxalate. With LOD of 6.7 nM and 0.3 mM, correspondingly, CQDs having additional chemical or product generated from pyrolytic processing of poly-ethylenimine and aminophenylboronic acid, have also demonstrated the ability to identify Cu (II) ions in water and a cellular matrices [98, 99]. It was discovered that following adding Cu(II) metallic ions to CQDs, the fluorescence strength progressively drops because Cu(II) binds to the carbonyl and carboxyl groups on the CQD interfaces. It was demonstrated that the fluorescence emission impact may result *via* the transmission of excited-state electrons [100]. Utilizing a "off-on" strategy, polyacrylamide sodium citrate and sodium citrate were utilized to easily synthesise CQDs for Pb (II) and pyrophosphate detection [101]. With an inner filter impact, the introduction of cumulative Pb (II) levels greatly reduced the bright-blue luminescence of CQDs, which was then restored using pyrophosphate. As a result, the created probe had LOD for Pb (II) and pyrophosphate of 4.6 nM and 54 nM, correspondingly. Additionally, the probe was successfully used with actual water matrix and human urine specimens.

Other harmful metals are Cr(VI) and CQDs made through the contamination of sucrose, ascorbic acid, citric acid, glucose and 2-hydroxyphenylboronic have been used to detect it in H_2O [102 - 109]. CQDs produced by injecting citric acid using L-reduced cysteamine, glutathione, glycine and glutamic acid had the highest quantum yields of 69, 78, 81, and 54%, respectfully. These improved CQDs have corresponding LOD values for Cr(VI) of 0.57 mM, 4.15 mM, 0.28 mM, and 5 nM. Depending on the reduction in the CQDs excitation wavelength, the measurement of Cr(VI) was established, and CQDs discharge lacking Cr(VI) may be seen at 425 nm. Furthermore, the energy acquired and electrons transferred from the ground phase to the excited phase caused the fluorescence intensity of the CQDs to progressively diminish when they were introduced to the Cr(VI) mixture [68]. As a result, the chromium ion could absorb the CQDs' excitation wavelength, turning them luminous so the Cr (VI) could efficiently extinguish them.

Luminous probes for metallic ions detection have been made from CQDs that have been co-doped with elements like phosphorus and nitrogen [110, 111]. Another efficient method for adjusting the PL characteristics is dosing, and loaded CQDs exhibit improved PL emissions having quantum efficiency >20% as a result of the generation of novel electronic properties that will adjust inherent energy transfer [112]. Citric acid-derived CQDs were treated using a variety of organic alloying elements, such as dextran, 4, 7, 10-trioxa-1, 13-tridecanediamne (TTDDA), L-cysteine, PEI, urea, and triethylenetetramine, for the purpose of detecting Cu (II) metal ions (TETA) [113 - 117]. According to reports, the N, S-CQDs made using L-cysteine possess the greatest fluorescence intensity output (82%), and the substance they produced demonstrated Cu (II) detection having a LOD of 2.1 mg/L for actual water laboratory testing [118]. Various P, N, S, and P, such as, ethane-diamine, ammonia, cysteine, thiourea, and 1-aminopropyl-3-methylimidazolium, were used to create CD derived from citrus that were then disclosed to have a LOD for Fe (III) ions in H_2O of 0.5 mM, 14 nM, 84 nM, 9.8 mM, and 13.68 mM [119 - 124]. The exceptional LOD of S-CQDs for Fe (III) ions has been found and might be attributed to the high quantum output (69%) of cysteine-loaded CQDs (N, S-CQDs) [119]. The efficient quenched of luminescence caused by Fe (III) is made possible by dosing using N and S heteroatoms, which supports the electron transport process and directs the precise interaction among N, S-CQDs and Fe (III) ions. Sulfur doped CQDs were utilised in colorimetric and florescent simultaneous readout methods to identify As(III). They were made by pyrolyzing citric acid and sodium thiosulphate. The addition of glutathione to the SCQDs substantially enhanced the sensitivity of the biosensor components against As (III). For treated river water, streams and rivers, and filtered water [125]. Having LOD of 0.201 mM and 0.234 mM, correspondingly, citric acid laced with diethylenetriamine (N-CQDs) yields a significant quantum efficiency of 82.4 percent when utilized to sense both Hg (II) and iodide ions. Additionally, N-CQDs demonstrated great stability in specimens of bovine urine and genuine lake water with increased salt contents [126]. With a LOD of 10.6 mM, citric acid and N-CQDs generated from L-lysine can identify selenite in the region of 15.6-42.80 mM. The architecture of the detector is based on the interface of the hydroxyl and carboxylate groups on the N-CQD interface between the ions of europium and selenite. The discharge of the europium ions on the N-CQD substrate was caused by the selenite ions' strong interaction propensity for them, and their blue luminescence was restored in detectable quantities [127].

Owing to their affordability and sustainability, sustainable techniques for the production of CQDs have subsequently been established. CQDs from the environmental path have gotten a lot of interest, and they are also being used more and more to detect water contamination. Thus, as precursors for the manufacture of ecofriendly CQDs, scientists have investigated a huge spectrum of biologically existing organic components and also waste resources (fruit extracts, veggies, food products, *etc*.). Sustainable CQDs have been claimed to be produced using a number of biological precursors, including millet, garlic, rose-heart radish, lyciifructus, egg white, papaya and chionanthusretusus. Above mentioned natural precursors-based green CQDs prepared and used for Fe (III) detection in H_2O and they displayed LOD of 46 nM, 49 mM, 0.28 mM, 0.2 mM, 0.18 mM, 0.030 mM, and 70 mM [128 - 134]. The green CQDs synthesized from egg white showed a quantum yield of 64%, which was due to surface groups and abundant surface defects in the core [129]. Doped green QDs were also utilized for sensing heavy metals. Foxtail millet and lyciumbarbarum N/S/P co-doped CQDs with a quantum yield of 21.2% were used to detect Fe (III) having superior sensitivity with a minimal LOD of 46 nM [128]. Recently, kitchen chimney oil waste-based green quantum dots were used for Fe (III) detection in broad spectrum 1 nM-600 μM [135]. These green CQDs were termed "tracer metal chemosensors" owing to their minimal LOD of 0.18 nM toward Fe (III) ions.

The fluorometric-based Pb (II) sensing method had (LOD-0.59 nM) and Lantana camara berries (LOD-9.64 nM) and was effective in real water specimens. Green

CQDs were made from Ocimum sanctum leaves [136, 137]. Lantana camara berries-based green CQDs had 33.15% quantum yield having a LOD of 200 nM for Pb(II) ion detection. Hydrothermal treatment of chocolate results in the formation of green CQDs, which were used for the detection of Pb ions in water upto a concentration of 12.7 nM. CQDs exhibited high sensitivity and selectivity for Pb(II). High selectivity of chocolate-based green CQDs was due to the strong affinity between Pb (II) and hydroxyl groups on the surface. Table sugar green CQDs were developed for colorimetric analysis of Pb (II) ions in water, and it was found that Pb (II) ions addition results in aggregation of green CQDs visible even through the naked eye [138]. The created sensor had a 67 nM LOD for Pb (II), and the instability of ecofriendly CQDs was substantiated by a reduction in absorption spectra as well as an expansion in particle diameter.

Environmental CQDs have been made from renewable fruit resources, including pineapple juice, lemon peel and jack fruit and utilized to detect Cr (VI) in water. The limit of detection for Cr (VI) by green CQDs form lemon peel, pineapple juice and jack fruit was 73 nM, 0-18 μM, 10 nM, respectively [139 - 141]. Besides fruit sources, green carbon quantum dots have also been prepared from waste, including degreased cotton, pig hair, pomelo and wool biomass [141 - 143]. The waste-derived green carbon quantum dots have excellent LOD for Cr (VI) below the WHO suggested range (50 ng/ml). For example, doped green CQDs made of pig and wool hair displayed a 25.6 percent quantum yield. With a linear spectrum of 50 nm–100 M and a LOD of 16.8 nM, ecofriendly CQDs identified Cr(VI). Due to a redox process involving Cr(VI), oxygen-containing surface groups on ecofriendly CQDs, and Cr(VI), fluorescence emission was quenched as a result of Cr(VI) detection. Surface passivating agents such as ammonium persulfate are also important for the post-modification of green carbon quantum dots as they improve the quantum yield of CQDs. Post modification of N,S- green CQDs using pomelo waste were utilized for detection of Cr(VI) in the 1-40 μM spectrum and were also effectively implied in real water sample evaluation for Cr(VI) [143]. Recently, the tulsi leaf-prepared green CQDs probe was used for selective sensing of Cr(VI), and it presented a good analytical detection range (1.6-50 μM) and LOD (229 nM).

For detection of Cu(II) in water, green CQDs derived from a number of natural sources (*e.g.*, prawn shells (LOD: 5 nM), peanut shells (LOD: 4.8 μM), kelp juice (LOD: 12.5 μM), and bamboo leaves (LOD: 115 nM) have been used [144 - 146]. Green quantum dots from prawn shells and peanut also showed satisfactory performance for real water specimens. Further, bamboo leaf-derived quantum dots were modified with polyethyleneimine and electrostatic adsorption was also utilized to effectively detect Cu(II) ions at a minimal pH 4.0 having a LOD of 115 nM [147]. Coal and camphor-derived green CQDs were used to sense Cu(II) ions

[148, 149]. Sustainable QDs made from camphor had a quantum output of 21.16%, however, their sensing ability was insufficiently focused. Sustainable CQDs made from coal had a very excellent LOD of 2 nM for copper (II) [149]. Owing to Cu (IIaffinity)'s for N, O-functional units upon the interface of coal CQDs and the comparatively slow metal-to-surface group interaction dynamics, the detector specificity was achieved [149]. The proportion metrics double signal probe created by conjugating the nettle juice-derived ecofriendly CQDs using fluorescein isothiocyanate was utilized to assess Cu (II) ions in actual water specimens [146]. For Cu (II) ions, the probe has a 9 nM LOD. Fig. (**4**) illustrates the process of heavy metal/metalloid screening using CQDs [33].

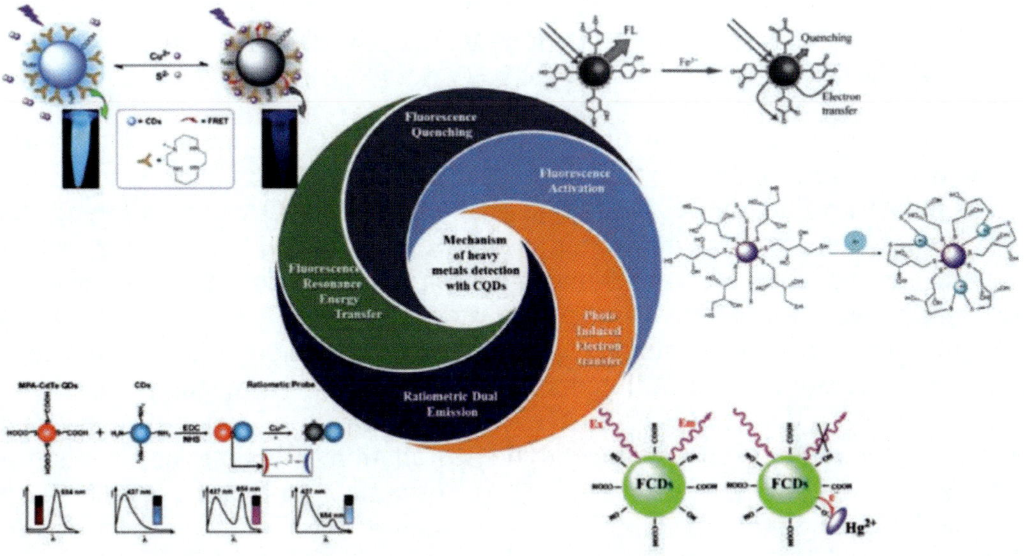

Fig. (4). Mechanism of heavy metal/metalloids detection with CQDs(Reproduced with permission from [33], copyright 2019, Elsevier).

Natural sources, including Jinhua bergamot, flour, elephant foot yam, mushrooms, lotus root, hongcaitai, which contains carbon, nitrogen, and sulfur elements, were used to derive green CQDs and used for Hg(II) detection having LOD of 0.5 nM, 5.5 nM, 4.13 nM, 18.7 nM, 0.06 μM, and 4.13 nM [150 - 156]. Sustainable CDs made using Jinhua bergamot showed a very significant quantum output of 50.78%, compared to merely 5.4 percent for CQDs made using flour. It was discovered that these biological precursors contained a variety of macromolecules, including minerals, proteins, carbs, and minerals. Therefore, owing to alterations in the structural framework and filling with various heterocyclic rings depending on the composition of the original precursors, the interface functioning and quantum output of ecofriendly CQDs changed.

Mixed natural and synthetic precursors have been used by researchers to obtain carbon quantum dots. Orange juice mixed ethylenediamine-based carbon quantum dots were reported for sensing Hg (II) ions in a sensing spectrum of 6-79 μM. Sustainable CQDs generated from mushrooms and vegetables were also used to detect Hg(II) in spiked and unspiked water specimens [156]. The 25 percent quantum output displayed by the mushroom-based ecofriendly QDs was similar to that of orange juice combined using ethylene diamine. These QDs have a LOD of 4.13 nM for Hg (II) ions as well, but dihydrolipoic acid was lowered after surface treatment with a synthetic linker. Attributed to the prevalence of carboxyl and hydroxyl functionalities on their exterior, ecofriendly synthetic CQDs can detect metal ions.

CARBON QUANTUM DOTS FOR REMOVAL AND PHOTOCATALYTIC DEGRADATION OF ORGANIC POLLUTANTS

Organic pollutants are one of the major environmental concerns as they pose serious health and environmental risks. These pollutants are mainly released from various industries like textile, paper, and pharmaceuticals. Organic pollutants, like dyes, phenols, and polycyclic aromatic hydrocarbons (PAHs), are known to be toxic and carcinogenic. Hence, removing these pollutants from wastewater before discharging them into the environment is crucial. In this regard, several methods, like biological treatment, advanced oxidation, and adsorption, have been developed to remove these pollutants from wastewater. In recent years, carbon quantum dots (CQDs) have been gaining attention as an effective adsorbent material for the removal and photocatalytic degradation of organic pollutants from wastewater [157]. The mechanism of organic pollutant removal and photocatalytic degradation by CQDs is mainly based on adsorption and photooxidation. The surface functional groups of CQDs, like carboxyl, amino, and hydroxyl groups, can adsorb organic pollutants by electrostatic interaction and hydrogen bonding. The adsorption capacity of CQDs is mainly determined by the functional groups present on their surfaces. The functional groups on the surface of CQDs can be tailored to increase the adsorption capacity of CQDs towards organic pollutants. For example, the presence of amino groups on the surface of CQDs enhances the adsorption of cationic dyes through electrostatic interaction.

Moreover, CQDs can also photo-catalytically degrade organic pollutants in the presence of ultraviolet (UV) light. Under UV light, CQDs can produce reactive oxygen species (ROS) that can oxidize and degrade organic pollutants. The photocatalytic activity of CQDs is mainly governed by the surface functional groups, size, and bandgap of CQDs. The bandgap of CQDs can be controlled by various methods like doping, which can enhance the photocatalytic activity of

CQDs. The ROS produced by CQDs can react with organic pollutants and break them down into simpler and less toxic products. The degradation efficiency of CQDs depends on various factors like the type and concentration of organic pollutants, the size and morphology of CQDs, and the intensity and duration of UV light exposure. Doped carbon quantum dots with nitrogen and sulfur was loaded with hexagonal mesoporous silica from inside. The N, S-CQDs were prepared from histidine and cysteine amino acids as the source through the hydrothermal method. Modified CQDs were used to remove methylene blue dye from an aqueous solution to effectively remove methylene blue dye (MB) from an aqueous solution. Modified CQDs maximum adsorption capacity of 370.4 mg/g and removal efficiency of 97%, for methylene blue, respectively, at 303 K. The adsorption isotherm studies were fitted with the Freundlich isotherm, and the dye removal kinetics of the adsorbent followed the pseudo-second-order model. Thermodynamic studies showed that the adsorption process had exothermic and spontaneous behavior. The efficient removal of MB dye was due to a small bandgap between HOMO and LUMO of N,S-CQDs, which increase π and π^* interactions and facilitates the dye adsorption process. Besides these interactions mentioned above, the surface area and polarity of the adsorbent materials are important parameters for the removal of pollutants in the adsorption process. Brunauer-Emmett-Teller (BET) surface area and pore volume measurements can be used for the determination of surface area and textural properties of CQDs. It has been found that CQDs have high as 1690 $m^2\ g^{-1}$ surface area, and hence they find extensive use in the adsorption of pollutants [158]. CQDs conjugated iron oxide adsorbents (Fe_3O_4, C11-Fe_3O_4 and CQDs/C11-Fe_3O_4) were prepared and used to remove benzo[a]pyrene (BaP) in wastewater samples [159]. From the BET surface measurement, the surface area Fe_3O_4, C11-Fe_3O_4 and CQDs/C11-Fe_3O_4 were 88.2, 163.5 and 289.8 $m^2\ g^{-1}$, respectively. The surface area of iron oxide increases in the presence of CQDs, and it was found that CQDs/C11-Fe_3O_4 showed 93.9% of adsorption capability for the BaP adsorption due to high surface adsorption sites. Besides the adsorption of organic pollutants, CQDs are also highly efficient photo-catalysts due to their good availability in transferring and trapping electrons [160]. Photons play an important role in the generation of electron-hole pairs on the surface of the photocatalyst, and reactive oxidative species are generated in the conduction band. Concurrently, hydroxyl radicals with strong oxidation capacity are produced due to the photo-induced holes. Fig. (**5**) shows the photocatalytic degradation mechanism of organic pollutants using a photocatalyst.

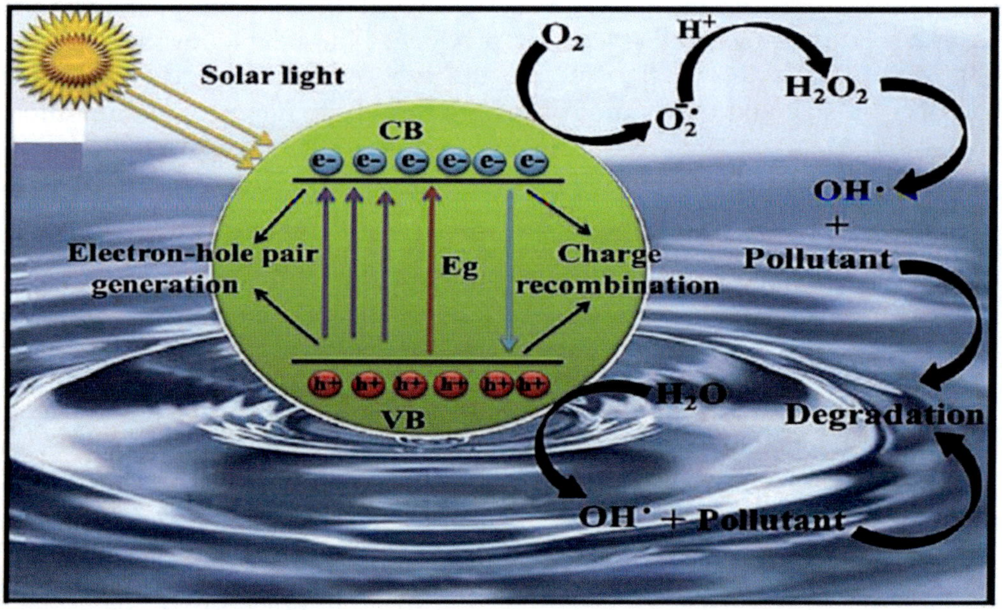

Fig. (5). A basic photo-catalytic mechanism in the degradation of organic pollutants. (Reproduced with permission from [161], copyright 2019, Elsevier).

Ratnayake *et al.* prepared a photo-catalyst by incorporating CQDs on the zirconia surface (ZrO_2) and used this material for photocatalytic degradation of methylene blue [162]. The photo-catalyst was added into a methylene blue solution and sonicated for 30 s. The mixture was kept under UV irradiation for 60 min, and it was found that 95% of methylene blue was degraded after 60 min of UV irradiation. However, only 34% of methylene blue was degraded in the absence of CQDs. Percentage degradation dye increases because CQDs are capable of avoiding the recombination of electron-hole pairs, which can significantly increase the degradation capability of the photo-catalysts [163]. Acid violet 43 and p-nitrophenol were degradation by CQDs/P25 photo-catalyst [164]. CQDs/P25 photo-catalyst was prepared by chemical adsorption of CQDs onto the surface of pyrogenic nanoparticles (P25), and a mixture of the photo-catalyst and acid violet 43 and p-nitrophenol samples were placed under a mercury lamp for some time. After that, the photocatalyst was removed from the mixture by centrifuging and analysis was done by using UV-Vis spectrophotometer. The results showed that 93% of acid violet 43 and 96% of p-nitrophenol were degraded.

In another approach, CQDs attached to a metal-organic framework were used as photo-catalyst for Rhodamine B (RhB) organic pollutants. CQDs were prepared from ascorbic acid by using microwave irradiation, and then photocatalyst composite (CQDs/NH2-MIL-125) was prepared by homogeneous dispersing of CQDs and NH_2-MIL-125 (metal-organic framework) into distilled water which was heated at 90 °C for 3 h [165]. The photo-catalyst prepared was dispersed in Rhodamine B (RhB) solution (10mg/L). The solution was stirred for 40 minutes in the dark room and then stirred under xenon lamp irradiation for 200min. After every 20 min, the solution samples were analyzed to evaluate the RhB and 100% degradation of RhB removal was achieved within 120 min. The photocatalytic activity was increased with an increasing amount of CQDs, and the result reveals that the CQDs can be considered as an enhancing luminescent material [166].

The incorporation of nanocomposite substances onto the surfaces of CQDs also improves the performance of photo-catalysis. This incorporation can be performed by electrostatic forces between π-π and hydrogen-bonding interactions. $La_2Ti_2O_7$ and CQDs composite was prepared and further used as photo-catalyst for degradation of Rhodamine B (RhB) [166]. It was found that after 90min of visible light irradiation 99% percent of RhB degraded. Photocatalytic property was raised because CQDs can convert visible light to shorter wavelengths, which can excite the $La_2Ti_2O_7$ nanosheets to generate electrons and holes. At the same time, photo-induced electrons from the conduction band of $La_2Ti_2O_7$, will be accepted by CQDs, resulting in the separation of electron-hole pairs. Hydroxyl radicals formed with dissolved oxygen and holes in the valance band of $La_2Ti_2O_7$ could oxidise the RhB molecules directly.

In another report, CQDs/TiO_2 composite nanofibers were sued for photocatalytic degradation of methylene blue (MB) dye [167]. TiO_2 nanofibers degraded about 71% of MB in 95 min, but CQDs/TiO_2 composite nanofibers degraded almost 100% of MB dye within the same period. Thus CQDs/TiO_2 composite nanofibers exhibit better photo-catalytic effects in comparison to TiO_2. CQDs absorb light and re-emit shorter wavelengths, which excites the TiO_2 nanofibers to generate electron-hole pairs. Thus, CQDs act as acceptors and transporters for photo-generated electrons [25]. The electron-hole pairs formed after the excitement of electrons were trapped by hydroxyl groups at the catalyst surfaces to yield OH• radicals essential for dye degradation [168]. Meanwhile, excited electrons react with dissolved oxygen molecules to yield superoxide radical anions O_2^- and the protonation process can generate hydroperoxyl radicals ($HO_2^•$) and hydroxyl radicals ($OH^•$). The $HO_2^•$ and $OH^•$ radicals act as strong oxidizing agents which decompose dye molecules into non-toxic gas like CO_2 as well as H_2O molecules [168]. The recycling process of photocatalytic activity has been illustrated in Fig. (**6**).

In the case of CQDs/$K_2Ti_6O_{13}$, the incorporation of potassium hexatitanate ($K_2Ti_6O_{13}$) nanotubes onto the surfaces of CQDs could not achieve the expected photo-catalytic degradation due to the rapid recombination rate of photo-generated electron-hole pairs [169]. $K_2Ti_6O_{13}$ has been reported to have good optical properties, but the incorporation of $K_2Ti_6O_{13}$ onto the surfaces of CQDs can generate a binary composite system that can significantly hinder the recombination rate of photo-excited charge carriers. Thus, fast recombination of charges and limited light utilisation properties are the main barriers of photo-catalyst materials. Hence, CQDs-based nanocomposites have been used for photo-catalysis because CQDs can be easily dispersed onto the surfaces of other nanocomposites, have excellent electron-transfer properties and CQDs can improve the light utilisation range of wide band-gap semiconductor photo-catalyst from UV-light to visible-light range.

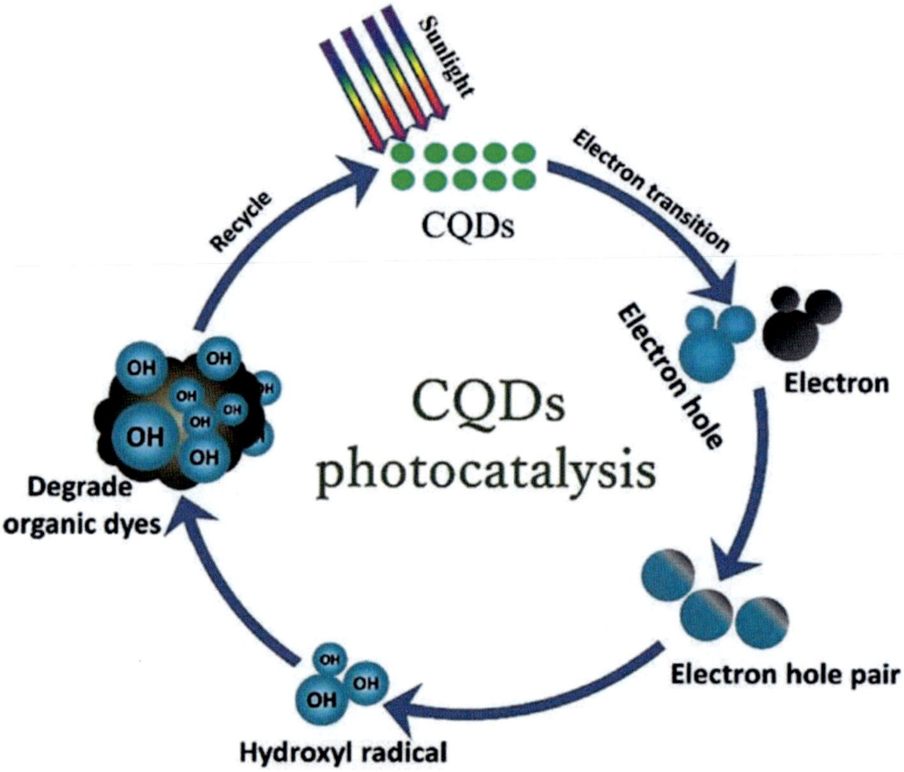

Fig. (6). CQDs are used as the photocatalyst in a conceptual depiction of the photocatalytic destruction of organic contaminants. (Reproduced with permission from [170], copyright 2019, Elsevier).

CONCLUSION

In this chapter, we summarize the recent efforts made for the fabrication of CQDs and their applications in wastewater remediation. Gelatin, cholesterol, glucose, graphite, plant sources, fruits, carbohydrates, citric acid, ascorbic acid, bio-waste lignin, and ammonium citrate are just a few of the raw materials that were utilized to create CQDs. Furthermore, because of its minimal price, accessibility, low cytotoxicity, and sustainably, the production of green CQDs using plant sources and carbon wastes needs to be expanded up. CQDs have been prepared using a variety of techniques, including heating, electrochemical process, hydrothermal, laser irradiation, microwave irradiation, and ultrasonication. However, because it involves minimal equipment, uses tiny power, and only demands one stage of processing, the hydrothermal process approach is frequently utilized to create CQDs.

CQDs have low toxicity, strong fluorescent effect, larger surface area and high photochemical stability. Due to these properties, it finds extensive applications in wastewater monitoring and remediation. In this chapter, we have summarized the utilization of CQDs for the extraction of toxic metal ions from wastewater. Different toxic metals, including Cd(II), Pb(II), Cu(II), U(VI), Hg(II) and many others, have been removed by CQDs and are reported in the chapter. Besides metal ion adsorption, CQDs and green carbon quantum dots-based sensors in heavy metal ion detection have been discussed in detail. For the enzymatic elimination of organic contaminants, including coloring agents and new organic contaminants, CQDs have been effectively tested. QDs were added to several nanocomposites, which altered the materials' characteristics and increased the effectiveness of removing diverse contaminants. The majority of research has documented the usage of CQDs in combination with additional semiconductor NPs for wastewater pollution remediation. Furthermore, research on the effective utilization CQDs for a photo-catalyst in photocatalytic activity is currently in its infancy. Consequently, it is reasonable to anticipate further scientific investigations and in-depth investigations of the optical characteristics of CQDs as photo-catalysts in the coming years. Although CQDs-based nanomaterials have made some significant advancements, additional research remains to be done to assure the manufacturing and use of comparable components in humongous operations.

REFERENCES

[1] Lodeiro, P.; Barriada, J.L.; Herrero, R.; Sastre de Vicente, M.E. The marine macroalga Cystoseira baccata as biosorbent for cadmium(II) and lead(II) removal: Kinetic and equilibrium studies. *Environ. Pollut.,* **2006**, *142*(2), 264-273.
[http://dx.doi.org/10.1016/j.envpol.2005.10.001] [PMID: 16360252]

[2] Martins, R.J.E.; Pardo, R.; Boaventura, R.A.R. Cadmium(II) and zinc(II) adsorption by the aquatic moss Fontinalis antipyretica: effect of temperature, pH and water hardness. *Water Res.,* **2004**, *38*(3), 693-699.
[http://dx.doi.org/10.1016/j.watres.2003.10.013] [PMID: 14723939]

[3] Kazantzis, G. Cadmium, osteoporosis and calcium metabolism. *Biometals,* **2004**, *17*(5), 493-498.
[http://dx.doi.org/10.1023/B:BIOM.0000045727.76054.f3] [PMID: 15688852]

[4] Zhitkovich, A. Chromium in drinking water: Sources, metabolism, and cancer risks. *Chem. Res. Toxicol.,* **2011**, *24*(10), 1617-1629.
[http://dx.doi.org/10.1021/tx200251t] [PMID: 21766833]

[5] Rengaraj, S.; Yeon, J.W.; Kim, Y.; Jung, Y.; Ha, Y.K.; Kim, W.H. Adsorption characteristics of Cu(II) onto ion exchange resins 252H and 1500H: Kinetics, isotherms and error analysis. *J. Hazard. Mater.,* **2007**, *143*(1-2), 469-477.
[http://dx.doi.org/10.1016/j.jhazmat.2006.09.064] [PMID: 17097805]

[6] Fagerlund, A.; Møretrø, T.; Heir, E.; Briandet, R.; Langsrud, S. Cleaning and disinfection of biofilms composed of Listeria monocytogenes and background microbiota from meat processing surfaces. *Appl. Environ. Microbiol.,* **2017**, *83*(17), e01046-17.
[http://dx.doi.org/10.1128/AEM.01046-17] [PMID: 28667108]

[7] Lineback, C.B.; Nkemngong, C.A.; Wu, S.T.; Li, X.; Teska, P.J.; Oliver, H.F. Hydrogen peroxide and sodium hypochlorite disinfectants are more effective against *Staphylococcus aureus* and *Pseudomonas aeruginosa* biofilms than quaternary ammonium compounds. *Antimicrob. Resist. Infect. Control,* **2018**, *7*(1), 154-163.
[http://dx.doi.org/10.1186/s13756-018-0447-5] [PMID: 30568790]

[8] Fukuzaki, S. Mechanisms of actions of sodium hypochlorite in cleaning and disinfection processes. *Biocontrol Sci.,* **2006**, *11*(4), 147-157.
[http://dx.doi.org/10.4265/bio.11.147] [PMID: 17190269]

[9] Jonnalagadda, S.B.; Nadupalli, S. Chlorine dioxide for bleaching, industrial applications and water treatment. *Indian Chemical Engineer,* **2014**, *56*(2), 123-136.
[http://dx.doi.org/10.1080/00194506.2014.881032]

[10] Xiang, M.; Bai, Y.; Li, Y.; Wei, S.; Shu, T.; Wang, H.; Li, P.; Yu, T.; Yu, L. An eco-friendly degumming process of flax roving without acid pickling and NaClO2-bleaching. *Process Biochem.,* **2020**, *93*, 77-84.
[http://dx.doi.org/10.1016/j.procbio.2020.03.018]

[11] Myers, J.N.; Steinecker, W.H.; Sandlin, Z.D.; Cox, J.A.; Gordon, G.; Pacey, G.E. Development of an automated on-line electrochemical chlorite ion sensor. *Talanta,* **2012**, *94*, 227-231.
[http://dx.doi.org/10.1016/j.talanta.2012.03.026] [PMID: 22608440]

[12] Zambrano-Monserrate, M.A.; Ruano, M.A.; Sanchez-Alcalde, L. Indirect effects of COVID-19 on the environment. *Sci. Total Environ.,* **2020**, *728*, 138813.
[http://dx.doi.org/10.1016/j.scitotenv.2020.138813] [PMID: 32334159]

[13] Sharma, E.; Vashisht, D.; Vashisht, A.; Vats, V.K.; Mehta, S.K.; Singh, K. Facile synthesis of sulfur and nitrogen codoped graphene quantum dots for optical sensing of Hg and Ag ions. *Chem. Phys. Lett.,* **2019**, *730*, 436-444.
[http://dx.doi.org/10.1016/j.cplett.2019.06.040]

[14] Shorie, M.; Kaur, H.; Chadha, G.; Singh, K.; Sabherwal, P. Graphitic carbon nitride QDs impregnated biocompatible agarose cartridge for removal of heavy metals from contaminated water samples. *J. Hazard. Mater.,* **2019**, *367*, 629-638.
[http://dx.doi.org/10.1016/j.jhazmat.2018.12.115] [PMID: 30654280]

[15] Namasivayam, C.; Ranganathan, K. Removal of Pb(II), Cd(II) and Ni(II) and mixture of metal ions by adsorption onto waste Fe(III)/ Cr(III) hydroxide and fixed bed studies. *Environ. Technol.,* **1995**, *16*, 851-860.

[16] O'Connell, D.W.; Birkinshaw, C.; O'Dwyer, T.F. Heavy metal adsorbents prepared from the modification of cellulose: A review. *Bioresour. Technol.,* **2008**, *99*(15), 6709-6724.
[http://dx.doi.org/10.1016/j.biortech.2008.01.036] [PMID: 18334292]

[17] Thakur, V; Sharma, E; Guleria, A; Sangar, S; Singh, K. Modification and management of lignocellulosic waste as an ecofriendly biosorbent for the application of heavy metal ions sorption.

Mater Today Proc, **2020**, *2020*(2), 756.
[http://dx.doi.org/10.1016/j.matpr.2020.02.756]

[18] Sabet, M.; Mahdavi, K. Green synthesis of high photoluminescence nitrogen-doped carbon quantum dots from grass *via* a simple hydrothermal method for removing organic and inorganic water pollutions. *Appl. Surf. Sci.,* **2019**, *463*, 283-291.
[http://dx.doi.org/10.1016/j.apsusc.2018.08.223]

[19] Das, R.; Bandyopadhyay, R.; Pramanik, P. Carbon quantum dots from natural resource: A review. *Mater. Today Chem.,* **2018**, *8*, 96-109.
[http://dx.doi.org/10.1016/j.mtchem.2018.03.003]

[20] Molaei, M.J. A review on nanostructured carbon quantum dots and their applications in biotechnology, sensors, and chemiluminescence. *Talanta,* **2019**, *196*, 456-478.
[http://dx.doi.org/10.1016/j.talanta.2018.12.042] [PMID: 30683392]

[21] Shorie, M.; Kaur, H.; Chadha, G.; Singh, K.; Sabherwal, P. Graphitic carbon nitride QDs impregnated biocompatible agarose cartridge for removal of heavy metals from contaminated water samples. *J. Hazard. Mater.,* **2019**, *367*, 629-638.
[http://dx.doi.org/10.1016/j.jhazmat.2018.12.115] [PMID: 30654280]

[22] Chen, Q.; Chen, L.; Qi, J.; Tong, Y.; Lv, Y.; Xu, C.; Ni, J.; Liu, W. Photocatalytic degradation of amoxicillin by carbon quantum dots modified $K_2Ti_6O_{13}$ nanotubes: Effect of light wavelength. *Chin. Chem. Lett.,* **2019**, *30*(6), 1214-1218.
[http://dx.doi.org/10.1016/j.cclet.2019.03.002]

[23] Xue, B.; Yang, Y.; Sun, Y.; Fan, J.; Li, X.; Zhang, Z. Photoluminescent lignin hybridized carbon quantum dots composites for bioimaging applications. *Int. J. Biol. Macromol.,* **2019**, *122*, 954-961.
[http://dx.doi.org/10.1016/j.ijbiomac.2018.11.018] [PMID: 30412762]

[24] Joshi, P.N.; Mathias, A.; Mishra, A.; Mathias, A. Synthesis of ecofriendly fluorescent carbon dots and their biomedical and environmental applications. *Mater. Technol.,* **2018**, *33*(10), 672-680.
[http://dx.doi.org/10.1080/10667857.2018.1492683]

[25] Liang, Z.; Yang, J.; Zhou, C.; Mo, Q.; Zhang, Y. Carbon quantum dots modified BiOBr microspheres with enhanced visible light photocatalytic performance. *Inorg. Chem. Commun.,* **2018**, *90*, 97-100.
[http://dx.doi.org/10.1016/j.inoche.2018.02.013]

[26] Zhou, L.; Qiao, M.; Zhang, L.; Sun, L.; Zhang, Y.; Liu, W. Green and efficient synthesis of carbon quantum dots and their luminescent properties. *J. Lumin.,* **2019**, *206*, 158-163.
[http://dx.doi.org/10.1016/j.jlumin.2018.10.057]

[27] Rani, U.A.; Ng, L.Y.; Ng, C.Y.; Mahmoudi, E. A review of carbon quantum dots and their applications in wastewater treatment. *Adv. Colloid Interface Sci.,* **2020**, *278*, 102124.
[http://dx.doi.org/10.1016/j.cis.2020.102124] [PMID: 32142942]

[28] Parthiban, V.; Panda, S.K.; Sahu, A.K. Highly fluorescent carbon quantum dots-Nafion as proton selective hybrid membrane for direct methanol fuel cells. *Electrochim. Acta,* **2018**, *292*, 855-864.
[http://dx.doi.org/10.1016/j.electacta.2018.09.193]

[29] Piri, M.; Sepehr, E.; Rengel, Z. Citric acid decreased and humic acid increased Zn sorption in soils. *Geoderma,* **2019**, *341*, 39-45.
[http://dx.doi.org/10.1016/j.geoderma.2018.12.027]

[30] Javed, M.; Saqib, A.N.S.; Ata-ur-Rehman, ; Ali, B.; Faizan, M.; Anang, D.A.; Iqbal, Z.; Abbas, S.M. Carbon quantum dots from glucose oxidation as a highly competent anode material for lithium and sodium-ion batteries. *Electrochim. Acta,* **2019**, *297*, 250-257.
[http://dx.doi.org/10.1016/j.electacta.2018.11.167]

[31] Kalaiyarasan, G.; Joseph, J. Cholesterol derived carbon quantum dots as fluorescence probe for the specific detection of hemoglobin in diluted human blood samples. *Mater. Sci. Eng. C,* **2019**, *94*, 580-586.

[http://dx.doi.org/10.1016/j.msec.2018.10.007] [PMID: 30423743]

[32] Salifairus, M.J.; Abd Hamid, S.B.; Soga, T.; Alrokayan, S.; Khan, H.; Rusop, M. Structural and optical properties of graphene from green carbon source *via* thermal chemical vapor deposition. *J. Mater. Res.*, **2016**, *31*, 1947-1956.
[http://dx.doi.org/10.1557/jmr.2016.200]

[33] Devi, P.; Rajput, P.; Thakur, A.; Kim, K.H.; Kumar, P. Recent advances in carbon quantum dot-based sensing of heavy metals in water. *Trends Analyt. Chem.*, **2019**, *114*, 171-195.
[http://dx.doi.org/10.1016/j.trac.2019.03.003]

[34] Namdari, P.; Negahdari, B.; Eatemadi, A. Synthesis, properties and biomedical applications of carbon-based quantum dots: An updated review. *Biomed. Pharmacother.*, **2017**, *87*, 209-222.
[http://dx.doi.org/10.1016/j.biopha.2016.12.108] [PMID: 28061404]

[35] Yang, P.; Zhu, Z.; Chen, M.; Chen, W.; Zhou, X. Microwave-assisted synthesis of xylan-derived carbon quantum dots for tetracycline sensing. *Opt. Mater.*, **2018**, *85*, 329-336.
[http://dx.doi.org/10.1016/j.optmat.2018.06.034]

[36] Coral Medina, J.D.; Woiciechowski, A.L.; Filho, A.Z.; Brar, S.K.; Magalhães Júnior, A.I.; Soccol, C.R. Energetic and economic analysis of ethanol, xylitol and lignin production using oil palm empty fruit bunches from a Brazilian factory. *J. Clean. Prod.*, **2018**, *195*, 44-55.
[http://dx.doi.org/10.1016/j.jclepro.2018.05.189]

[37] Zhong, Q.; Chen, Y.; Su, A.; Wang, Y. Synthesis of catalytically active carbon quantum dots and its application for colorimetric detection of glutathione. *Sens. Actuators B Chem.*, **2018**, *273*, 1098-1102.
[http://dx.doi.org/10.1016/j.snb.2018.07.026]

[38] Sakho, E.M.; Oluwafemi, O.S.; Perumbilavil, S.; Philip, R.; Kala, M.S.; Thomas, S.; Kalarikkal, N. Rapid and facile synthesis of graphene oxide quantum dots with good linear and nonlinear optical properties. *J. Mater. Sci. Mater. Electron.*, **2016**, *27*(10), 10926-10933.
[http://dx.doi.org/10.1007/s10854-016-5204-z]

[39] Algarra, M.; González-Calabuig, A.; Radotić, K.; Mutavdzic, D.; Ania, C.O.; Lázaro-Martínez, J.M.; Jiménez-Jiménez, J.; Rodríguez-Castellón, E.; del Valle, M. Enhanced electrochemical response of carbon quantum dot modified electrodes. *Talanta,* **2018**, *178*, 679-685.
[http://dx.doi.org/10.1016/j.talanta.2017.09.082] [PMID: 29136880]

[40] Vieira, K.O.; Bettini, J.; de Oliveira, L.F.C.; Ferrari, J.L.; Schiavon, M.A.; Schiavon, M.A. Synthesis of multicolor photoluminescent carbon quantum dots functionalized with hydrocarbons of different chain lengths. *N. Carbon Mater.*, **2017**, *32*(4), 327-337.
[http://dx.doi.org/10.1016/S1872-5805(17)60126-4]

[41] Sui, Y.; Wu, L.; Zhong, S.; Liu, Q. Carbon quantum dots/TiO_2 nanosheets with dominant (001) facets for enhanced photocatalytic hydrogen evolution. *Appl. Surf. Sci.*, **2019**, *480*, 810-816.
[http://dx.doi.org/10.1016/j.apsusc.2019.03.028]

[42] He, M.; Guo, X.; Huang, J.; Shen, H.; Zeng, Q.; Wang, L. Mass production of tunable multicolor graphene quantum dots from an energy resource of coke by a one-step electrochemical exfoliation. *Carbon,* **2018**, *140*, 508-520.
[http://dx.doi.org/10.1016/j.carbon.2018.08.067]

[43] Wang, Y.; Hu, A. Carbon quantum dots: synthesis, properties and applications. *J. Mater. Chem. C Mater. Opt. Electron. Devices,* **2014**, *2*(34), 6921.
[http://dx.doi.org/10.1039/C4TC00988F]

[44] Tian, J.; Leng, Y.; Zhao, Z.; Xia, Y.; Sang, Y.; Hao, P.; Zhan, J.; Li, M.; Liu, H. Carbon quantum dots/hydrogenated TiO_2 nanobelt heterostructures and their broad spectrum photocatalytic properties under UV, visible, and near-infrared irradiation. *Nano Energy,* **2015**, *11*, 419-427.
[http://dx.doi.org/10.1016/j.nanoen.2014.10.025]

[45] Kaur, M.; Kaur, M.; Sharma, V.K. Nitrogen-doped graphene and graphene quantum dots: A review

onsynthesis and applications in energy, sensors and environment. *Adv. Colloid Interface Sci.,* **2018**, *259*, 44-64.
[http://dx.doi.org/10.1016/j.cis.2018.07.001] [PMID: 30032930]

[46] Yang, W.; Yang, H.; Ding, W.; Zhang, B.; Zhang, L.; Wang, L.; Yu, M.; Zhang, Q. High quantum yield ZnO quantum dots synthesizing *via* an ultrasonication microreactor method. *Ultrason. Sonochem.,* **2016**, *33*, 106-117.
[http://dx.doi.org/10.1016/j.ultsonch.2016.04.020] [PMID: 27245962]

[47] Ratnayake, S.P.; Mantilaka, M.M.M.G.P.G.; Sandaruwan, C.; Dahanayake, D.; Murugan, E.; Kumar, S.; Amaratunga, G.A.J.; de Silva, K.M.N. Carbon quantum dots-decorated nano-zirconia: A highly efficient photocatalyst. *Appl. Catal. A Gen.,* **2019**, *570*, 23-30.
[http://dx.doi.org/10.1016/j.apcata.2018.10.022]

[48] Liang, J.Y.; Han, L.; Liu, S.G.; Ju, Y.J.; Gao, X.; Li, N.B.; Luo, H.Q. Green fluorescent carbon quantum dots as a label-free probe for rapid and sensitive detection of hematin. *Spectrochim. Acta A Mol. Biomol. Spectrosc.,* **2019**, *212*, 167-172.
[http://dx.doi.org/10.1016/j.saa.2019.01.001] [PMID: 30639601]

[49] Gai, W.; Zhao, D.L.; Chung, T.S. Thin film nanocomposite hollow fiber membranes comprising Na^+-functionalized carbon quantum dots for brackish water desalination. *Water Res.,* **2019**, *154*, 54-61.
[http://dx.doi.org/10.1016/j.watres.2019.01.043] [PMID: 30771707]

[50] Czajczyńska, D.; Nannou, T.; Anguilano, L.; Krzyżyńska, R.; Ghazal, H.; Spencer, N.; Jouhara, H. Potentials of pyrolysis processes in the waste management sector. *Energy Procedia,* **2017**, *123*, 387-394.
[http://dx.doi.org/10.1016/j.egypro.2017.07.275]

[51] Collard, F.X.; Blin, J. A review on pyrolysis of biomass constituents: Mechanisms and composition of the products obtained from the conversion of cellulose, hemicelluloses and lignin. *Renew. Sustain. Energy Rev.,* **2014**, *38*, 594-608.
[http://dx.doi.org/10.1016/j.rser.2014.06.013]

[52] Liu, G.; Jia, H.; Li, N.; Li, X.; Yu, Z.; Wang, J.; Song, Y. High-fluorescent carbon dots (CDs) originated from China grass carp scales (CGCS) for effective detection of Hg(II) ions. *Microchem. J.,* **2019**, *145*, 718-728.
[http://dx.doi.org/10.1016/j.microc.2018.11.044]

[53] Tian, P.; Tang, L.; Teng, K.S.; Lau, S.P. Graphene quantum dots from chemistry to applications. *Mater. Today Chem.,* **2018**, *10*, 221-258.
[http://dx.doi.org/10.1016/j.mtchem.2018.09.007]

[54] Pan, D.; Zhang, J.; Li, Z.; Wu, M. Hydrothermal route for cutting graphene sheets into blue-luminescent graphene quantum dots. *Adv. Mater.,* **2010**, *22*(6), 734-738.
[http://dx.doi.org/10.1002/adma.200902825] [PMID: 20217780]

[55] Liang, Q.; Ma, W.; Shi, Y.; Li, Z.; Yang, X. Easy synthesis of highly fluorescent carbon quantum dots from gelatin and their luminescent properties and applications. *Carbon,* **2013**, *60*, 421-428.
[http://dx.doi.org/10.1016/j.carbon.2013.04.055]

[56] Liu, Y.; Li, W.; Wu, P.; Ma, C.; Wu, X.; Xu, M.; Luo, S.; Xu, Z.; Liu, S. Hydrothermal synthesis of nitrogen and boron co-doped carbon quantum dots for application in acetone and dopamine sensors and multicolor cellular imaging. *Sens. Actuators B Chem.,* **2019**, *281*, 34-43.
[http://dx.doi.org/10.1016/j.snb.2018.10.075]

[57] Wong, K.T.; Jang, S.B.; Saravanan, P.; Nah, I.W.; Park, S.; Choi, J.; Park, C.; Kim, Y.; Yoon, Y.; Jang, M. Critical insight on the hydrothermal effects toward exfoliation of g-C3N4 and simultaneous *in-situ* deposition of carbon quantum dots. *Appl. Surf. Sci.,* **2019**, *471*, 703-713.
[http://dx.doi.org/10.1016/j.apsusc.2018.12.064]

[58] Arumugam, N.; Kim, J. Synthesis of carbon quantum dots from Broccoli and their ability to detect silver ions. *Mater. Lett.,* **2018**, *219*, 37-40.

[http://dx.doi.org/10.1016/j.matlet.2018.02.043]

[59] Singh, A.; Mohapatra, P.K.; Kalyanasundaram, D.; Kumar, S. Self-functionalized ultrastable water suspension of luminescent carbon quantum dots. *Mater. Chem. Phys.,* **2019**, *225*, 23-27. [http://dx.doi.org/10.1016/j.matchemphys.2018.12.031]

[60] Singh, R.K.; Kumar, R.; Singh, D.P.; Savu, R.; Moshkalev, S.A. Progress in microwave-assisted synthesis of quantum dots (graphene/carbon/semiconducting) for bioapplications: a review. *Mater. Today Chem.,* **2019**, *12*, 282-314. [http://dx.doi.org/10.1016/j.mtchem.2019.03.001]

[61] Kaur, M.; Kaur, M.; Sharma, V.K. Nitrogen-doped graphene and graphene quantum dots: A review onsynthesis and applications in energy, sensors and environment. *Adv. Colloid Interface Sci.,* **2018**, *259*, 44-64. [http://dx.doi.org/10.1016/j.cis.2018.07.001] [PMID: 30032930]

[62] Liang, Z.; Yang, J.; Zhou, C.; Mo, Q.; Zhang, Y. Carbon quantum dots modified BiOBr microspheres with enhanced visible light photocatalytic performance. *Inorg. Chem. Commun.,* **2018**, *90*, 97-100. [http://dx.doi.org/10.1016/j.inoche.2018.02.013]

[63] Lim, H.; Liu, Y.; Kim, H.Y.; Son, D.I. Facile synthesis and characterization of carbon quantum dots and photovoltaic applications. *Thin Solid Films,* **2018**, *660*, 672-677. [http://dx.doi.org/10.1016/j.tsf.2018.04.019]

[64] Rajabi, H.R.; Khani, O.; Shamsipur, M.; Vatanpour, V. High-performance pure and Fe^{3+}-ion doped ZnS quantum dots as green nanophotocatalysts for the removal of malachite green under UV-light irradiation. *J. Hazard. Mater.,* **2013**, *250-251*, 370-378. [http://dx.doi.org/10.1016/j.jhazmat.2013.02.007] [PMID: 23474411]

[65] Aghamali, A.; Khosravi, M.; Hamishehkar, H.; Modirshahla, N.; Behnajady, M.A. Synthesis and characterization of high efficient photoluminescent sunlight driven photocatalyst of N-Carbon Quantum Dots. *J. Lumin.,* **2018**, *201*, 265-274. [http://dx.doi.org/10.1016/j.jlumin.2018.04.061]

[66] Hemmateenejad, B.; Shadabipour, P.; Khosousi, T.; Shamsipur, M. Chemometrics investigation of the light-free degradation of methyl green and malachite green by starch-coated CdSe quantum dots. *J. Ind. Eng. Chem.,* **2015**, *27*, 384-390. [http://dx.doi.org/10.1016/j.jiec.2015.01.018]

[67] Algarra, M.; González-Calabuig, A.; Radotić, K.; Mutavdzic, D.; Ania, C.O.; Lázaro-Martínez, J.M.; Jiménez-Jiménez, J.; Rodríguez-Castellón, E.; del Valle, M. Enhanced electrochemical response of carbon quantum dot modified electrodes. *Talanta,* **2018**, *178*, 679-685. [http://dx.doi.org/10.1016/j.talanta.2017.09.082] [PMID: 29136880]

[68] Athika, M.; Prasath, A.; Duraisamy, E.; Sankar Devi, V.; Selva Sharma, A.; Elumalai, P. Carbon-quantum dots derived from denatured milk for efficient chromium-ion sensing and supercapacitor applications. *Mater. Lett.,* **2019**, *241*, 156-159. [http://dx.doi.org/10.1016/j.matlet.2019.01.064]

[69] Xie, X.; Yang, Y.; Xiao, Y.H.; Huang, X.; Shi, Q.; Zhang, W.D. Enhancement of photoelectrochemical activity of Fe_2O_3 nanowires decorated with carbon quantum dots. *Int. J. Hydrogen Energy,* **2018**, *43*(14), 6954-6962. [http://dx.doi.org/10.1016/j.ijhydene.2018.02.099]

[70] Ramar, V.; Moothattu, S.; Balasubramanian, K. Metal free, sunlight and white light based photocatalysis using carbon quantum dots from Citrus grandis: A green way to remove pollution. *Sol. Energy,* **2018**, *169*, 120-127. [http://dx.doi.org/10.1016/j.solener.2018.04.040]

[71] Alarfaj, N.; El-Tohamy, M.; Oraby, H. CA 19-9 pancreatic tumor marker fluorescence immunosensing detection *via* immobilized carbon quantum dots conjugated gold nanocomposite. *Int. J. Mol. Sci.,* **2018**, *19*(4), 1162.

[http://dx.doi.org/10.3390/ijms19041162] [PMID: 29641488]

[72] Zhang, Z.; Wu, L.; Wang, P.; Zhang, Y.; Wan, S.; Guo, X.; Jin, W.; Zhang, J. Carbon quantum dots modified $La_2Ti_2O_7$ nanosheets for visible light photocatalysis. *Mater. Lett.,* **2018**, *230*, 72-75. [http://dx.doi.org/10.1016/j.matlet.2018.07.086]

[73] Azimi, A.; Azari, A.; Rezakazemi, M.; Ansarpour, M. Removal of heavy metals from industrial wastewaters: A review. *ChemBioEng Rev.,* **2017**, *4*(1), 37-59. [http://dx.doi.org/10.1002/cben.201600010]

[74] Abdelsalam, H.; Teleb, N.H.; Yahia, I.S.; Zahran, H.Y.; Elhaes, H.; Ibrahim, M.A. First principles study of the adsorption of hydrated heavy metals on graphene quantum dots. *J. Phys. Chem. Solids,* **2019**, *130*, 32-40. [http://dx.doi.org/10.1016/j.jpcs.2019.02.014]

[75] Dou, J.; Gan, D.; Huang, Q.; Liu, M.; Chen, J.; Deng, F.; Zhu, X.; Wen, Y.; Zhang, X.; Wei, Y. Functionalization of carbon nanotubes with chitosan based on MALI multicomponent reaction for Cu^{2+} removal. *Int. J. Biol. Macromol.,* **2019**, *136*, 476-485. [http://dx.doi.org/10.1016/j.ijbiomac.2019.06.112] [PMID: 31220490]

[76] Zhang, X.; Huang, Q.; Deng, F.; Huang, H.; Wan, Q.; Liu, M.; Wei, Y. Mussel-inspired fabrication of functional materials and their environmental applications: Progress and prospects. *Appl. Mater. Today,* **2017**, *7*, 222-238. [http://dx.doi.org/10.1016/j.apmt.2017.04.001]

[77] Huang, Q.; Liu, M.; Chen, J.; Wan, Q.; Tian, J.; Huang, L.; Jiang, R.; Wen, Y.; Zhang, X.; Wei, Y. Facile preparation of MoS_2 based polymer composites *via* mussel inspired chemistry and their high efficiency for removal of organic dyes. *Appl. Surf. Sci.,* **2017**, *419*, 35-44. [http://dx.doi.org/10.1016/j.apsusc.2017.05.006]

[78] Huang, S.; Jiang, S.; Pang, H.; Wen, T.; Asiri, A.M.; Alamry, K.A.; Alsaedi, A.; Wang, X.; Wang, S. Dual functional nanocomposites of magnetic $MnFe_2O_4$ and fluorescent carbon dots for efficient U(VI) removal. *Chem. Eng. J.,* **2019**, *368*, 941-950. [http://dx.doi.org/10.1016/j.cej.2019.03.015]

[79] Gan, D.; Liu, M.; Huang, H.; Chen, J.; Dou, J.; Wen, Y.; Huang, Q.; Yang, Z.; Zhang, X.; Wei, Y. Facile preparation of functionalized carbon nanotubes with tannins through mussel-inspired chemistry and their application in removal of methylene blue. *J. Mol. Liq.,* **2018**, *271*, 246-253. [http://dx.doi.org/10.1016/j.molliq.2018.08.079]

[80] Zainal Abidin, N.H.; Wongso, V.; Hui, K.C.; Cho, K.; Sambudi, N.S.; Ang, W.L.; Saad, B. The effect of functionalization on rice-husks derived carbon quantum dots properties and cadmium removal. *J. Water Process Eng.,* **2020**, *38*, 101634. [http://dx.doi.org/10.1016/j.jwpe.2020.101634]

[81] Zhang, X.; Huang, Q.; Liu, M.; Tian, J.; Zeng, G.; Li, Z.; Wang, K.; Zhang, Q.; Wan, Q.; Deng, F.; Wei, Y. Preparation of amine functionalized carbon nanotubes *via* a bioinspired strategy and their application in Cu^{2+} removal. *Appl. Surf. Sci.,* **2015**, *343*, 19-27. [http://dx.doi.org/10.1016/j.apsusc.2015.03.081]

[82] Lei, Y.; Cui, Y.; Huang, Q.; Dou, J.; Gan, D.; Deng, F.; Liu, M.; Li, X.; Zhang, X.; Wei, Y. Facile preparation of sulfonic groups functionalized Mxenes for efficient removal of methylene blue. *Ceram. Int.,* **2019**, *45*(14), 17653-17661. [http://dx.doi.org/10.1016/j.ceramint.2019.05.331]

[82] Mallakpour, S.; Behranvand, V. Synthesis of mesoporous recycled poly(ethylene terephthalate)/MWNT/carbon quantum dot nanocomposite from sustainable materials using ultrasonic waves: Application for methylene blue removal. *J. Clean. Prod.,* **2018**, *190*, 525-537. [http://dx.doi.org/10.1016/j.jclepro.2018.04.120]

[83] Mashkani, M.; Mehdinia, A.; Jabbari, A.; Bide, Y.; Nabid, M.R. Preconcentration and extraction of lead ions in vegetable and water samples by N-doped carbon quantum dot conjugated with Fe_3O_4 as a

green and facial adsorbent. *Food Chem.*, **2018**, *239*, 1019-1026. [http://dx.doi.org/10.1016/j.foodchem.2017.07.042] [PMID: 28873517]

[84] Huang, Q.; Zhao, J.; Liu, M.; Chen, J.; Zhu, X.; Wu, T.; Tian, J.; Wen, Y.; Zhang, X.; Wei, Y. Preparation of polyethylene polyamine@tannic acid encapsulated MgAl-layered double hydroxide for the efficient removal of copper (II) ions from aqueous solution. *J. Taiwan Inst. Chem. Eng.*, **2018**, *82*, 92-101. [http://dx.doi.org/10.1016/j.jtice.2017.10.019]

[85] Yao, W.; Wang, X.; Liang, Y.; Yu, S.; Gu, P.; Sun, Y.; Xu, C.; Chen, J.; Hayat, T.; Alsaedi, A.; Wang, X. Synthesis of novel flower-like layered double oxides/carbon dots nanocomposites for U(VI) and ^{241}Am(III) efficient removal: Batch and EXAFS studies. *Chem. Eng. J.*, **2018**, *332*, 775-786. [http://dx.doi.org/10.1016/j.cej.2017.09.011]

[86] Rahmanian, O.; Dinari, M.; Abdolmaleki, M.K. Carbon quantum dots/layered double hydroxide hybrid for fast and efficient decontamination of Cd(II): The adsorption kinetics and isotherms. *Appl. Surf. Sci.*, **2018**, *428*, 272-279. [http://dx.doi.org/10.1016/j.apsusc.2017.09.152]

[87] Mahmoud, M.E.; Fekry, N.A.; Abdelfattah, A.M. Removal of uranium (VI) from water by the action of microwave-rapid green synthesized carbon quantum dots from starch-water system and supported onto polymeric matrix. *J. Hazard. Mater.*, **2020**, *397*, 122770. [http://dx.doi.org/10.1016/j.jhazmat.2020.122770] [PMID: 32388094]

[88] Song, Z.; Chen, X.; Gong, X.; Gao, X.; Dai, Q.; Nguyen, T.T.; Guo, M. Luminescent carbon quantum dots/nanofibrillated cellulose composite aerogel for monitoring adsorption of heavy metal ions in water. *Opt. Mater.*, **2020**, *100*, 109642. [http://dx.doi.org/10.1016/j.optmat.2019.109642]

[89] Devi, P.; Rajput, P.; Thakur, A.; Kim, K.H.; Kumar, P. Recent advances in carbon quantum dot-based sensing of heavy metals in water. *Trends Analyt. Chem.*, **2019**, *114*, 171-195. [http://dx.doi.org/10.1016/j.trac.2019.03.003]

[90] Qu, K.; Wang, J.; Ren, J.; Qu, X. Carbon dots prepared by hydrothermal treatment of dopamine as an effective fluorescent sensing platform for the label-free detection of iron(III) ions and dopamine. *Chemistry,* **2013**, *19*(22), 7243-7249. [http://dx.doi.org/10.1002/chem.201300042] [PMID: 23576265]

[91] Zhou, L.; Lin, Y.; Huang, Z.; Ren, J.; Qu, X. Carbon nanodots as fluorescence probes for rapid, sensitive, and label-free detection of Hg^{2+} and biothiols in complex matrices. *Chem. Commun.*, **2012**, *48*(8), 1147-1149. [http://dx.doi.org/10.1039/C2CC16791C] [PMID: 22159407]

[92] Liu, R.; Li, H.; Kong, W.; Liu, J.; Liu, Y.; Tong, C.; Zhang, X.; Kang, Z. Ultra-sensitive and selective Hg2+ detection based on fluorescent carbon dots. *Mater. Res. Bull.*, **2013**, *48*(7), 2529-2534. [http://dx.doi.org/10.1016/j.materresbull.2013.03.015]

[93] Zhang, R.; Chen, W. Nitrogen-doped carbon quantum dots: Facile synthesis and application as a "turn-off" fluorescent probe for detection of Hg^{2+} ions. *Biosens. Bioelectron.*, **2014**, *55*, 83-90. [http://dx.doi.org/10.1016/j.bios.2013.11.074] [PMID: 24365697]

[94] Guo, Y.; Wang, Z.; Shao, H.; Jiang, X. Hydrothermal synthesis of highly fluorescent carbon nanoparticles from sodium citrate and their use for the detection of mercury ions. *Carbon,* **2013**, *52*, 583-589. [http://dx.doi.org/10.1016/j.carbon.2012.10.028]

[95] Xu, S.; Liu, Y.; Yang, H.; Zhao, K.; Li, J.; Deng, A. Fluorescent nitrogen and sulfur co-doped carbon dots from casein and their applications for sensitive detection of Hg^{2+} and biothiols and cellular imaging. *Anal. Chim. Acta,* **2017**, *964*, 150-160. [http://dx.doi.org/10.1016/j.aca.2017.01.037] [PMID: 28351631]

[96] Pourreza, N.; Ghomi, M. Green synthesized carbon quantum dots from Prosopis juliflora leaves as a

dual off-on fluorescence probe for sensing mercury (II) and chemet drug. *Mater. Sci. Eng. C,* **2019**, *98*, 887-896.
[http://dx.doi.org/10.1016/j.msec.2018.12.141] [PMID: 30813094]

[97] Zhang, S.; Wang, Q.; Tian, G.; Ge, H. A fluorescent turn-off/on method for detection of $Cu^{2+}$and oxalate using carbon dots as fluorescent probes in aqueous solution. *Mater. Lett.,* **2014**, *115*, 233-236.
[http://dx.doi.org/10.1016/j.matlet.2013.10.086]

[98] Rong, M.C.; Zhang, K.X.; Wang, Y.R.; Chen, X. The synthesis of B, N-carbon dots by a combustion method and the application of fluorescence detection for Cu 2+. *Chin. Chem. Lett.,* **2017**, *28*(5), 1119-1124.
[http://dx.doi.org/10.1016/j.cclet.2016.12.009]

[99] Wang, J.; Sheng Li, R.; Zhi Zhang, H.; Wang, N.; Zhang, Z.; Huang, C.Z. Highly fluorescent carbon dots as selective and visual probes for sensing copper ions in living cells *via* an electron transfer process. *Biosens. Bioelectron.,* **2017**, *97*, 157-163.
[http://dx.doi.org/10.1016/j.bios.2017.05.035] [PMID: 28595076]

[100] Xu, X.; He, L.; Long, Y.; Pan, S.; Liu, H.; Yang, J.; Hu, X. S-doped carbon dots capped ZnCdTe quantum dots for ratiometric fluorescence sensing of guanine. *Sens. Actuators B Chem.,* **2019**, *279*, 44-52.
[http://dx.doi.org/10.1016/j.snb.2018.09.102]

[101] Liu, Y.; Zhou, Q.; Yuan, Y.; Wu, Y. Hydrothermal synthesis of fluorescent carbon dots from sodium citrate and polyacrylamide and their highly selective detection of lead and pyrophosphate. *Carbon,* **2017**, *115*, 550-560.
[http://dx.doi.org/10.1016/j.carbon.2017.01.035]

[102] Chang, M.M.F.; Ginjom, I.R.; Ngu-Schwemlein, M.; Ng, S.M. Synthesis of yellow fluorescent carbon dots and their application to the determination of chromium(III) with selectivity improved by pH tuning. *Mikrochim. Acta,* **2016**, *183*(6), 1899-1907.
[http://dx.doi.org/10.1007/s00604-016-1819-2]

[103] Fang, L.; Zhang, L.; Chen, Z.; Zhu, C.; Liu, J.; Zheng, J. Ammonium citrate derived carbon quantum dot as on-off-on fluorescent sensor for detection of chromium(VI) and sulfites. *Mater. Lett.,* **2017**, *191*, 1-4.
[http://dx.doi.org/10.1016/j.matlet.2016.12.098]

[104] Gong, X.; Liu, Y.; Yang, Z.; Shuang, S.; Zhang, Z.; Dong, C. An *on-off-on* fluorescent nanoprobe for recognition of chromium(VI) and ascorbic acid based on phosphorus/nitrogen dual-doped carbon quantum dot. *Anal. Chim. Acta,* **2017**, *968*, 85-96.
[http://dx.doi.org/10.1016/j.aca.2017.02.038] [PMID: 28395778]

[105] Vaz, R.; Bettini, J.; Júnior, J.G.F.; Lima, E.D.S.; Botero, W.G.; Santos, J.C.C.; Schiavon, M.A. High luminescent carbon dots as an eco-friendly fluorescence sensor for Cr(VI) determination in water and soil samples. *J. Photochem. Photobiol. Chem.,* **2017**, *346*, 502-511.
[http://dx.doi.org/10.1016/j.jphotochem.2017.06.047]

[106] Yang, M.; Meng, X.; Li, B.; Ge, S.; Lu, Y. N, S co-doped carbon dots with high quantum yield: tunable fluorescence in liquid/solid and extensible applications. *J. Nanopart. Res.,* **2017**, *19*(6), 217.
[http://dx.doi.org/10.1007/s11051-017-3914-7]

[107] Guo, Y.; Chen, Y.; Cao, F.; Wang, L.; Wang, Z.; Leng, Y. Hydrothermal synthesis of nitrogen and boron doped carbon quantum dots with yellow-green emission for sensing Cr(VI), anti-counterfeiting and cell imaging. *RSC Advances,* **2017**, *7*(76), 48386-48393.
[http://dx.doi.org/10.1039/C7RA09785A]

[108] Zhang, Y.; Fang, X.; Zhao, H.; Li, Z. A highly sensitive and selective detection of Cr(VI) and ascorbic acid based on nitrogen-doped carbon dots. *Talanta,* **2018**, *181*, 318-325.
[http://dx.doi.org/10.1016/j.talanta.2018.01.027] [PMID: 29426518]

[109] Singh, V.K.; Singh, V.; Yadav, P.K.; Chandra, S.; Bano, D.; Kumar, V.; Koch, B.; Talat, M.; Hasan,

S.H. Bright-blue-emission nitrogen and phosphorus-doped carbon quantum dots as a promising nanoprobe for detection of Cr(VI) and ascorbic acid in pure aqueous solution and in living cells. *New J. Chem.*, **2018**, *42*(15), 12990-12997.
[http://dx.doi.org/10.1039/C8NJ02126K]

[110] Omer, K.M.; Tofiq, D.I.; Ghafoor, D.D. Highly photoluminescent label free probe for Chromium (II) ions using carbon quantum dots co-doped with nitrogen and phosphorous. *J. Lumin.*, **2019**, *206*, 540-546.
[http://dx.doi.org/10.1016/j.jlumin.2018.10.100]

[111] Qi, H.; Teng, M.; Liu, M.; Liu, S.; Li, J.; Yu, H.; Teng, C.; Huang, Z.; Liu, H.; Shao, Q.; Umar, A.; Ding, T.; Gao, Q.; Guo, Z. Biomass-derived nitrogen-doped carbon quantum dots: highly selective fluorescent probe for detecting Fe^{3+} ions and tetracyclines. *J. Colloid Interface Sci.*, **2019**, *539*, 332-341.
[http://dx.doi.org/10.1016/j.jcis.2018.12.047] [PMID: 30594008]

[112] Madrakian, T.; Maleki, S.; Gilak, S.; Afkhami, A. Turn-off fluorescence of amino-functionalized carbon quantum dots as effective fluorescent probes for determination of isotretinoin. *Sens. Actuators B Chem.*, **2017**, *247*, 428-435.
[http://dx.doi.org/10.1016/j.snb.2017.03.071]

[113] Salinas-Castillo, A.; Ariza-Avidad, M.; Pritz, C.; Camprubí-Robles, M.; Fernández, B.; Ruedas-Rama, M.J.; Megia-Fernández, A.; Lapresta-Fernández, A.; Santoyo-Gonzalez, F.; Schrott-Fischer, A.; Capitan-Vallvey, L.F. Carbon dots for copper detection with down and upconversion fluorescent properties as excitation sources. *Chem. Commun.*, **2013**, *49*(11), 1103-1105.
[http://dx.doi.org/10.1039/c2cc36450f] [PMID: 23283251]

[114] Zong, J.; Yang, X.; Trinchi, A.; Hardin, S.; Cole, I.; Zhu, Y.; Li, C.; Muster, T.; Wei, G. Carbon dots as fluorescent probes for "off–on" detection of Cu^{2+} and l-cysteine in aqueous solution. *Biosens. Bioelectron.*, **2014**, *51*, 330-335.
[http://dx.doi.org/10.1016/j.bios.2013.07.042] [PMID: 23994615]

[115] Liu, Q.; Zhang, N.; Shi, H.; Ji, W.; Guo, X.; Yuan, W.; Hu, Q. One-step microwave synthesis of carbon dots for highly sensitive and selective detection of copper ions in aqueous solution. *New J. Chem.*, **2018**, *42*(4), 3097-3101.
[http://dx.doi.org/10.1039/C7NJ05000C]

[116] Wang, Y.; Chang, Q.; Hu, S. Carbon dots with concentration-tunable multicolored photoluminescence for simultaneous detection of Fe^{3+} and Cu^{2+} ions. Sensor. Actuator. *Biol. Chem.*, **2017**, *253*, 928-933.

[117] Yang, R.; Guo, X.; Jia, L.; Zhang, Y. A fluorescent "on-off-on" assay for selective recognition of Cu(II) and glutathione based on modified carbon nanodots, and its application to cellular imaging. *Mikrochim. Acta,* **2017**, *184*(4), 1143-1150.
[http://dx.doi.org/10.1007/s00604-017-2076-8]

[118] Amjadi, M.; Manzoori, J.L.; Hallaj, T.; Azizi, N. Sulfur and nitrogen co-doped carbon quantum dots as the chemiluminescence probe for detection of Cu^{2+} ions. *J. Lumin.*, **2017**, *182*, 246-251.
[http://dx.doi.org/10.1016/j.jlumin.2016.10.021]

[119] Wu, P.; Li, W.; Wu, Q.; Liu, Y.; Liu, S. Hydrothermal synthesis of nitrogen-doped carbon quantum dots from microcrystalline cellulose for the detection of Fe^{3+} ions in an acidic environment. *RSC Advances,* **2017**, *7*(70), 44144-44153.
[http://dx.doi.org/10.1039/C7RA08400E]

[120] Wang, C.; Hu, T.; Wen, Z.; Zhou, J.; Wang, X.; Wu, Q.; Wang, C. Nitrogen-doped carbon nanoparticles derived from silkworm excrement as on-off-on fluorescent sensors to detect Fe(III) and Biothiols J. *Colloid Interface Sci.*, **2018**, *521*, 33-41.
[http://dx.doi.org/10.1016/j.jcis.2018.03.021]

[121] Lv, P.; Yao, Y.; Li, D.; Zhou, H.; Naeem, M.A.; Feng, Q.; Huang, J.; Cai, Y.; Wei, Q. Self-assembly of nitrogen-doped carbon dots anchored on bacterial cellulose and their application in iron ion

detection. *Carbohydr. Polym.,* **2017**, *172*, 93-101.
[http://dx.doi.org/10.1016/j.carbpol.2017.04.086] [PMID: 28606552]

[122] Miao, X.; Yan, X.; Qu, D.; Li, D.; Tao, F.F.; Sun, Z. Red emissive sulfur, nitrogen codoped carbon dots and their application in ion detection and theraonostics. *ACS Appl. Mater. Interfaces,* **2017**, *9*(22), 18549-18556.
[http://dx.doi.org/10.1021/acsami.7b04514] [PMID: 28508626]

[123] Guo, T.; Cao, F.; Li, Y. Solid phase synthesis of nitrogen and phosphor co-doped carbon quantum dots for sensing Fe^{3+} and the enhanced photocatalytic degradation of dyes. Sensor. Actuator. *Biol. Chem.,* **2018**, *255*, 1105-1111.

[124] Xie, Z.; Sun, X.; Jiao, J.; Xin, X. Ionic liquid-functionalized carbon quantum dots as fluorescent probes for sensitive and selective detection of iron ion and ascorbic acid. *Colloids Surf. A Physicochem. Eng. Asp.,* **2017**, *529*, 38-44.
[http://dx.doi.org/10.1016/j.colsurfa.2017.05.069]

[125] Gupta, A.; Verma, N.C.; Khan, S.; Nandi, C.K. Carbon dots for naked eye colorimetric ultrasensitive arsenic and glutathione detection. *Biosens. Bioelectron.,* **2016**, *81*, 465-472.
[http://dx.doi.org/10.1016/j.bios.2016.03.018] [PMID: 27015150]

[126] He, J.; Zhang, H.; Zou, J.; Liu, Y.; Zhuang, J.; Xiao, Y.; Lei, B. Carbon dots-based fluorescent probe for "off-on" sensing of Hg(II) and I. *Biosens. Bioelectron.,* **2016**, *79*, 531-535.
[http://dx.doi.org/10.1016/j.bios.2015.12.084] [PMID: 26748370]

[127] Amin, N.; Afkhami, A.; Madrakian, T. Construction of a novel "Off-On" fluorescence sensor for highly selective sensing of selenite based on europium ions induced crosslinking of nitrogen-doped carbon dots. *J. Lumin.,* **2018**, *194*, 768-777.
[http://dx.doi.org/10.1016/j.jlumin.2017.09.048]

[128] Chen, Z.; Zhao, Z.; Wang, Z.; Zhang, Y.; Sun, X.; Hou, L.; Yuan, C. Foxtail millet-derived highly fluorescent multi-heteroatom doped carbon quantum dots towards fluorescent inks and smart nanosensors for selective ion detection. *New J. Chem.,* **2018**, *42*(9), 7326-7331.
[http://dx.doi.org/10.1039/C8NJ01072B]

[129] Zhang, Z.; Sun, W.; Wu, P. Highly photoluminescent carbon dots derived from egg white: facile and green synthesis, photoluminescence properties, and multiple applications. *ACS Sustain. Chem.& Eng.,* **2015**, *3*(7), 1412-1418.
[http://dx.doi.org/10.1021/acssuschemeng.5b00156]

[130] Wang, N.; Wang, Y.; Guo, T.; Yang, T.; Chen, M.; Wang, J. Green preparation of carbon dots with papaya as carbon source for effective fluorescent sensing of Iron (III) and *Escherichia coli. Biosens. Bioelectron.,* **2016**, *85*, 68-75.
[http://dx.doi.org/10.1016/j.bios.2016.04.089] [PMID: 27155118]

[131] Sun, C.; Zhang, Y.; Wang, P.; Yang, Y.; Wang, Y.; Xu, J.; Wang, Y.; Yu, W.W. Synthesis of nitrogen and sulfur co-doped carbon dots from garlic for selective detection of Fe3+. *Nanoscale Res. Lett.,* **2016**, *11*(1), 110-116.
[http://dx.doi.org/10.1186/s11671-016-1326-8] [PMID: 26924814]

[132] Sun, X.; He, J.; Yang, S.; Zheng, M.; Wang, Y.; Ma, S.; Zheng, H. Green synthesis of carbon dots originated from Lycii Fructus for effective fluorescent sensing of ferric ion and multicolor cell imaging. *J. Photochem. Photobiol. B,* **2017**, *175*, 219-225.
[http://dx.doi.org/10.1016/j.jphotobiol.2017.08.035] [PMID: 28915491]

[133] Liu, W.; Diao, H.; Chang, H.; Wang, H.; Li, T.; Wei, W. Green synthesis of carbon dots from rose-heart radish and application for Fe^{3+} detection and cell imaging. Sensor. Actuator. *Biol. Chem.,* **2017**, *241*, 190-198.

[134] Atchudan, R.; Edison, T.N.J.I.; Chakradhar, D.; Perumal, S.; Shim, J-J.; Lee, Y.R. Facile green synthesis of nitrogen-doped carbon dots using Chionanthus retusus fruit extract and investigation of their suitabilityfor metal ion sensing and biological applications. Sensor. Actuator. *Biol. Chem.,* **2017**,

246, 497-509.

[135] Das, P.; Ganguly, S.; Maity, P.P.; Bose, M.; Mondal, S.; Dhara, S.; Das, A.K.; Banerjee, S.; Das, N.C. Waste chimney oil to nanolights: A low cost chemosensor for tracer metal detection in practical field and its polymer composite for multidimensional activity. *J. Photochem. Photobiol. B,* **2018**, *180*, 56-67.
[http://dx.doi.org/10.1016/j.jphotobiol.2018.01.019] [PMID: 29413702]

[136] Kumar, A.; Chowdhuri, A.R.; Laha, D.; Mahto, T.K.; Karmakar, P.; Sahu, S.K. Green synthesis of carbon dots from Ocimum sanctum for effective fluorescent sensing of Pb2+ ions and live cell imaging. Sensor. Actuator. *Biol. Chem.,* **2017**, *242*, 679-686.

[137] Bandi, R.; Dadigala, R.; Gangapuram, B.R.; Guttena, V. Green synthesis of highly fluorescent nitrogen – Doped carbon dots from Lantana camara berries for effective detection of lead(II) and bioimaging. *J. Photochem. Photobiol. B,* **2018**, *178*, 330-338.
[http://dx.doi.org/10.1016/j.jphotobiol.2017.11.010] [PMID: 29178994]

[138] Ansi, V.; Renuka, N. Table sugar derived Carbon dot : A naked eye sensor for toxic Pb^{2+} ions. Sensor. Actuator. *Biol. Chem.,* **2018**, *264*, 67-75.

[139] Tyagi, A.; Tripathi, K.M.; Singh, N.; Choudhary, S.; Gupta, R.K. Green synthesis of carbon quantum dots from lemon peel waste: applications in sensing and photocatalysis. *RSC Advances,* **2016**, *6*(76), 72423-72432.
[http://dx.doi.org/10.1039/C6RA10488F]

[140] Rajendran, K.; Rajendiran, N. Bluish green emitting carbon quantum dots synthesized from jackfruit (*Artocarpus heterophyllus*) and its sensing applications of Hg (II) and Cr (VI) ions. *Mater. Res. Express,* **2018**, *5*(2), 024008-024014.
[http://dx.doi.org/10.1088/2053-1591/aaae4b]

[141] Sharma, S.; Umar, A.; Mehta, S.K.; Kansal, S.K. Fluorescent spongy carbon nanoglobules derived from pineapple juice: A potential sensing probe for specific and selective detection of chromium (VI) ions. *Ceram. Int.,* **2017**, *43*(9), 7011-7019.
[http://dx.doi.org/10.1016/j.ceramint.2017.02.127]

[142] Wang, S.; Niu, H.; He, S.; Cai, Y. One-step fabrication of high quantum yield sulfur- and nitrogen-doped carbon dots for sensitive and selective detection of Cr(VI). *RSC Advances,* **2016**, *6*(109), 107717-107722.
[http://dx.doi.org/10.1039/C6RA21059G]

[143] Shen, J.; Shang, S.; Chen, X.; Wang, D.; Cai, Y. Highly fluorescent N, S-co-doped carbon dots and their potential applications as antioxidants and sensitive probes for Cr (VI) detection. Sensor. Actuator. *Biol. Chem.,* **2017**, *248*, 92-100.

[144] Liu, Y.; Zhao, Y.; Zhang, Y. One-step green synthesized fluorescent carbon nanodots from bamboo leaves for copper(II) ion detection. Sensor. Actuator. *Biol. Chem.,* **2014**, *196*, 647-652.

[145] Gedda, G.; Lee, C-Y.; Lin, Y-C.; Wu, H-f. Green synthesis of carbon dots from prawn shells for highly selective and sensitive detection of copper ions. Sensor. Actuator. *Biol. Chem.,* **2016**, *224*, 396-403.

[146] Ma, X.; Dong, Y.; Sun, H.; Chen, N. Highly fluorescent carbon dots from peanut shells as potential probes for copper ion: The optimization and analysis of the synthetic process. *Mater. Today Chem.,* **2017**, *5*, 1-10.
[http://dx.doi.org/10.1016/j.mtchem.2017.04.004]

[147] Zhu, X.; Jin, H.; Gao, C.; Gui, R.; Wang, Z. Ratiometric, visual, dual-signal fluorescent sensing and imaging of pH/copper ions in real samples based on carbon dots-fluorescein isothiocyanate composites. *Talanta,* **2017**, *162*, 65-71.
[http://dx.doi.org/10.1016/j.talanta.2016.10.015] [PMID: 27837885]

[148] Gaddam, R.R.; Vasudevan, D.; Narayan, R.; Raju, K.V.S.N. Controllable synthesis of biosourced

blue-green fluorescent carbon dots from camphor for the detection of heavy metal ions in water. *RSC Advances,* **2014**, *4*(100), 57137-57143.
[http://dx.doi.org/10.1039/C4RA10471D]

[149] Hu, C.; Yu, C.; Li, M.; Wang, X.; Yang, J.; Zhao, Z.; Eychmüller, A.; Sun, Y.P.; Qiu, J. Chemically tailoring coal to fluorescent carbon dots with tuned size and their capacity for Cu(II) detection. *Small,* **2014**, *10*(23), 4926-4933.
[http://dx.doi.org/10.1002/smll.201401328] [PMID: 25048718]

[150] Qin, X.; Lu, W.; Asiri, A.M.; Al-Youbi, A.O.; Sun, X. Microwave-assisted rapid green synthesis of photoluminescent carbon nanodots from flour and their applications for sensitive and selective detection of mercury(II) ions. Sensor. Actuator. *Biol. Chem.,* **2013**, *184*, 156-162.

[151] Yu, J.; Song, N.; Zhang, Y-K.; Zhong, S-X.; Wang, A-J.; Chen, J. Green preparation of carbon dots by Jinhua bergamot for sensitive and selective fluorescent detection of Hg^{2+} and Fe^{3+}. Sensor. Actuator. *Biol. Chem.,* **2015**, *214*, 29-35.

[152] Choudhary, R.; Madhuri, R.; Sharma, P.K. Detection of Hg^{2+} Ion using fluorescent carbon dots derived from elephant foot yum *via* Green-Chemistry. AIP Publishing, **2017**; p. 050010.
[http://dx.doi.org/10.1063/1.4980243]

[153] Guo, Y.; Yang, L.; Li, W.; Wang, X.; Shang, Y.; Li, B. Carbon dots doped with nitrogen and sulfur and loaded with copper(II) as a *turn-on* fluorescent probe for cystein, glutathione and homocysteine. *Mikrochim. Acta,* **2016**, *183*(4), 1409-1416.
[http://dx.doi.org/10.1007/s00604-016-1779-6]

[154] Li, Z.; Zhang, Y.; Niu, Q.; Mou, M.; Wu, Y.; Liu, X.; Yan, Z.; Liao, S. A fluorescence probe based on the nitrogen-doped carbon dots prepared from orange juice for detecting Hg $^{2+}$ in water. *J. Lumin.,* **2017**, *187*, 274-280.
[http://dx.doi.org/10.1016/j.jlumin.2017.03.023]

[155] Li, L-S.; Jiao, X-Y.; Zhang, Y.; Cheng, C.; Huang, K.; Xu, L. Green synthesis of fluorescent carbon dots from Hongcaitai for selective detection of hypochlorite and mercuric ions and cell imaging. Sensor. Actuator. *Biol. Chem.,* **2018**, *263*, 426-435.

[156] Venkateswarlu, S.; Viswanath, B.; Reddy, A.S.; Yoon, M. Fungus-derived photoluminescent carbon nanodots for ultrasensitive detection of Hg^{2+} ions and photoinduced bactericidal activity. Sensor. Actuator. *Biol. Chem.,* **2018**, *258*, 172-183.

[157] Teymoorian, T.; Hashemi, N.; Mousazadeh, M.H.; Entezarian, Z. N, S doped carbon quantum dots inside mesoporous silica for effective adsorption of methylene blue dye. *SN Applied Sciences,* **2021**, *3*(3), 305-312.
[http://dx.doi.org/10.1007/s42452-021-04287-z]

[158] Ren, X.; Zhang, F.; Guo, B.; Gao, N.; Zhang, X. Synthesis of N-doped micropore carbon quantum dots with high quantum yield and dual-wavelength photoluminescence emission from biomass for cellular imaging. *Nanomaterials,* **2019**, *9*(4), 495.
[http://dx.doi.org/10.3390/nano9040495] [PMID: 30939724]

[159] Yang, D.; Tammina, S.K.; Li, X.; Yang, Y. Enhanced removal and detection of benzo[a]pyrene in environmental water samples using carbon dots-modified magnetic nanocomposites. *Ecotoxicol. Environ. Saf.,* **2019**, *170*, 383-390.
[http://dx.doi.org/10.1016/j.ecoenv.2018.11.138] [PMID: 30550968]

[160] Devi, P.; Rajput, P.; Thakur, A.; Kim, K.H.; Kumar, P. Recent advances in carbon quantum dot-based sensing of heavy metals in water. *Trends Analyt. Chem.,* **2019**, *114*, 171-195.
[http://dx.doi.org/10.1016/j.trac.2019.03.003]

[161] Sharma, S.; Dutta, V.; Singh, P.; Raizada, P.; Rahmani-Sani, A.; Hosseini-Bandegharaei, A.; Thakur, V.K. Carbon quantum dot supported semiconductor photocatalysts for efficient degradation of organic pollutants in water: A review. *J. Clean. Prod.,* **2019**, *228*, 755-769.
[http://dx.doi.org/10.1016/j.jclepro.2019.04.292]

[162] Ratnayake, S.P.; Mantilaka, M.M.M.G.P.G.; Sandaruwan, C.; Dahanayake, D.; Murugan, E.; Kumar, S.; Amaratunga, G.A.J.; de Silva, K.M.N. Carbon quantum dots-decorated nano-zirconia: A highly efficient photocatalyst. *Appl. Catal. A Gen.,* **2019**, *570*, 23-30. [http://dx.doi.org/10.1016/j.apcata.2018.10.022]

[163] Guo, Y.; Cao, F.; Li, Y. Solid phase synthesis of nitrogen and phosphor co-doped carbon quantum dots for sensing Fe3+ and the enhanced photocatalytic degradation of dyes. *Sens. Actuators B Chem.,* **2018**, *255*, 1105-1111. [http://dx.doi.org/10.1016/j.snb.2017.08.104]

[164] Zheng, F.; Wang, Z.; Chen, J.; Li, S. Synthesis of carbon quantum dot-surface modified P25 nanocomposites for photocatalytic degradation of p-nitrophenol and acid violet 43. *RSC Advances,* **2014**, *4*(58), 30605-30609. [http://dx.doi.org/10.1039/C4RA02707H]

[165] Wang, Q.; Wang, G.; Liang, X.; Dong, X.; Zhang, X. Supporting carbon quantum dots on NH2-MI--125 for enhanced photocatalytic degradation of organic pollutants under a broad spectrum irradiation. *Appl. Surf. Sci.,* **2019**, *467-468*, 320-327. [http://dx.doi.org/10.1016/j.apsusc.2018.10.165]

[166] Zhang, Z.; Wu, L.; Wang, P.; Zhang, Y.; Wan, S.; Guo, X.; Jin, W.; Zhang, J. Carbon quantum dots modified $La_2Ti_2O_7$ nanosheets for visible light photocatalysis. *Mater. Lett.,* **2018**, *230*, 72-75. [http://dx.doi.org/10.1016/j.matlet.2018.07.086]

[167] Saud, P.S.; Pant, B.; Alam, A.M.; Ghouri, Z.K.; Park, M.; Kim, H.Y. Carbon quantum dots anchored TiO_2 nanofibers: Effective photocatalyst for waste water treatment. *Ceram. Int.,* **2015**, *41*(9), 11953-11959. [http://dx.doi.org/10.1016/j.ceramint.2015.06.007]

[168] Lu, K.Q.; Quan, Q.; Zhang, N.; Xu, Y.J. Multifarious roles of carbon quantum dots in heterogeneous photocatalysis. *J. Energy Chemistry,* **2016**, *25*(6), 927-935. [http://dx.doi.org/10.1016/j.jechem.2016.09.015]

[169] Chen, Q.; Chen, L.; Qi, J.; Tong, Y.; Lv, Y.; Xu, C.; Ni, J.; Liu, W. Photocatalytic degradation of amoxicillin by carbon quantum dots modified K2Ti6O13 nanotubes: Effect of light wavelength. *Chin. Chem. Lett.,* **2019**, *30*(6), 1214-1218. [http://dx.doi.org/10.1016/j.cclet.2019.03.002]

[170] Cheng, Y.; Bai, M.; Su, J.; Fang, C.; Li, H.; Chen, J.; Jiao, J. Synthesis of fluorescent carbon quantum dots from aqua mesophase pitch and their photocatalytic degradation activity of organic dyes. *J. Mater. Sci. Technol.,* **2019**, *35*(8), 1515-1522. [http://dx.doi.org/10.1016/j.jmst.2019.03.039]

CHAPTER 9

Application of Carbonaceous Quantum dots in Energy Storage

Vaishali Thakur[1,*] and **Ekta Sharma**[1]

[1] *Department of Chemistry, School of Basic and Applied Sciences, Maharaja Agrasen University, Baddi, India*

Abstract: Carbon quantum dots (CQDs) are a type of carbon nanomaterial that has lately received attention as a potential replacement for standard semiconductor quantum dots (QDs). CQDs feature a quasi-spherical structure and amorphous to nanocrystalline carbon cores with diameters of 10-20 nm. Based on the carbon core, CQDs are further classified as graphene quantum dots (GQDs), carbon nanodots (CNDs), and polymer dots (PDs). CQDs exhibit unique electrical and optical properties due to their bigger edge effects and quantum confinement; better than graphene oxide nanosheets, they can also be easily split into electrons and holes due to their high dielectric constant and extinction coefficient. CQDs are crucial in the sector of energy storage and transformation because CQDs offer the advantageous properties of low toxicity, environmental friendliness, low cost, photostability, favourable charge transfer with increased electronic conductivity, and comparably simple synthesis processes. Due to their superior crystal structure and surface properties, CQD nanocomposites often helped to shorten charge transfer paths and maintain electrode material cycle stability. CQDs provide cost-effective and environmentally friendly nanocomposites used for supplying high energy density and stable electrodes for energy storage applications. This chapter provides a summary of the role that CQDs play in energy transmit technologies, including solar cells, supercapacitors, lithium-ion batteries, and hydrogen and oxygen evolution reactions.

Keywords: CQDs, Hydrogen and oxygen evolution reactions, Lithium-ion batteries, Supercapacitor.

INTRODUCTION

CQDs are classified as zero-dimensional carbon-dominated nanomaterials when compared to spherical, tubular, and sheet fillers, as well as nanomaterials from the graphene family. CQDs, being a structure made up of several layers of graphene oxide (GO) nanosheets, provide many of the benefits of GO while also having their own distinct qualities. CQDs exhibit unique electrical and optical properties

* **Corresponding author Vaishali Thakur:** Department of Chemistry, School of Basic and Applied Sciences, Maharaja Agrasen University, Baddi, India; Email: vaishalithakur137@gmail.com

Kulvinder Singh & Vineet Kumar (Eds.)

due to their bigger edge effects and quantum confinement; better than GO nanosheets, they can also be easily split into electrons and holes due to their high dielectric constant and extinction coefficient. Furthermore, long-term *in-vivo* testing results on CQDs showed that it is far less cytotoxic than GO sheets that are only a few micrometres in size; this suggests that it would be safer for *in-vivo* biological study. Amorphous to nanocrystalline carbon cores of CQDs are smaller than 10–20 nm in diameter and feature a quasi-spherical structure. CQDs are further subdivided into graphene quantum dots (GQDs), carbon nanodots (CNDs), and polymer dots (PDs) based on the various carbon cores [1 - 4]. A form of carbon NMs, preferably called carbon quantum dots (CQDs), has recently gained recognition as a potent alternative to conventional semiconductor quantum dots (QDs). CQDs possess several benefits over conventional semiconductors, including minimal toxicity, low price, luminescence, advantageous electrostatic interactions with increased electron mobility, and comparably simple synthesis techniques [1 - 4]. Quantum dot (QD)-based materials have been used in energy production and development to enhance the kinetic and thermodynamic features of electrochemical reactions. However, QDs' strong reactivity, chemical inertness, substance agglomeration, and limited electrical conductivity pose the biggest obstacles to their broad implementation. In this study, we merged the most recent findings on QD-carbonaceous-based nanohybrids used in rechargeable energy-storing equipment, including Sodium-ion batteries (SIBs), Perovskite, solar cells, Lithium-ion batteries (LIBs), supercapacitors, and other devices. Changes in photostability and optical properties of semiconductor QDs, such as adsorption and emission wavelength, are influenced by particle size. Semiconductor QDs, on the other hand, have several downsides, such as high toxicity, owing to the usage of heavy metals (like lead (Pb), and mercury (Hg) in their construction [2, 5 - 10]. Since then, researchers have become interested in CQDs because of their unique properties, which include good optoelectronic properties, luminescence, simple surface modification, robust biocompatibility, a sizable high surface area, and low cytotoxicity. CQDs can either be amorphous or nanocrystals and have a quasi-spherical architecture with carbon centers that are less than 10–20 nm in width. Based on the carbon cores, CQDs are further split into GQDs, carbon nanodots (CNDs), and polymer dots. In the graphene family, GQDs and CQDs are 0D NMs that share characteristics of both graphene and carbon dots, such as anisotropy and horizontal lengths of less than 20 nm [11, 12]. This chapter goes into detail regarding the developments and energy applications of CQDs. The part after that discussed some of the energy uses for CQDs, like solar panels, super-capacitors, and LIBs [13 - 16]. Additionally, the Shockley-Queisser performance barrier for silicon-based solar cells may be successfully exceeded by the simultaneous hot carrier gathering method of CQDs [10, 17]. The exceptional bioavailability, chemical stability, and less hazardous properties of graphene are also carried through into CQDs. CQDs, in particular, make excellent nonzero band gap and luminescence semiconductor materials because of their large number of functional units after grafting. Due to their higher conductance, high effective surface area, simplicity in loading and customizing, and superior conductance, the original CQDs might be used as a conducting reagent to the electrode in energy storage systems like supercapacitors [1, 3, 18 - 20].

CQDs in the Energy Storage and Conversion Applications

CQDs were formerly thought to have only a few applications, but now that their variety of special characteristics has come to light, they are being used more frequently than first thought. The attributes of CQDS in many energy-related sectors, including supercapacitors, photovoltaic cells, solar cells, batteries, water electrolysis, photodiodes, light-emitting diodes, and other fields, are covered in this section.

CQDs in Supercapacitors

Supercapacitors are energy-storing systems that, in contrast to the battery, may sustain maximum power production for prolonged durations. Owing to their unique benefits, like extended life, environmentally friendly nature, and significant-high - power density, supercapacitors have drawn considerable attention. Supercapacitors can be replaced with an electrode made of NMs that has a higher effective capacitance and extended duration durability [21]. Inflatable electrodes for supercapacitors have been shown to be quite utilized for storing power quickly and delivering both extreme power densities and rapid recharging speeds. In contrast to continued attempts, more study is necessary to create innovative porosity, conductive, and large specific interface area carbon composites having excellent specific capacity and extended battery durability. CQDs have been utilized in power storage systems due to their high conductance and shape- and dimension characteristics [10, 14, 22 - 24]. It stands for the argument that manufacturing nanohybrid frameworks utilizing sulfide/metal oxide and CQDs could be a viable method for creating powerful and reliable supercapacitors [21]. Arul *et al.* produced CQD-MnO nanostructures using a degraded recycling resource using an eco-benign method to achieve high energy density and electronic conductivity [25]. As per the morphological investigation, CQD-MnO had a larger surface region and superior electrical conductance than virgin MnO, which was represented by the extremely conducting CQDs. These nanohybrids showed a particular capacity of 189 F g^{-1} when used as the electrode in symmetrical superconductors, demonstrating an extended cyclic lifetime owing to strong coulombic efficacy and quick current-voltage responsiveness discovered through electrochemical studies. While improving the electrochemical properties of energy storage devices, CQDs may also improve the electrochemical characteristics of metal oxide-sulfide composites. Binesh *et al.* developed manganese oxide-graphene (MnO -CDGs) nanohybrids employing an easy, green, and one-pot manufacturing method [26]. The enhanced electrical characteristics and increased electrolytic reactivity of transition metal sulphides make them more appropriate for use in energy storage devices from an electrical standpoint. These characteristics aid in the creation of extremely efficient and dependable energy storage technologies. For example, in 2020, Quan *et al.* and colleagues combined CQDs with a CuS nanostructure to generate an unique porous structure nanoflower-like nanocomposite (CuS@CQDs) [27]. The solvothermal manufacturing technique and the implantation combined methodology (ICM) were used separately to synthesize the CuS@CQDs nanocomposite (SSM). With relatively extreme power intensity, compact size, efficient charging-discharging processes, and prolonged lifetimes, supercapacitors and micro-supercapacitors (SCs/MSCs) are becoming more and more used as energy storage systems [13, 28 - 31]. SCs are classified into two classes: pseudo capacitors and electrochemical double-layer capacitors, based on their unique high-energy storage mechanisms. SCs using CQDs as the electrode elements are typical of the electrochemical double-layer variety, having charge absorption over the substrate of the electrode acting as the pathway. On a gold finger electrode, electrodeposition by Xue *et al.* resulted in an MSC with a specified capacity of 311.3 mF cm^{-2}, a rate efficiency of up to 1000 V s^{-1}, and an electronic conductivity of 95.2% following 5000 cycles in comparison to its original specified capacity [32]. Traditional pseudo-capacitors are SCs constructed of CQDs loaded by compounds like N and S. Wu *et al.* created SCs using a carbonized ZIF-8 framework, a conductive matrix made of CNTs, and a strongly pseudocapacitive substance made of N-doped CQDs [33, 34]. In 1 M H_2SO_4 electrolytes, their SCs displayed outstanding cycle endurance and a significant particular capacitance of 540 F g^{-1} at 0.5 A g^{-1}. To increase the conductivity of an electrode made of carbonaceous material with a very large surface area, Fan *et al.* included GQDs into the electrode. With a specific capacity of 388 F g^{-1} at 1 A g^{-1}, the electrical double layer SC in a two-electrode system demonstrates exceptional rate capability (τ_0 = 0.68 s) and cyclability (100 percent of its initial value after 10 000 cycles). The SC displays a 60% capacitor preservation at current densities of 100 A g^{-1}. Despite having a class of substances having a strong pseudocapacitance, transition metal compounds are hard to utilize as electrodes on their own due to their poor electrical properties, high integrity, rate efficiency, and cyclability [35]. To create asymmetrical SCs, Wang *et al.* created a multilayer NiO@Co_3O_4 hollow sphere using activated carbon as the anode and GQDs as the cathode. The asymmetrical SCs have an impressive cyclability, a stated inductance of 1361 F g^{-1}, and a power density of 38.44 W h kg^{-1} (76.41 percent retention after 3000 cycles) [36].

CQDs in Photovoltaics

Due to their unique optoelectronic properties, up-conversion site fluorescence emission characteristics, band gap processability, and reduced application expenses, the integration of CQDs over NMs presented the opportunity for scenic charge transport extraction abilities for the improvement of solar cell effectiveness. The use of CQDs is possible in a variety of high-performance photovoltaic (pv) technologies, including dye-sensitized solar panels and perovskite solar cells, to name a few. CQDs are frequently utilized as a photo-catalyst, electron/hole transmission control procedure, or absorber coating additives for solar systems due to their huge, high surface area, high electron predilection, and outstanding capacity to absorb UV-vis energy and transform it to visible light [13, 14, 37 - 39].

CQDs in Dye-sensitized-solar Cells (DSSCs) and Quantum-dot-sensitized Solar Cells (QDSSCs)

In order to create effective and profitable light-harvesting materials, CQDs have recently been used in dye-sensitized solar cells (DSSCs) and quantum dot-sensitized solar cells (QDSSCs). For instance, Yuanyuan *et al.* created extremely effective solar cells by combining biomass-converted CQDs (LPP-CQDs), which were produced *via* the hydrothermal process from low-cost soybean powders, with long persistence phosphors (LPP). It was discovered that the LPP-CQDs architecture that was produced could capture enough light to emit monochromatic green light even in complete darkness when used in a solar cell [40]. The LPP-CQDs sensitizer-equipped device attained efficiencies as high as 7.97%. Additionally, it claimed that these CQDs could function in any weather and were capable of converting photons to electricity both during the day and at night. For instance, using N-doped CQDs (NCQDs) as a sensitizer, Rama *et al.* created effective DSSCs based on the ZnO photoanode. A greater power conversion efficiency (PCE) of 1.18% was achieved by the ZnO-NCQDs device as compared to the ZnO device alone (0.88%). The higher short-circuit current density (JSC), which was redirected from the increased light absorption and efficient electron-hole separation, was the primary factor in the increased device efficiencies [41]. N-CQDs can extend the absorption spectrum, act as a donor material, and carry out the crucial charge transfer kinetics task [42]. Adding fluorescent CQDs to the dye sensitizer significantly increased the meso-TiO_2 photoanode's charge-collecting efficiency, electron transport speed, and electron diffusion length [43]. The scientists suggested that CQDs could act as a dual-purpose light-harvesting material and enable the electron electron-transport channel in the dye-sensitized N719-TiO_2 screen based on optical and electrochemical studies. However, CQDs-modified photoanode revealed significantly higher absorption than those of pure N719-TiO_2 film. The CQDs were also utilised in the QDSSCs as a sensitizer to improve their photovoltaic performance, much like the DSSCs. Zhao *et al.* and colleagues created up and down-convention luminescences Er-doped CQDs (E-CQDs) for the first time using a hydrothermal method, and they applied it as a sensitizer in QDSSCs application. They used sequential ionic layer adsorption and reaction to apply various layers of CdS QDs and E-CQDs over a TiO_2 photoanode (SILAR). Due to better light absorption and electron transport, the CdS/Er-

CQDs/TiO_2 photoanode achieved efficiencies of 0.43%, which were greater than those of the CdS/TiO_2 photoanode [44].

CQDs in Perovskite Solar Cells (PSCs)

PSCs have emerged in interest for years because of their impressive efficiency and inexpensive development method. Just as important as creating extremely effective perovskite films are getting the correct chemicals to integrate with PSCs to enhance efficiency. For instance, sophisticated NMs with adjustable band gaps and chemical resistance include CQDs and tiny graphene sheets. GQDs are adaptive electron accumulators and donors when photoexcited [45 - 53]. Along with DSSCs and QDSSCs, the usage of CQDs in PSCs as a hole transport material (HTM) and for the regulation of charge recombination, in which the CQDs were utilised to adjust the interfacial characteristics, is well documented. Since the CQDs concept was first put forth, numerous efforts have been made to better understand how they work and improve the efficiency of perovskite solar cells (PSC). For instance, Paulo *et al.* proposed the first report of CQDs as a hole transport material (HTM) in 2016, the perovskite methylammonium lead iodide (MAPI). According to the electrochemical measurements, the proper levels of the CQDs' highest occupied molecular orbital (HOMO) and lowest unoccupied molecular orbital (LUMO) contributed to the efficient hole transfer and prevented the leakage of electrons from the perovskite to the CQDs [54]. In a different study, Kim *et al.* showed that nickel oxide (NiO), a p-type metal oxide, injected into CQDs, could function as an effective HTM in PSC [55]. It was shown in their research that the oxygen and nitrogen-rich surfaces of the CQDs helped to enhance the electronic characteristics of the NiO and properly match with the perovskite and conducting glass work-functions, resulting in cascade type charge transport in planar p-i-n type PSCs. In conclusion, the presence of CQDs in the perovskite light coating not only efficiently reduces the quantity of intergranular, eliminates electron trapping, and lowers the crossover process, but also enhances electron transport in the perovskite layer [56 - 59].

CQDs in Lithium-ion Batteries

Due to their unique qualities, such as their lightweight, economic efficiency, high energy density, and extended cycle life, lithium-ion batteries (LIB) are a smart energy storage technology that has increased commercial demand for various consumer power sources and electric or hybrid cars [60 - 62]. Due to self-aggregation, uncontrolled volume expansion in the electrode material, dissolution, formation of a solid electrolyte interface layer over the electrode, and rapidly elevated charge transfer resistances during cycles, many LIBs electrode materials, however, suffer from a rapid capacity fading during the charge-discharge process and poor rate performance. Low coulombic efficiency, electrolyte depletion, and safety problems are frequent with regard to the LIBs electrode materials [63 - 65]. In order to improve the LIB's next-generation efficiency, these types of problems in the electrode materials have recently been solved by combining CQDs with LIB electrode materials and altering the surface states and internal structures of the electrode materials. Because CQDs boost the conductivity of electrode materials, the active surface sites over the electrode materials grow, increasing their surface area and wettability, and in addition, it helps control the volume expansion of the electrode material during the charging-discharging process [66]. As an anode material for LIBs, Prasath *et al.* used a nanocomposite of bismuth oxide (Bi_2O_3) and CQDs [67]. Compared to either Bi_2O_3 or CQDs, these nanocomposites had better electrical conductivity and surface area. The coin cell made with Bi_2O_3-CQDs as a result, demonstrated a better discharge capacity of 1500 mAh g-1 at 0.2C and preserved almost 100% of its coulombic efficiency even in the 2500th cycle. In a different study, Zhang *et al.* found that heteroatom doping into CDs might be a potential method for increasing the electrochemical efficiency and surface area of carbon-based electrocatalysts. Jing *et al.* noted comparable outcomes when coating Mn_3O_4 particles on the surface of CQDs [68]. According to their research, the Mn_3O_4-CQDs nanocomposite that was produced after the CQDs coating was carried out using an alternating voltage electrochemical route changed structurally from a particle to a nano-octahedra structure, which in turn significantly improved the surface area and electrochemical properties. In comparison to just the Mn_3O_4 particles that were visible from the charge-discharge profile, the nanocomposite's discharge capacity was considerably improved by the higher surface area.

CQDs in Hydrogen Evolution Reaction

Among other energy sources, hydrogen is one of the best renewable energy alternatives because it has a high energy density and produces no carbon. In this case, we have two options for producing hydrogen: photocatalytic water splitting and electrocatalytic water splitting. With the aid of sunlight falling on the catalyst in the case of photocatalytic hydrogen evolution, and with the aid of voltage given to the catalyst in the case of electrocatalytic hydrogen evolution, hydrogen is produced [69 - 71]. CQDs are being used as a catalyst in photocatalytic and electrochemical processes to produce hydrogen. It contains all the necessary characteristics for both processes, including a high surface area and a number of active edges, including COOH, OH, CHO, and NH_2. As a result, they can help to coordinate with metal to form the metal-carbon dots composite, defect sites, rapid electron transfer, tunable photoluminescence (PL), and form holes and electrons by absorbing UV and visible light. They can also help to increase the wettability of the catalyst. In addition to these special qualities, they are economical and manageable in size [4]. Here, we've covered a few recent articles on composite materials based on CQDs. The metal-free photocatalyst was created by Elsayed *et al.* by implanting the nitrogen-doped CQDs (NCQDs) polymer (PS-PEGCOOH) using a straightforward chemical process [72]. The generated NCQDs covered in PS-PEGCOOH heterostructure material exhibit a 5-fold increase in hydrogen evolution efficiency in visible light when compared to PS-PEGCOOH. Because NCQDs are included, the catalyst in heterostructures has a higher hydrogen evolution efficiency, which benefits PS-charge PEGCOOH's separation and photocatalytic efficiency. Additionally, it has been shown that when NCQDs are implanted over PS-PEGCOOH, the produced charge practically doubles in recombination as compared to PS-PEGCOOH, which has not been implanted.

CQDs in Oxygen Evolution Reaction (OER)

One of the two components of the hydrogen fuel cell that produces hydrogen is the oxygen evolution process. Because OER requires a high overpotential to evaluate oxygen from water, where a covalent bond is generated by the transfer of four electrons, the oxygen evolution process is slow [73, 74]. Here, well-known catalysts based on iridium and ruthenium may readily split the water molecule and create the O-O bond. However, there are certain drawbacks to these catalysts, including high cost and scarcity, which restrict their widespread use and commercial potential [75, 76]. While transition metal-based catalysts provide a non-precious alternative catalyst for oxygen evolution, their limited activity prevents them from being fully utilised for large-scale oxygen synthesis. Since CQDs may not only increase the electronic conductivity of nanocomposites but also shield them from oxidation during water splitting and metal oxide corrosion, they have demonstrated good catalytic activity towards the evolution of oxygen in recent years [77]. The hydrothermal approach was employed by Chang *et al.* to create the single atomically anchored iron-on graphene quantum dots. The obtained material was then used as an electrocatalyst for electrocatalytic OER [76].

CONCLUSION

CQDs are crucial in the sector of energy transmission and storage. As realistic solutions to some of the most pressing problems in energy production, on-demand availability, pollution cleanup, and sophisticated production while cutting costs and minimizing environmental effects, novel energy transmission and storage technologies have been rapidly gaining support. Due to their distinctive and inherent properties, CQDs have a number of essential applications in the future field of energy storage and conversion. Researchers and the scientific community are beginning to understand the significance of using more energy and converting it in a greener way. It is possible for CQDs to have different electrical and chemical structures depending on their size, shape, surface functional groups, and heteroatom doping. Be aware that doping and modification increase electrical conductivity to the extent that this enhances their physiochemical properties. The insertion of CQDs into nanomaterials improves the ability of extraction of solar cells to become more effective due to its astounding optoelectronic features that can separate charge carriers that have low light absorption and tuneable bandgap. Due to their huge surface area and fast electron mobility, CQDs are frequently utilised as photo-sensitizers or active layers for photovoltaic systems. This allows them to absorb light throughout a broad spectrum. The CQDs modified electrodes, on the other hand, excel as counter electrodes for Li/Na ion batteries because they reduce volume expansion, increase ion diffusion rate, improve the electronic double layer kinetics, and enhance interfacial electrical conductivity. Furthermore, by combining CQDs with conducting polymers, carbonates, and activated carbon materials, the energy density, specific capacitance, and conductivity of supercapacitors might be greatly enhanced. Therefore, it is the responsibility of CQDs to provide more active sites in such structures and locations (either by creating new ones or activating existing ones).

REFERENCES

[1] Sharma, E.; Thakur, V.; Sangar, S.; Singh, K. *Recent progress on heterostructures of photocatalysts for environmental remediation*; *Mater.Today Proc*; Elsevier Ltd, **2020**, pp. 584-593.

[2] Shorie, M.; Kaur, H.; Chadha, G.; Singh, K.; Sabherwal, P. Graphitic carbon nitride QDs impregnated biocompatible agarose cartridge for removal of heavy metals from contaminated water samples. *J. Hazard. Mater.*, **2019**, *367*, 629-638. [http://dx.doi.org/10.1016/j.jhazmat.2018.12.115] [PMID: 30654280]

[3] Sharma, E.; Vashisht, D.; Thakur, V.; Vashisht, A.; Mehta, S.K.; Singh, K. Graphene quantum dots functionalized with bovine serum albumin for sensing of hypochlorite ions. *Mater. Chem. Phys.*, **2021**, *273*, 125088. [http://dx.doi.org/10.1016/j.matchemphys.2021.125088]

[4] Rasal, A.S.; Yadav, S.; Yadav, A.; Kashale, A.A.; Manjunatha, S.T.; Altaee, A.; Chang, J.Y. Carbon quantum dots for energy applications: A review. *ACS Appl. Nano Mater.*, **2021**, *4*(7), 6515-6541. [http://dx.doi.org/10.1021/acsanm.1c01372]

[5] Rani, A.; Singh, K.; Patel, A.S.; Chakraborti, A.; Kumar, S.; Ghosh, K.; Sharma, P. Visible Light Driven Photocatalysis of Organic Dyes Using SnO2 Decorated MoS2 Nanocomposites. *Chem. Phys. Lett. Electron Mater Lett*, **2019**.

[6] Vashisht, D.; Sharma, E.; Kaur, M.; Vashisht, A.; Mehta, S.K.; Singh, K. Spectrochim. Acta - Part A Mol. Biomol. Spectrosc. *Spectrochim. Acta*, **2019**.

[7] Sangar, S.; Sharma, S.; Vats, V.K.; Mehta, S.K.; Singh, K. Biosynthesis of silver nanocrystals, their kinetic profile from nucleation to growth and optical sensing of mercuric ions. *J. Clean. Prod.*, **2019**, *228*, 294-302. [http://dx.doi.org/10.1016/j.jclepro.2019.04.238]

[8] Singh, K.; Mehta, S.K. Utilization of ZnO nanoflowers as efficient electrochemical catalyst for the oxidation of hydrazine. *Sens. Lett.*, **2015**, *13*(11), 1002-1006.

[http://dx.doi.org/10.1166/sl.2015.3574]

[9] Zhen, Chen.; Cuihua, Xue.; Wei, Shi.; Fen-Tair, Luo.; Sarah, Green.; Jian, Chen.; Haiying, Liu. Return to issueprevsmall correspondence selective and sensitive fluorescent sensors for metal ions based on manipulation of side-chain compositions of Poly(p-phenyleneethynylene)s *Anal. Chem.,* **2004**, *76,* 6513-6518.

[10] Liu, Q.; Sun, J.; Gao, K.; Chen, N.; Sun, X.; Ti, D.; Bai, C.; Cui, R.; Qu, L. Graphene quantum dots for energy storage and conversion: from fabrication to applications. *Mater. Chem. Front.,* **2020**, *4*(2), 421-436.
[http://dx.doi.org/10.1039/C9QM00553F]

[11] Li, Z.; Yu, H.; Bian, T.; Zhao, Y.; Zhou, C.; Shang, L.; Liu, Y.; Wu, L.Z.; Tung, C.H.; Zhang, T. Highly luminescent nitrogen-doped carbon quantum dots as effective fluorescent probes for mercuric and iodide ions. *J. Mater. Chem. C Mater. Opt. Electron. Devices,* **2015**, *3*(9), 1922-1928.
[http://dx.doi.org/10.1039/C4TC02756F]

[12] Yang, S.; Sun, J.; Li, X.; Zhou, W.; Wang, Z.; He, P.; Ding, G.; Xie, X.; Kang, Z.; Jiang, M. Large-scale fabrication of heavy doped carbon quantum dots with tunable-photoluminescence and sensitive fluorescence detection. *J. Mater. Chem. A Mater. Energy Sustain.,* **2014**, *2*(23), 8660-8667.
[http://dx.doi.org/10.1039/c4ta00860j]

[13] Li, X.; Rui, M.; Song, J.; Shen, Z.; Zeng, H. Carbon and graphene quantum dots for optoelectronic and energy devices: A review. *Adv. Funct. Mater.,* **2015**, *25*(31), 4929-4947.
[http://dx.doi.org/10.1002/adfm.201501250]

[14] Riaz, R.; Ali, M.; Sahito, I.A.; Arbab, A.A.; Maiyalagan, T.; Anjum, A.S.; Ko, M.J.; Jeong, S.H. Self-assembled nitrogen-doped graphene quantum dots (N-GQDs) over graphene sheets for superb electro-photocatalytic activity. *Appl. Surf. Sci.,* **2019**, *480,* 1035-1046.
[http://dx.doi.org/10.1016/j.apsusc.2019.02.228]

[15] Guo, Q.; Ma, Z.; Zhou, C.; Ren, Z.; Yang, X. Single molecule photocatalysis on TiO_2 surfaces. *Chem. Rev.,* **2019**, *119*(20), 11020-11041.
[http://dx.doi.org/10.1021/acs.chemrev.9b00226] [PMID: 31503466]

[16] Lijuan, K.; Yongqiang, Y.; Ruiyi, L.; Zaijun, L. Phenylalanine-functionalized graphene quantum dot-silicon nanoparticle composite as an anode material for lithium ion batteries with largely enhanced electrochemical performance. *Electrochim. Acta,* **2016**, *198,* 144-155.
[http://dx.doi.org/10.1016/j.electacta.2016.03.034]

[17] Sharma, E.; Vashisht, D.; Vashisht, A.; Vats, V.K.; Mehta, S.K.; Singh, K. Facile synthesis of sulfur and nitrogen codoped graphene quantum dots for optical sensing of Hg and Ag ions. *Chem. Phys. Lett.,* **2019**, *730,* 436-444.
[http://dx.doi.org/10.1016/j.cplett.2019.06.040]

[18] Guo, X.; Wang, C.F.; Yu, Z.Y.; Chen, L.; Chen, S. Facile access to versatile fluorescent carbon dots toward light-emitting diodes. *Chem. Commun.,* **2012**, *48*(21), 2692-2694.
[http://dx.doi.org/10.1039/c2cc17769b] [PMID: 22306963]

[19] Wang, X.; Cao, L.; Yang, S.T.; Lu, F.; Meziani, M.J.; Tian, L.; Sun, K.W.; Bloodgood, M.A.; Sun, Y.P. bandgap-like strong fluorescence in functionalized carbon nanoparticles. *Angew. Chem. Int. Ed.,* **2010**, *49*(31), 5310-5314.
[http://dx.doi.org/10.1002/anie.201000982]

[20] Singh, K.; Mehta, S.K. Luminescent ZnO quantum dots as an efficient sensor for free chlorine detection in water. *Analyst,* **2016**, *141*(8), 2487-2492.
[http://dx.doi.org/10.1039/C5AN02599K]

[21] Liu, W.W.; Feng, Y.Q.; Yan, X-B.; Chen, J.T.; Xue, Q.J. Superior micro-supercapacitors based on graphene quantum dots. *Adv. Funct. Mater.,* **2013**, *23*(33), 4111-4122.
[http://dx.doi.org/10.1002/adfm.201203771]

[22] Singh, K.; Kumar, S.; Agarwal, K.; Soni, K.; Ramana Gedela, V.; Ghosh, K. Three-dimensional graphene with MoS 2 nanohybrid as potential energy storage/transfer device. *Sci. Rep.,* **2017**, 7.

[23] Haque, E.; Kim, J.; Malgras, V.; Reddy, K.R.; Ward, A.C.; You, J.; Bando, Y.; Hossain, M.S.A.; Yamauchi, Y. Recent advances in graphene quantum dots: Synthesis, properties, and applications. *Small Methods,* **2018**, *2*(10), 1800050.
[http://dx.doi.org/10.1002/smtd.201800050]

[24] Huang, Y.; Shi, T.; Zhong, Y.; Cheng, S.; Jiang, S.; Chen, C.; Liao, G.; Tang, Z. Graphene-quantu--dots induced NiCo2S4 with hierarchical-like hollow nanostructure for supercapacitors with enhanced electrochemical performance. *Electrochim. Acta,* **2018**, *269*, 45-54.
[http://dx.doi.org/10.1016/j.electacta.2018.02.145]

[25] Prasath, A.; Athika, M.; Duraisamy, E.; Sharma, A.S.; Elumalai, P. Carbon□quantum□dot□derived nanostructured MnO $_2$ and its symmetrical supercapacitor performances. *ChemistrySelect,* **2018**, *3*(30), 8713-8723.
[http://dx.doi.org/10.1002/slct.201801950]

[26] Unnikrishnan, B.; Wu, C.W.; Chen, I.W.P.; Chang, H.T.; Lin, C.H.; Huang, C.C. Carbon dot-mediated synthesis of manganese oxide decorated graphene nanosheets for supercapacitor application. *ACS Sustain. Chem.& Eng.,* **2016**, *4*(6), 3008-3016.
[http://dx.doi.org/10.1021/acssuschemeng.5b01700]

[27] Wang, Q.; Huang, J.; Zhou, J.; Liu, Z.; Geng, Y.; Liang, Z.; Du, Y.; Tian, X. Different nanostructured tungsten oxides synthesized by facile solvothermal route for chlorine gas sensing. *Sens. Actuators B Chem.,* **2018**, *275*, 306-311.
[http://dx.doi.org/10.1016/j.snb.2018.08.047]

[28] Ganganboina, A.B.; Chowdhury, A.D.; Doong, R. Nano assembly of N-doped graphene quantum dots anchored Fe3O4/halloysite nanotubes for high performance supercapacitor. *Electrochim. Acta,* **2017**, *245*, 912-923.
[http://dx.doi.org/10.1016/j.electacta.2017.06.002]

[29] Tang, P.; Han, L.; Zhang, L.; Wang, S.; Feng, W.; Xu, G.; Zhang, L. Controlled construction of hierarchical nanocomposites consisting of mno $_2$ and pedot for high-performance supercapacitor applications. *ChemElectroChem,* **2015**, *2*(7), 949-957.
[http://dx.doi.org/10.1002/celc.201500025]

[30] Zhang, S.; Zhu, J.; Qing, Y.; Wang, L.; Zhao, J.; Li, J.; Tian, W.; Jia, D.; Fan, Z. supercapacitors. *Adv. Funct. Mater.,* **2018**, *28*(52), 1805898.

[31] Jia, H.; Cai, Y.; Lin, J.; Liang, H.; Qi, J.; Cao, J.; Feng, J.; Fei, W.D. High-performance flexible quasi-solid-state supercapacitors realized by molybdenum dioxide@ nitrogen-doped carbon and copper cobalt sulfide tubular nanostructures. *Adv. Sci.,* **2018**, **•••**, 5.

[32] Xu, Y.; Jia, X.H.; Yin, X.B.; He, X.W.; Zhang, Y.K. Carbon quantum dot stabilized gadolinium nanoprobe prepared *via* a one-pot hydrothermal approach for magnetic resonance and fluorescence dual-modality bioimaging. *Anal. Chem.,* **2014**, *86*(24), 12122-12129.
[http://dx.doi.org/10.1021/ac503002c] [PMID: 25383762]

[33] Wu, Z.L.; Gao, M.X.; Wang, T.T.; Wan, X.Y.; Zheng, L.L.; Huang, C.Z. A general quantitative pH sensor developed with dicyandiamide N-doped high quantum yield graphene quantum dots. *Nanoscale,* **2014**, *6*(7), 3868-3874.
[http://dx.doi.org/10.1039/C3NR06353D] [PMID: 24589665]

[34] Z, W.; W, L.; J, C.; Y., C. A Graphene Quantum Dot-Based Method for the Highly Sensitive and Selective Fluorescence Turn on Detection of Biothiols. *Science,* **2014**, *119*, 538-543.

[35] Fan, X.; Feng, Y.; Su, Y.; Zhang, L.; Lv, Y. A green solid-phase method for preparation of carbon nitride quantum dots and their applications in chemiluminescent dopamine sensing. *RSC Advances,* **2015**, *5*(68), 55158-55164.

[http://dx.doi.org/10.1039/C5RA05397H]

[36] Wang, H.; Wang, B.; Bian, Y.; Dai, L.; Appl, A.C.S. Enhancing photocatalytic activity of graphitic carbon nitride by codoping with P and C for efficient hydrogen generation. *ACS Appl. Mater. Interfaces,* **2017**, *9*(26), 21730-21737.
[http://dx.doi.org/10.1021/acsami.7b02445]

[37] Xing, G.; Wu, B.; Chen, S.; Chua, J.; Yantara, N.; Mhaisalkar, S.; Mathews, N.; Sum, T.C. Interfacial electron transfer barrier at compact TiO_2 /$CH_3NH_3PbI_3$ heterojunction. *Small,* **2015**, *11*(29), 3606-3613.
[http://dx.doi.org/10.1002/smll.201403719] [PMID: 25824264]

[38] Liu, X.; Tsai, K.W.; Zhu, Z.; Sun, Y.; Chueh, C.C.; Jen, A.K.Y. Fluoroalkyl-substituted fullerene/perovskite heterojunction for efficient and ambient stable perovskite solar cells. *Adv. Mater. Interfaces,* **2016**, 3.

[39] Mahmoudi, T.; Wang, Y.; Hahn, Y.B. Stability enhancement in perovskite solar cells with perovskite/silver–graphene composites in the active layer. *ACS Energy Lett.,* **2019**, *4*(1), 235-241.
[http://dx.doi.org/10.1021/acsenergylett.8b02201]

[40] Meng, Y.; Zhang, Y.; Sun, W.; Wang, M.; He, B.; Chen, H.; Tang, Q. Biomass converted carbon quantum dots for all-weather solar cells. *Electrochim. Acta,* **2017**, *257*, 259-266.
[http://dx.doi.org/10.1016/j.electacta.2017.10.086]

[41] Rama Krishna, C.; Im, Y.; Kang, M. Nitrogen doped carbon quantum dots as a green luminescent sensitizer to functionalize ZnO nanoparticles for enhanced photovoltaic conversion devices. In: *Mater. Res. Bull*; Elsvier, **2017**; 94, pp. 399-407.
[http://dx.doi.org/10.1016/j.materresbull.2017.06.040]

[42] Yang, Q.; Yang, W.; Zhang, Y.; Ge, W.; Yang, X.; Yang, P. Enhanced stretchable and sensitive strain sensor *via* controlled strain distribution. *Nanomaterials,* **2020**, *10*(2), 218.

[43] Shi, Y.; Na, Y.; Su, T.; Li, L.; Yu, J.; Fan, R.; Yang, Y. Fluorescent carbon quantum dots incorporated into dye-sensitized TiO_2 photoanodes with dual contributions. *ChemSusChem,* **2016**, *9*(12), 1498-1503.
[http://dx.doi.org/10.1002/cssc.201600067] [PMID: 27218888]

[44] Zhao, C.; Zhang, X.; Shu, X.; Liu, X.; Fang, D.; Song, Y.; Wang, J. Er-doped carbon dots broadening light absorption range and accelerating electron transport for enhancing photovoltaic performance of CdS quantum dots sensitized cells. *Opt. Mater.,* **2018**, *84*, 242-251.
[http://dx.doi.org/10.1016/j.optmat.2018.07.016]

[45] Tai, M.; Zhao, X.; Wei, H.; Wang, G.; Hao, F.; Li, X.; Yin, X.; Zhou, Y.; Han, J.; Wei, Y.; Jiang, K.; Lin, H. Laser-induced flash-evaporation printing $CH_3NH_3PbI_3$ thin films for high-performance planar solar cells. *ACS Appl. Mater. Interfaces,* **2018**, *10*(31), 26206-26212.
[http://dx.doi.org/10.1021/acsami.8b05918] [PMID: 30003774]

[46] AlSalka, Y.; Granone, L.I.; Ramadan, W.; Hakki, A.; Dillert, R.; Bahnemann, D.W. Iron-based photocatalytic and photoelectrocatalytic nano-structures: Facts, perspectives, and expectations. *Appl. Catal. B,* **2019**, *244*, 1065-1095.
[http://dx.doi.org/10.1016/j.apcatb.2018.12.014]

[47] Sung, H.; Ahn, N.; Jang, M.S.; Lee, J.K.; Yoon, H.; Park, N.G.; Choi, M. Transparent conductive oxide-free graphene-based perovskite solar cells with over 17% efficiency. In: *Adv. Energy Mater*; Wiley, **2016**; 6, p. (3)1501873.

[48] Q, L.; B, G.; Z, R.; B, Z.; G., JR. Strong two-photon-induced fluorescence from photostable, biocompatible nitrogen-doped graphene quantum dots for cellular and deep-tissue imaging *Nano Lett.,* **2013**, *13*, 2436-2441.

[49] Liu, Y.; Liao, M.; He, X.; Liu, X.; Kou, X.; Xiao, D. one-step synthesis of highly luminescent nitrogen-doped carbon dots for selective and sensitive detection of Mercury(II) Ions and cellular

imaging. *Anal. Sci.,* **2015**, *31*(10), 971-977.
[http://dx.doi.org/10.2116/analsci.31.971] [PMID: 26460360]

[50] Wang, Y.; Mahmoudi, T.; Rho, W.Y.; Yang, H.Y.; Seo, S.; Bhat, K.S.; Ahmad, R.; Hahn, Y.B. Ambient-air-solution-processed efficient and highly stable perovskite solar cells based on CH3NH3PbI3−xClx-NiO composite with Al2O3/NiO interfacial engineering. *Nano Energy,* **2017**, *40*, 408-417.
[http://dx.doi.org/10.1016/j.nanoen.2017.08.047]

[51] Shen, B.; Lang, J.; Guo, R.; Zhang, X.; Yan, X. Engineering the electrochemical capacitive properties of microsupercapacitors based on graphene quantum Dots/MnO_2 using ionic liquid gel electrolytes. *ACS Appl. Mater. Interfaces,* **2015**, *7*(45), 25378-25389.
[http://dx.doi.org/10.1021/acsami.5b07909] [PMID: 26502031]

[52] Yang, Z.; Xie, J.; Arivazhagan, V.; Xiao, K.; Qiang, Y.; Huang, K.; Hu, M.; Cui, C.; Yu, X.; Yang, D. Efficient and highly light stable planar perovskite solar cells with graphene quantum dots doped PCBM electron transport layer. *Nano Energy,* **2017**, *40*, 345-351.
[http://dx.doi.org/10.1016/j.nanoen.2017.08.008]

[53] Wang, H.C.; Li, Y.; Yang, M.J. Fast response thin film SnO2 gas sensors operating at room temperature. *Sens. Actuators B Chem.,* **2006**, *119*(2), 380-383.
[http://dx.doi.org/10.1016/j.snb.2005.12.037]

[54] Paulo, S.; Palomares, E.; Martinez-Ferrero, E. Graphene and carbon quantum dot-based materials in photovoltaic devices: From synthesis to applications. *Nanomaterials,* **2016**, *6*(9), 157.

[55] Kim, J.K.; Nguyen, D.N.; Lee, J.H.; Kang, S.; Kim, Y.; Kim, S.S.; Kim, H.K. Carbon quantum dot-incorporated nickel oxide for planar p-i-n type perovskite solar cells with enhanced efficiency and stability. *J. Alloys Compd.,* **2020**, *818*, 152887.
[http://dx.doi.org/10.1016/j.jallcom.2019.152887]

[56] Xie, J.; Yu, X.; Sun, X.; Huang, J.; Zhang, Y.; Lei, M.; Huang, K.; Xu, D.; Tang, Z.; Cui, C.; Yang, D. Improved performance and air stability of planar perovskite solar cells *via* interfacial engineering using a fullerene amine interlayer. *Nano Energy,* **2016**, *28*, 330-337.
[http://dx.doi.org/10.1016/j.nanoen.2016.08.048]

[57] Ke, W.; Fang, G.; Liu, Q.; Xiong, L.; Qin, P.; Tao, H.; Wang, J.; Lei, H.; Li, B.; Wan, J.; Yang, G.; Yan, Y. Low-temperature solution-processed tin oxide as an alternative electron transporting layer for efficient perovskite solar cells. *J. Am. Chem. Soc.,* **2015**, *137*(21), 6730-6733.
[http://dx.doi.org/10.1021/jacs.5b01994] [PMID: 25987132]

[58] Tan, H.; Jain, A.; Voznyy, O.; Lan, X.; De Arquer, F.P.G.; Fan, J.Z.; Quintero-Bermudez, R.; Yuan, M.; Zhang, B.; Zhao, Y.; Fan, F.; Li, P.; Quan, L.N.; Zhao, Y.; Lu, Z.H.; Yang, Z.; Hoogland, S.; Sargent, E.H. Efficient and stable solution-processed planar perovskite solar cells *via* contact passivation. *Science,* **2017**, *355*(6326), 722-726.

[59] Shen, D.; Zhang, W.; Xie, F.; Li, Y.; Abate, A.; Wei, M. Graphene quantum dots decorated TiO2 mesoporous film as an efficient electron transport layer for high-performance perovskite solar cells. *J. Power Sources,* **2018**, *402*, 320-326.
[http://dx.doi.org/10.1016/j.jpowsour.2018.09.056]

[60] Evarts, E.C. Lithium batteries: To the limits of lithium. *Nature,* **2015**, *526*(7575), S93-S95.
[http://dx.doi.org/10.1038/526S93a]

[61] Diouf, B.; Pode, R. Potential of lithium-ion batteries in renewable energy. *Renew. Energy,* **2015**, *76*, 375-380.
[http://dx.doi.org/10.1016/j.renene.2014.11.058]

[62] Z.John, Zhang; Ramadass, P. Lithium-ion battery systems and technology. In: *Encyclopedia of Sustainability Science and Technology*; Meyers, R.A., Ed.; Springer: New York, NY, **2012**.

[63] Pender, J.P.; Jha, G.; Youn, D.H.; Ziegler, J.M.; Andoni, I.; Choi, E.J.; Heller, A.; Dunn, B.S.; Weiss,

P.S.; Penner, R.M.; Mullins, C.B. Electrode degradation in lithium-ion batteries. *ACS Nano,* **2020**, *14*(2), 1243-1295.
[http://dx.doi.org/10.1021/acsnano.9b04365] [PMID: 31895532]

[64] Hausbrand, R.; Jaegermann, W. Encycl. Interfacial Chem. Surf. Sci. Electrochem. **2018**, 232-245.

[65] Pinson, M.B.; Bazant, M.Z. Theory of SEI formation in rechargeable batteries: capacity fade, accelerated aging and lifetime prediction. *J. Electrochem. Soc.,* **2013**, *160*(2), A243-A250.
[http://dx.doi.org/10.1149/2.044302jes]

[66] Song, T.B.; Huang, Z.H.; Niu, X.Q.; Liu, J.; Wei, J.S.; Chen, X.B.; Xiong, H.M. Applications of carbon dots in next□generation lithium□ion batteries. *ChemNanoMat,* **2020**, *6*(10), 1421-1436.
[http://dx.doi.org/10.1002/cnma.202000355]

[67] Prasath, A.; Athika, M.; Duraisamy, E.; Selva Sharma, A.; Sankar Devi, V.; Elumalai, P. Carbon quantum dot-anchored bismuth oxide composites as potential electrode for lithium-ion battery and supercapacitor applications. *ACS Omega,* **2019**, *4*(3), 4943-4954.
[http://dx.doi.org/10.1021/acsomega.8b03490] [PMID: 31459678]

[68] Jing, M.; Wang, J.; Hou, H.; Yang, Y.; Zhang, Y.; Pan, C.; Chen, J.; Zhu, Y.; Ji, X. Carbon quantum dot coated Mn_3O_4 with enhanced performances for lithium-ion batteries. *J. Mater. Chem. A Mater. Energy Sustain.,* **2015**, *3*(32), 16824-16830.
[http://dx.doi.org/10.1039/C5TA03610K]

[69] Tachibana, Y.; Vayssieres, L.; Durrant, J.R. Artificial photosynthesis for solar water-splitting. *Nature Photonics,* **2012**, *5*, 511-518.

[70] Fujishima, A.; Honda, K. Electrochemical photolysis of water at a semiconductor electrode. *Nature,* **1972**, *238*(5358), 37-38.
[http://dx.doi.org/10.1038/238037a0] [PMID: 12635268]

[71] Han, M.; Zhu, S.; Lu, S.; Song, Y.; Feng, T.; Tao, S.; Liu, J.; Yang, B. Recent progress on the photocatalysis of carbon dots: Classification, mechanism and applications. *Nano Today,* **2018**, *19*, 201-218.
[http://dx.doi.org/10.1016/j.nantod.2018.02.008]

[72] Elsayed, M.H.; Jayakumar, J.; Abdellah, M.; Mansoure, T.H.; Zheng, K.; Elewa, A.M.; Chang, C.L.; Ting, L.Y.; Lin, W.C.; hua, Yu; Wang, W.H.; Chung, C.C.; Chou, H.H. Visible-light-driven hydrogen evolution using nitrogen-doped carbon quantum dotimplanted polymer dots as metal-free photocatalysts *Appl. Catal. B,* **2021**, 283.

[73] Anantharaj, S.; Ede, S.R.; Karthick, K.; Sam Sankar, S.; Sangeetha, K.; Karthik, P.E.; Kundu, S. Precision and correctness in the evaluation of electrocatalytic water splitting: Revisiting activity parameters with a critical assessment. *Energy Environ. Sci.,* **2018**, *11*(4), 744-771.
[http://dx.doi.org/10.1039/C7EE03457A]

[74] Suen, N.T.; Hung, S.F.; Quan, Q.; Zhang, N.; Xu, Y.J.; Chen, H.M. Electrocatalysis for the oxygen evolution reaction: Recent development and future perspectives. *Chem. Soc. Rev.,* **2017**, *46*(2), 337-365.
[http://dx.doi.org/10.1039/C6CS00328A] [PMID: 28083578]

[75] Natter, H.; Hempelmann, R. Electrochim. Acta. Pergamon, **2003**; pp. 51-61.

[76] Chang, C.Y.; Kashale, A.A.; Lee, C.M.; Chu, S.L.; Lin, Y.F.; Chen, I-W.P. Single atomically anchored iron on graphene quantum dots for a highly efficient oxygen evolution reaction. *Mater. Today Energy,* **2021**, *20*, 100693.
[http://dx.doi.org/10.1016/j.mtener.2021.100693]

[77] Tian, L.; Li, Z.; Wang, P.; Zhai, X.; Wang, X.; Li, T. Carbon quantum dots for advanced electrocatalysis. *J. Energy Chem.,* **2021**, *55*, 279-294.
[http://dx.doi.org/10.1016/j.jechem.2020.06.057]

CHAPTER 10

Future Prospect of Carbonaceous Quantum Dots

Devika Vashisht[1], **Aseem Vashisht**[2], **Shweta Sharma**[3] and **Surinder Kumar Mehta**[1,4,*]

[1] *Department of Chemistry and Centre of Advanced Studies in Chemistry, Panjab University, Chandigarh, India*

[2] *Punjab State Council for Science and technology Sector - 26 , Chandigarh, India*

[3] *Institute of Forensic Science and Criminology, Panjab University, Chandigarh, India*

[4] *University of Ladakh, Leh, India*

Abstract: Carbonaceous quantum dots (CQDs) are carbon nanomaterials offering significant contributions in the fields of sensing, biosensing, optoelectronics, and energy storage. CQDs materials with edge defects are suitable candidates for LED emitters, water splitting, optoelectronic and photodetectors devices. The role of attached functional groups and edge effects on flexible energy storage devices is discussed in this chapter. The future prospects and underlying challenges for CQD-based material for wastewater remediation, in addition to factors such as their sustainability, durability, performance and economics in the context of industrial scale-up, have also been explored in this chapter.

Keywords: Carbon dots, Energy devices, Photodetectors, Quantum dots, Wastewater remediation.

INTRODUCTION

Carbonaceous quantum dots (CQDs) are nanoparticles made up of carbon that are of extremely small size (1-10 nm). CQDs were first accidental discovery by Xu *et al.* [1] in 2004. Sun *et al.* [2] called these fluorescent nanomaterials as "carbon quantum dots" later in 2006. The CQDs have intriguing features over traditional semiconductor quantum dots (SQDs). Due to the inherent features, the possibility of using CQDs has been highlighted in countless applications in biological, material, chemical and physical sciences due to the features like biocompatibility, chemical inertness, low cost and photo stability [3 - 8]. In the recent past, numerous reviews have been reported, which are majorly concentrated on the synthesis and applications of CQDs in various fields. In the application perspective, surface passivation and functionalization for the regulation of

* **Corresponding author Surinder Kumar Mehta:** Department of Chemistry and Centre of Advanced Studies in Chemistry, Panjab University, Chandigarh, India and University of Ladakh, Leh, India; E-mail: skmehta@pu.ac.in

Kulvinder Singh & Vineet Kumar (Eds.)

physiochemical characteristics have also been presented. The majority of CQDs reviews and chapters focus on the chemical applications rather than the exciting futuristic application of CQDs in energy and wastewater remediation.

We have already discussed the properties of CQDs in terms of structure and induced defects in their structures in previous chapters. This chapter will focus on the futuristic aspects of CQDs. This discussion is expanded to include the underlying issues as well as possible future directions.

CQDs may be synthesized by employing "top-down", "bottom-up," and "waste-derived" techniques. CQDs are made from a variety of sources, including graphite powder, bulk carbon, carbon fibers and carbon nanotubes, using a top-down technique. Some of them are Arc-discharge [1], Laser ablation [2], Plasma reactor [9], Ultrasonic synthesis [10], combustion [11] and electrochemical method [12, 13]. Carbonization or Pyrolysis of organic precursors is the root of a bottom-up strategy. Microwave [14 - 16], hydrothermal [17], pyrolysis [18] and MOF template-assisted [19] are some of the bottom-up techniques. Waste-derived procedures are methods that use waste resources like leftover fruits and vegetables as a source of carbon. The waste-derived synthesis directs the environmentally responsible use of natural resources. Numerous reports on the synthesis of CQDs from waste such as watermelon peel, sweet pepper, pomelo peel, dairy waste, orange juice and cabbage have been published [20 - 24].

According to their chemical structure, the majority of CQDs contain an amorphous base with a sizable variety of functional groups. Some investigations, however, revealed the presence of sp2 carbon within the base. The electronic structure of CQDs is explained on the basis of molecular orbital (MO) theory. Every possible absorption transition among these bonding, anti-bonding and non-bonding orbitals lies in the UV-vis range. Due to a higher degree of surface oxidation, top-down-produced CQDs are known to have a higher number of defects than bottom-up-created CQDs. Surface defects increase as surface oxidation increases. The fluorescence of CQDs is caused by these surface imperfections, which are referred to as molecular states. Based on the chemical and electronic structure of CQDs, a range of applications has been explored. The futuristic applications of CQDs are energy devices and wastewater treatment, which have been discussed further in this chapter.

Photodetectors

UV Photodetectors

Based on the response wavelength, CQDs-based photodetectors can be divided into three categories: Visible light detectors, UV detectors and near mid-infrared detectors [25 - 28]. UV photodetectors are commonly reported because $\pi \rightarrow \pi^*$ transitions produce absorption in the deep-UV region. They are majorly utilized in the transmission of signals and medical tools. In the last few decades, eco-friendly CQD detectors are more popular than the semiconductor or nitride-based detectors [29]. Due to the enormous binding energy of 60 meV, UV detectors made of high bandgap oxide materials show potential [30, 31]. Heterostructures of ZnO nanorods with CQDs/ GQDs have been studied under dark and UV light of 365 nm, revealing that ZnO-GQD has higher UV detectivity than ZnO-CQDs. The photodesorption of adsorbed oxygen under UV irradiation by trapping "hot" electrons is the process that encompasses oxide-based UV detection. However, in ZnO-CQD or ZnO-GQD-based heterostructures, the hydrophilic interface of the CQDs/GQDs extensively gathered oxygen moieties.

Broadband Photodetectors

These photodetectors often rely on substances that absorb and emit in the severe ultraviolet (UV), visible, and near-infrared (NIR) areas. CQDs may successfully simulate the absorption in the deep UV to NIR region by using the right designs and doping. This makes them great candidates for broadband photodetectors that are both efficient and affordable. Fig. (**1**) illustrates the mechanism underlying broadband absorption and emission. Deep UV absorption occurs when localized electrons in sp^2 hybridized carbon (C=C) bonds undergo a transition (v_3---c_3). On the other hand, the emission happens while the excited electron goes through an interband transmission (c_3---c_2) prior to radiative combination (c_2---v_2). Partially conjugated π-electrons give rise to visible light absorption (v_2----c_2) [32].

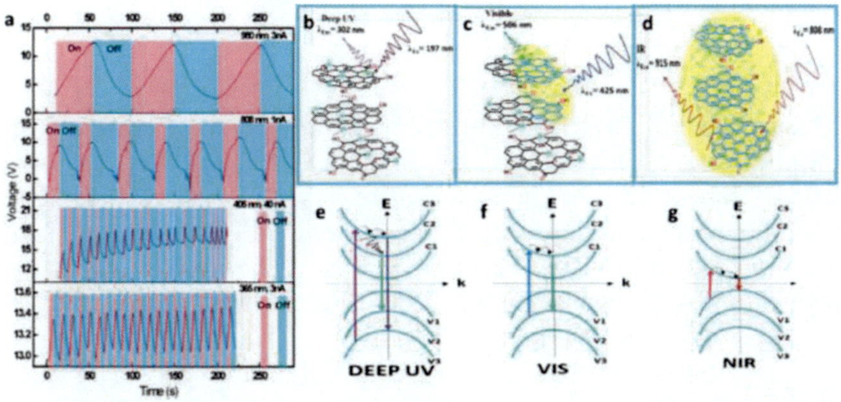

Fig. (1). N-GQD-based photodetectors: (**a**) The N-GQDs' reaction to different light streams (980 nm, 808 nm, 405 nm, 365 nm). The process by which N-GQDs emit wideband light is shown in (**b-d**). (**b**) Deep UV PL absorption, absorbance, and emissions for the electron in C=C. (**c**) The layered architecture of conjugated electrons absorbs visible light; (**d**) bulkier electrons are responsible for NIR absorption. (e-g) The band to-band transitions and vibrations relaxation-related emission process [32].

Solar Cells

Heterostructured Solar Cells

CQD-based solid-state heterostructured solar cells have proven to work quite well. After incorporating these quantum dots into heterostructures, optical absorbance, sensibility, and electron transit all significantly improved. A core-shell design of Si Nw and CQDs has been reported by Xie *et al.* [33] to investigate the photovoltaic features by varying the CQDs layer thickness. Under AM 1.5 irradiation, the Si/CQDs core-shell heterostructured device had a power conversion efficacy (PCE) of 9.10%. The charge particles produced in the Si/CQDs heterojunction during light irradiation were divided by the strong built-in potential field. Electrons were gathered at the bottom electrode, but the energy barrier brought on by band offsets prevented supplied electrons from passing to the Au electrode. As the HOMO band of CQDs allows the defects from Si to readily access the Au electrode. By acting as an electron-blocking barrier, the CQD layer increased the device's open circuit voltage. On similar lines, a solar cell device was fabricated based on the same working principle using Si/GQD heterojunction. The device was stable for over six months and provided 6.63% of PCE. Due to the band structure of CQDs, they improved open circuit voltage when incorporated into specific heterostructured devices. The advanced features of CQDs, such as efficient photo-absorptivity and their tendency to act as effective sensitizers, lead to this improvement. Gupta *et al.* [34] investigated the CQD/TiO_x composite as an electron transit planar heterojunction and discovered 19% efficiency. Similarly, Guo *et al.* [35] highlighted the passivation of $MAPbI_3$grain boundaries, achieving a PCE of 19.38%.

Dye-sensitized Solar Cells

Dye-sensitized solar cells (DSSCs) are photovoltaic systems that utilise dyes to absorb sunlight. The solutions can be processed through the spin coating and inkjet printing. Yan *et al.* [36] reported a very low-efficiency DSSC based on GQDs and TiO_2 surfaces. Therefore, nitrogen-doped CQDs were utilized to offer covalent binding and modify the band energy configuration in order to improve physical absorption. A conversion efficiency of 0.13% was obtained by optimized doping.

Light-emitting Diodes

Due to their high efficacy, excellent quantum yield, low hazardous composition, and improved stability, CQDs have a great ability to substitute phosphor-based white light emitting diodes (LEDs). These promising features have led researchers to develop methods to fabricate phosphor-based and electro-chemiluminescence based LEDs.

Chen *et al.* [37] reported broad yellow light fluorescence from CQDs generated by one-step pyrolysis of N-acetylcysteine due to the lack of stability of traditional LEDs. By integrating yellow light-producing CQDs into the blue GaN LEDs, the authors created white light-emitting diodes (WLEDs) as per CEI standards. Guo *et al.* [38] used a sensational way to make WLEDs by integrating UV LEDs with CQDs produced from photonic crystals. Despite the fact that the coordinates were near to those of white light, UV light leakage was a significant disadvantage. Sun *et al.* [39] adopted a different method for designing and fabricating a solid-state LED using a combination of starch and CQDs. The aggregation-induced quenching (AIQ) phenomenon has been identified as the most significant constraint to the development of solid-state CQD-based LEDs. To summarize, further investigations and research inputs are required to optimize the proper combination of material with CQDs to get better-performing LED devices.

Wearable and Flexible Electronic Devices

Flexible devices such as supercapacitors, photo sensors, non-volatile memories and LEDs have been reported due to excellent compatibility of flexible substrate processes with CQDs. Anchoring of CQDs under optimized conditions to conventional supercapacitors based on nanostructural materials has emerged as an efficient way for supercapacitor performance. Lv *et al.* [40] conducted a comparative analysis to examine the distinctions in performance among MnO_2 electrodes and CQDs decorated MnO_2 electrodes. A different approach was employed (Fig. **2**) by Jian *et al.* [41] to get solid-state supercapacitors where CQDs and Polypyrrole composite were prepared by electrochemical deposition.

To improve the relative conductance and energy storage density, oxygen and nitrogen co-doped GQDs were constructed with a carbon network. Utilizing an electrical double layer and pseudo capacitance, the carbon network structure's strong electrical conductivity and large surface region allowed for effective activation. Apart from flexible supercapacitors, researchers have looked into CQD compatibility in flexible memory systems, optoelectronics, and photovoltaics.

Hydrogen Fuels

The synthesis of hydrogen fuel to meet global energy needs has remained a concern. Production of hydrogen fuel through eco-benign CQDs is a viable option for providing clean energy. The tunable absorption range makes CQDs promising candidates for the development of photoelectrodes for photoelectrochemical (PEC) hydrogen generation using water splitting. Numerous structures having

CQDs have been reported with Si, ZnO, Cu_2S and TiO_2 nano-rods for PEC water splitting.

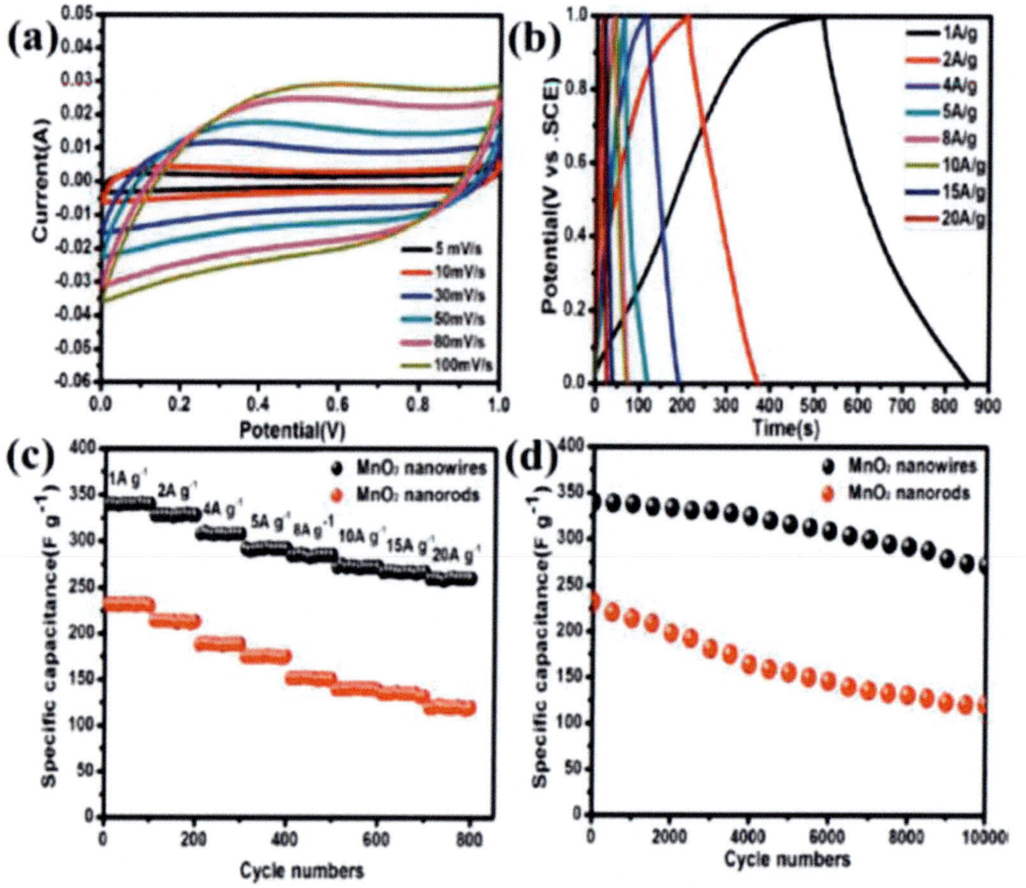

Fig. (2). (**a**) CV lines and (**b**) GCD lines of the MnO_2/CQD nanowires. Cycling efficacy of MnO_2/CQD nanowires and MnO_2nanorods and (**c**) specific capacitance at several current densities, (**d**) deviation of specified capacitance for 10,000charge–discharge cycles at 1 A g^{-1} [40].

Various CQDs decorated heterojunctions have been employed for PEC water oxidation. Yu and his co-workers [42] have reported TiO_2/CQDs heterojunction for PEC water splitting. PEC generated an optimum photocurrent of 1.2 mA/ cm^2 at 0 V Vs. Ag/AgCl in 100 mW/ cm^2 imitation sunlight. Liang *et al.* [43] developed TiO_2/CQDs photoanodes with a photocurrent density of 3.0 mA/cm^2 at 1.23 V in imitation sunlight in a similar study. The absorption spectrum of the $BiVO_4$-CQDs integrated photoanode, which is manufactured using an ultrathin layer of CQDs between $BiVO_4$cores, is enhanced between UV to visible to NIR. Hence, the introduction of CQDs not only increases the solar spectrum absorption but also act as an electron transfer medium for HER. Si nanowire (SiNw) supported with Au CQDs has also been reported as an efficient, scalable and stable photocathode for HER. SiNw-AuCQDs were four times more efficient than pristine SiNw. The photocurrent was optimized to 1.7 mA/ $cm^{2,}$ and the conversion efficiency (photon to current) was 0.8 percent.

Wastewater Treatment and Purification

This section encompasses carbonaceous materials for remediation of wastewater, drinking and groundwater by eliminating contaminants and metallic ions as adsorbents. In addition, current advancements and challenges exhibited by the CQDs for purifying and treating wastewater have been highlighted. In addition to energy-based applications, CQDs are intriguing prospects for applications like biosensing, drug delivery, gene transfer, and bioimaging.

Li *et al.* [44] used a single-step alkali-assisted electrochemical preparation procedure to make CQDs with strong up conversion PL characteristics. Almost 100% photodegradation ability of nanocomposites SiO_2/CQDs or TiO_2/CQDs was found after 25 minutes of irradiation with a 300 W halogen lamp. For the photodegradation of benzene and methanol in the air under visible light irradiation, researchers used a ZnO/CQDs nanocomposite synthesized by a one-step hydrothermal procedure. CQDs-coated nanocomposites of Bi_2WO_6 were used by Qian *et al.* [45] for photocatalytic elimination of VOCs under visible and UV light irradiation.

Metal sulphides have also been investigated for the improvement of CQDs. Li *et al.* [46] used a hydrothermal approach to make CQDs/CdS photocatalysts, which showed photodegradation of the dye Rhodamine B with 90%efficiency. Kaur and co-workers [47] reported CQDs/ZnS nanocatalysts for photodegradation of Alizarin red S (ARS) dye with 89% degradation after 250 minutes.

In another report [48], degradation of noxious gases benzene and methanol were effectively degraded under visible light irradiation using Fe_2O_3/CQDs nanohybrid, which depicted a perfect combination of magnetic response and fluorescence.

To protect the environment and human health, wastewater, including industrial water, must be treated. As this section shows, numerous research organizations around the world have achieved significant progress in recent years in terms of reducing pollution in the environment. However, several obstacles remain, including a lack of infrastructure and resource utilization, as well as concerns about other nanomaterials potential toxicity.

CONCLUSION AND FUTURE PERSPECTIVE

In concluding lines, this chapter elaborated on the benefits of CQDs, their energy solutions and wastewater remediation features such as solar cells, and optoelectronic devices such as flexible electronics, LEDs and photodetectors. Based on the review of the literature during the last 5-6 years on various advanced features of CQDs, major improvements in the existing device design are predicted to be made in order to expand their applications. Despite the high level of interest in research, the development of robust and efficient prototypes for energy applications still requires a significant amount of work.

Asymmetric electrodes in CQDs-based hetero structure detectors exhibit crucial terms for maintaining electron-hole flow. Heterostructure solar cells based on functionalized CQDs exhibited electron transporting. Fluorescent CQDs have also been employed for flexible LEDs. Surface passivation of heterojunctions decorated with CQDs has been demonstrated for PEC water splitting.

More work is needed to establish reliable experimental confirmation for their characterization and optical properties in order to comprehend the impact of dimensions, surface chemistry, and doping. To fully comprehend the interactions of hybrid materials made from CQDs with their equivalents, a thorough investigation is required. In applications such as electronic sensors and photodetectors, the functional groups that exist on the interface of CQDs is a key problem. Recombination eliminates the advantages; therefore, surface imperfections should be minimized. Further, the use of CQDs to enhance the functioning of existing photodetectors has also been discussed in this chapter. CQDs' success is obviously limited, but they offer enormous potential for improvement in order to broaden their applications in a variety of disciplines.

REFERENCES

[1] Xu, X.; Ray, R.; Gu, Y.; Ploehn, H.J.; Gearheart, L.; Raker, K.; Scrivens, W.A. Electrophoretic analysis and purification of fluorescent single-walled carbon nanotube fragments. *J. Am. Chem. Soc.*, **2004**, *126*(40), 12736-12737.
[http://dx.doi.org/10.1021/ja040082h] [PMID: 15469243]

[2] Sun, Y.P.; Zhou, B.; Lin, Y.; Wang, W.; Fernando, K.A.S.; Pathak, P.; Meziani, M.J.; Harruff, B.A.; Wang, X.; Wang, H.; Luo, P.G.; Yang, H.; Kose, M.E.; Chen, B.; Veca, L.M.; Xie, S.Y. Quantum-sized carbon dots for bright and colorful photoluminescence. *J. Am. Chem. Soc.*, **2006**, *128*(24), 7756-7757.
[http://dx.doi.org/10.1021/ja062677d] [PMID: 16771487]

[3] Wang, Y.; Li, X.; Song, J.; Xiao, L.; Zeng, H.; Sun, H. All-inorganic colloidal perovskite quantum dots: A new class of lasing materials with favorable characteristics. *Adv. Mater.*, **2015**, *27*(44), 7101-7108.
[http://dx.doi.org/10.1002/adma.201503573] [PMID: 26448638]

[4] Wang, Y.; Li, X.; Zhao, X.; Xiao, L.; Zeng, H.; Sun, H. Nonlinear absorption and low-threshold multiphoton pumped stimulated emission from all-inorganic perovskite nanocrystals. *Nano Lett.*, **2016**,

16(1), 448-453.
[http://dx.doi.org/10.1021/acs.nanolett.5b04110] [PMID: 26652773]

[5] Hardman, R. A toxicologic review of quantum dots: toxicity depends on physicochemical and environmental factors. *Environ. Health Perspect.,* **2006**, *114*(2), 165-172.
[http://dx.doi.org/10.1289/ehp.8284] [PMID: 16451849]

[6] Geys, J.; Nemmar, A.; Verbeken, E.; Smolders, E.; Ratoi, M.; Hoylaerts, M.F.; Nemery, B.; Hoet, P.H.M. Acute toxicity and prothrombotic effects of quantum dots: Impact of surface charge. *Environ. Health Perspect.,* **2008**, *116*(12), 1607-1613.
[http://dx.doi.org/10.1289/ehp.11566] [PMID: 19079709]

[7] Huang, X.; Zhang, F.; Zhu, L.; Choi, K.Y.; Guo, N.; Guo, J.; Tackett, K.; Anilkumar, P.; Liu, G.; Quan, Q.; Choi, H.S.; Niu, G.; Sun, Y.P.; Lee, S.; Chen, X. Effect of injection routes on the biodistribution, clearance, and tumor uptake of carbon dots. *ACS Nano,* **2013**, *7*(7), 5684-5693.
[http://dx.doi.org/10.1021/nn401911k] [PMID: 23731122]

[8] Yang, S.T.; Wang, X.; Wang, H.; Lu, F.; Luo, P.G.; Cao, L.; Meziani, M.J.; Liu, J.H.; Liu, Y.; Chen, M.; Huang, Y.; Sun, Y.P. Carbon dots as nontoxic and high-performance fluorescence imaging agents. *J. Phys. Chem. C,* **2009**, *113*(42), 18110-18114.
[http://dx.doi.org/10.1021/jp9085969] [PMID: 20357893]

[9] Jiang, H.; Chen, F.; Lagally, M.G.; Denes, F.S. New strategy for synthesis and functionalization of carbon nanoparticles. *Langmuir,* **2010**, *26*(3), 1991-1995.
[http://dx.doi.org/10.1021/la9022163] [PMID: 20038107]

[10] Li, H.; Liu, R.; Liu, Y.; Huang, H.; Yu, H.; Ming, H.; Lian, S.; Lee, S-T.; Kang, Z. Carbon quantum dots/Cu2O composites with protruding nanostructures and their highly efficient (near) infrared photocatalytic behavior. *J. Mater. Chem.,* **2012**, *22*(34), 17470-17475.
[http://dx.doi.org/10.1039/c2jm32827e]

[11] Ye, R.; Xiang, C.; Lin, J.; Peng, Z.; Huang, K.; Yan, Z.; Cook, N.P.; Samuel, E.L.G.; Hwang, C.C.; Ruan, G.; Ceriotti, G.; Raji, A.R.O.; Martí, A.A.; Tour, J.M. Coal as an abundant source of graphene quantum dots. *Nat. Commun.,* **2013**, *4*(1), 2943.
[http://dx.doi.org/10.1038/ncomms3943] [PMID: 24309588]

[12] Zhou, J.; Booker, C.; Li, R.; Zhou, X.; Sham, T.K.; Sun, X.; Ding, Z. An electrochemical avenue to blue luminescent nanocrystals from multiwalled carbon nanotubes (MWCNTs). *J. Am. Chem. Soc.,* **2007**, *129*(4), 744-745.
[http://dx.doi.org/10.1021/ja0669070] [PMID: 17243794]

[13] Lu, J.; Yang, J.; Wang, J.; Lim, A.; Wang, S.; Loh, K.P. One-pot synthesis of fluorescent carbon graphene by the exfoliation of graphite in ionic liquids. *ACS Nano,* **2009**, *3*(8), 2367-2375.
[http://dx.doi.org/10.1021/nn900546b] [PMID: 19702326]

[14] Zhu, H.; Wang, X.; Li, Y.; Wang, Z.; Yang, F.; Yang, X. Microwave synthesis of fluorescent carbon nanoparticles with electrochemiluminescence properties. *Chem. Commun.,* **2009**, (34), 5118-5120.
[http://dx.doi.org/10.1039/b907612c] [PMID: 20448965]

[15] Tang, L.; Ji, R.; Cao, X.; Lin, J.; Jiang, H.; Li, X.; Teng, K.S.; Luk, C.M.; Zeng, S.; Hao, J.; Lau, S.P. Deep ultraviolet photoluminescence of water-soluble self-passivated graphene quantum dots. *ACS Nano,* **2012**, *6*(6), 5102-5110.
[http://dx.doi.org/10.1021/nn300760g] [PMID: 22559247]

[16] Zhai, X.; Zhang, P.; Liu, C.; Bai, T.; Li, W.; Dai, L.; Liu, W. Highly luminescent carbon nanodots by microwave-assisted pyrolysis. *Chem. Commun.,* **2012**, *48*(64), 7955-7957.
[http://dx.doi.org/10.1039/c2cc33869f] [PMID: 22763501]

[17] Zhang, B.; Liu, C.; Liu, Y. A novel one-step approach to synthesize fluorescent carbon nanoparticles. *Eur. J. Inorg. Chem.,* **2010**, *2010*(28), 4411-4414.
[http://dx.doi.org/10.1002/ejic.201000622]

[18] Bourlinos, A.B.; Stassinopoulos, A.; Anglos, D.; Zboril, R.; Karakassides, M.; Giannelis, E.P. Surface functionalized carbogenic quantum dots. *Small,* **2008**, *4*(4), 455-458. [http://dx.doi.org/10.1002/smll.200700578] [PMID: 18350555]

[19] Krysmann, M.J.; Kelarakis, A.; Dallas, P.; Giannelis, E.P. Formation mechanism of carbogenic nanoparticles with dual photoluminescence emission. *J. Am. Chem. Soc.,* **2012**, *134*(2), 747-750. [http://dx.doi.org/10.1021/ja204661r] [PMID: 22201260]

[20] Alam, A.M.; Park, B.Y.; Ghouri, Z.K.; Park, M.; Kim, H.Y. Synthesis of carbon quantum dots from cabbage with down- and up-conversion photoluminescence properties: Excellent imaging agent for biomedical applications. *Green Chem.,* **2015**, *17*(7), 3791-3797. [http://dx.doi.org/10.1039/C5GC00686D]

[21] Luo, P.G.; Sahu, S.; Yang, S.T.; Sonkar, S.K.; Wang, J.; Wang, H.; LeCroy, G.E.; Cao, L.; Sun, Y.P. Carbon *quantum* dots for optical bioimaging. *J. Mater. Chem. B Mater. Biol. Med.,* **2013**, *1*(16), 2116-2127. [http://dx.doi.org/10.1039/c3tb00018d] [PMID: 32260843]

[22] Yin, B.; Deng, J.; Peng, X.; Long, Q.; Zhao, J.; Lu, Q.; Chen, Q.; Li, H.; Tang, H.; Zhang, Y.; Yao, S. Green synthesis of carbon dots with down- and up-conversion fluorescent properties for sensitive detection of hypochlorite with a dual-readout assay. *Analyst,* **2013**, *138*(21), 6551-6557. [http://dx.doi.org/10.1039/c3an01003a] [PMID: 23982153]

[23] Zhou, J.; Sheng, Z.; Han, H.; Zou, M.; Li, C. Facile synthesis of fluorescent carbon dots using watermelon peel as a carbon source. *Mater. Lett.,* **2012**, *66*(1), 222-224. [http://dx.doi.org/10.1016/j.matlet.2011.08.081]

[24] Sahu, S.; Behera, B.; Maiti, T.K.; Mohapatra, S. Simple one-step synthesis of highly luminescent carbon dots from orange juice: Application as excellent bio-imaging agents. *Chem. Commun.,* **2012**, *48*(70), 8835-8837. [http://dx.doi.org/10.1039/c2cc33796g] [PMID: 22836910]

[25] Sinha, R.; Roy, N.; Mandal, T.K. Growth of carbon dot-decorated zno nanorods on a graphite-coated paper substrate to fabricate a flexible and self-powered schottky diode for uv detection. *ACS Appl. Mater. Interfaces,* **2020**, *12*(29), 33428-33438. [http://dx.doi.org/10.1021/acsami.0c10484] [PMID: 32573201]

[26] Bhattacharjee, L.; Mohanta, K.; Batabyal, S.K.; Saha, A.; Bhattacharjee, R.R.; Bhattacharjee, R.R. Reversible light-responsive solventless-liquid switch: polarization-induced dynamic surface ordering–disordering in liquid-like carbon quantum dots. *J. Phys. Chem. Lett.,* **2020**, *11*(12), 4726-4733. [http://dx.doi.org/10.1021/acs.jpclett.0c01275] [PMID: 32464062]

[27] Chen, H.; Wen, K.; Chen, J.; Xing, W.; Wu, X.; Shi, Q.; Peng, A.; Huang, H. Ultra-stable tellurium-doped carbon quantum dots for cell protection and near-infrared photodynamic application. *Sci. Bull.,* **2020**, *65*(18), 1580-1586. [http://dx.doi.org/10.1016/j.scib.2020.05.021] [PMID: 36738076]

[28] Wang, R.C.; Lu, J.T.; Lin, Y.C. High-performance nitrogen doped carbon quantum dots: Facile green synthesis from waste paper and broadband photodetection by coupling with ZnO nanorods. *J. Alloys Compd.,* **2020**, *813*, 152201. [http://dx.doi.org/10.1016/j.jallcom.2019.152201]

[29] Rai, S.C.; Wang, K.; Ding, Y.; Marmon, J.K.; Bhatt, M.; Zhang, Y.; Zhou, W.; Wang, Z.L. piezo-phototronic effect enhanced uv/visible photodetector based on fully wide band gap type-II ZnO/ZnS core/shell nanowire array. *ACS Nano,* **2015**, *9*(6), 6419-6427. [http://dx.doi.org/10.1021/acsnano.5b02081] [PMID: 26039323]

[30] Nie, B.; Hu, J.G.; Luo, L.B.; Xie, C.; Zeng, L.H.; Lv, P.; Li, F.Z.; Jie, J.S.; Feng, M.; Wu, C.Y.; Yu, Y.Q.; Yu, S.H. Monolayer graphene film on ZnO nanorod array for high-performance Schottky junction ultraviolet photodetectors. *Small,* **2013**, *9*(17), 2872-2879.

[http://dx.doi.org/10.1002/smll.201203188] [PMID: 23495044]

[31] Peng, L.; Hu, L.; Fang, X. Low-dimensional nanostructure ultraviolet photodetectors. *Adv. Mater.*, **2013**, *25*(37), 5321-5328.
[http://dx.doi.org/10.1002/adma.201301802] [PMID: 24089350]

[32] Tang, L.; Ji, R.; Li, X.; Bai, G.; Liu, C.P.; Hao, J.; Lin, J.; Jiang, H.; Teng, K.S.; Yang, Z.; Lau, S.P. Deep ultraviolet to near-infrared emission and photoresponse in layered N-doped graphene quantum dots. *ACS Nano,* **2014**, *8*(6), 6312-6320.
[http://dx.doi.org/10.1021/nn501796r] [PMID: 24848545]

[33] Xie, C.; Nie, B.; Zeng, L.; Liang, F.X.; Wang, M.Z.; Luo, L.; Feng, M.; Yu, Y.; Wu, C.Y.; Wu, Y.; Yu, S.H. Core-shell heterojunction of silicon nanowire arrays and carbon quantum dots for photovoltaic devices and self-driven photodetectors. *ACS Nano,* **2014**, *8*(4), 4015-4022.
[http://dx.doi.org/10.1021/nn501001j] [PMID: 24665986]

[34] Gupta, V.; Chaudhary, N.; Srivastava, R.; Sharma, G.D.; Bhardwaj, R.; Chand, S. Luminscent graphene quantum dots for organic photovoltaic devices. *J. Am. Chem. Soc.*, **2011**, *133*(26), 9960-9963.
[http://dx.doi.org/10.1021/ja2036749] [PMID: 21650464]

[35] Guo, Q.; Yuan, F.; Zhang, B.; Zhou, S.; Zhang, J.; Bai, Y.; Fan, L.; Hayat, T.; Alsaedi, A.; Tan, Z. Passivation of the grain boundaries of $CH_3NH_3PbI_3$ using carbon quantum dots for highly efficient perovskite solar cells with excellent environmental stability. *Nanoscale,* **2019**, *11*(1), 115-124.
[http://dx.doi.org/10.1039/C8NR08295B] [PMID: 30525161]

[36] Yan, X.; Cui, X.; Li, B.; Li, L. Large, solution-processable graphene quantum dots as light absorbers for photovoltaics. *Nano Lett.*, **2010**, *10*(5), 1869-1873.
[http://dx.doi.org/10.1021/nl101060h] [PMID: 20377198]

[37] Chen, Q.L.; Wang, C.F.; Chen, S. One-step synthesis of yellow-emitting carbogenic dots toward white light-emitting diodes. *J. Mater. Sci.*, **2013**, *48*(6), 2352-2357.
[http://dx.doi.org/10.1007/s10853-012-7016-8]

[38] Guo, X.; Wang, C.F.; Yu, Z.Y.; Chen, L.; Chen, S. Facile access to versatile fluorescent carbon dots toward light-emitting diodes. *Chem. Commun.*, **2012**, *48*(21), 2692-2694.
[http://dx.doi.org/10.1039/c2cc17769b] [PMID: 22306963]

[39] Sun, M.; Qu, S.; Hao, Z.; Ji, W.; Jing, P.; Zhang, H.; Zhang, L.; Zhao, J.; Shen, D. Towards efficient solid-state photoluminescence based on carbon-nanodots and starch composites. *Nanoscale,* **2014**, *6*(21), 13076-13081.
[http://dx.doi.org/10.1039/C4NR04034A] [PMID: 25247822]

[40] Lv, H.; Gao, X.; Xu, Q.; Liu, H.; Wang, Y.G.; Xia, Y. Carbon quantum dot-induced MnO_2 nanowire formation and construction of a binder-free flexible membrane with excellent superhydrophilicity and enhanced supercapacitor performance. *ACS Appl. Mater. Interfaces,* **2017**, *9*(46), 40394-40403.
[http://dx.doi.org/10.1021/acsami.7b14761] [PMID: 29072448]

[41] Jian, X.; Yang, H.; Li, J.; Zhang, E.; Cao, L.; Liang, Z. Flexible all-solid-state high-performance supercapacitor based on electrochemically synthesized carbon quantum dots/polypyrrole composite electrode. *Electrochim. Acta,* **2017**, *228*, 483-493.
[http://dx.doi.org/10.1016/j.electacta.2017.01.082]

[42] Yu, X.; Liu, R.; Zhang, G.; Cao, H. Carbon quantum dots as novel sensitizers for photoelectrochemical solar hydrogen generation and their size-dependent effect. *Nanotechnology,* **2013**, *24*(33), 335401-335407.
[http://dx.doi.org/10.1088/0957-4484/24/33/335401] [PMID: 23892324]

[43] Liang, Z.; Hou, H.; Fang, Z.; Gao, F.; Wang, L.; Chen, D.; Yang, W. Hydrogenated TiO_2 nanorod arrays decorated with carbon quantum dots toward efficient photoelectrochemical water splitting. *ACS Appl. Mater. Interfaces,* **2019**, *11*(21), 19167-19175.
[http://dx.doi.org/10.1021/acsami.9b04059] [PMID: 31058485]

[44] Li, H.; He, X.; Kang, Z.; Huang, H.; Liu, Y.; Liu, J.; Lian, S.; Tsang, C.H.A.; Yang, X.; Lee, S.T. Water-soluble fluorescent carbon quantum dots and photocatalyst design. *Angew. Chem. Int. Ed.,* **2010**, *49*(26), 4430-4434.
[http://dx.doi.org/10.1002/anie.200906154] [PMID: 20461744]

[45] Qian, X.; Yue, D.; Tian, Z.; Reng, M.; Zhu, Y.; Kan, M.; Zhang, T.; Zhao, Y. Carbon quantum dots decorated Bi2WO6 nanocomposite with enhanced photocatalytic oxidation activity for VOCs. *Appl. Catal. B,* **2016**, *193*, 16-21.
[http://dx.doi.org/10.1016/j.apcatb.2016.04.009]

[46] Li, L.; Wu, G.; Yang, G.; Peng, J.; Zhao, J.; Zhu, J.J. Focusing on luminescent graphene quantum dots: current status and future perspectives. *Nanoscale,* **2013**, *5*(10), 4015-4039.
[http://dx.doi.org/10.1039/c3nr33849e] [PMID: 23579482]

[47] Kaur, S.; Sharma, S.; Kansal, S.K. Synthesis of ZnS/CQDs nanocomposite and its application as a photocatalyst for the degradation of an anionic dye, ARS. *Superlattices Microstruct.,* **2016**, *98*, 86-95.
[http://dx.doi.org/10.1016/j.spmi.2016.08.011]

[48] Zhang, H.; Ming, H.; Lian, S.; Huang, H.; Li, H.; Zhang, L.; Liu, Y.; Kang, Z.; Lee, S.T. Fe2O3/carbon quantum dots complex photocatalysts and their enhanced photocatalytic activity under visible light. *Dalton Trans.,* **2011**, *40*(41), 10822-10825.
[http://dx.doi.org/10.1039/c1dt11147g] [PMID: 21935522]

SUBJECT INDEX

A

Absorption 65, 83, 94, 101, 115, 123, 182, 194, 197
  broadband 194
  drug 83
  fluorescent acceptor's 123
  iron 65
  solar spectrum 197
  spectroscopy 115
  spectrum 94, 101, 182, 197
Acid 23, 28, 29, 31, 39, 65, 70, 115, 128, 150, 151, 154, 155, 156, 159, 161, 163
  ascorbic 29, 65, 150, 155, 161, 163
  citric 23, 28, 31, 39, 70, 115, 151, 154, 155, 156, 163
  dihydrolipoic 159
  glutamic 155
  pyrolysis, citric 128
Activity 8, 9
  light-determined electron transfer 9
  thrombin 8
AFM analysis 58
Aggregation-induced quenching (AIQ) 196
Applications 96, 103
  photocatalyst 96
  photocatalytic 96
  solar energy transmission 103
Arginine-CQDs gel retardation test 70, 71
Atomic force microscopy (AFM) 54, 62, 64

B

Band gaps 2, 5, 55, 100, 183
  electronic energy 5

C

Cancer 79, 82, 85
  cells, ovarian 85
  nanotechnology 79
  therapy 82
Carbonaceous 25, 26, 121, 178
  QDs-based sensors 121
  quantum 25, 26, 178
Carbonaceous quantum dots 57, 96
  nitrogen-containing 57
  in photocatalysis 96
Carbon nanomaterials 79, 82, 178, 192
Carbon quantum dots 21, 26, 27, 29, 31, 32, 33, 79, 86, 146, 150, 152, 153, 159, 178, 179
  dots in gene therapy 86
Chemical ablation 39
Chemiluminescence resonance energy transfer (CRET) 123, 126
Chemotherapy 49, 82
Chromatography 26, 27, 28
Chromophores 118
Cinematography 82
Combustion process 28
Conductivity, electric 20, 33
Conversion, light-energy 46
CQD-based 8, 10, 97
  biosensors 8
  hybrid photocatalysts 97
  LEDs 10
CQD(s) 45, 73, 82, 96, 155
  characterization techniques 73
  oxidized 45
  photoinduced 96
  synthesise 155
  theranostic 82
Cytotoxicity assay 48

D

Detection method 117
Devices 12, 182, 195, 199
  optoelectronic 12, 199
  sensitizer-equipped 182
  solar cell 195
Diseases 82, 84, 145

infectious 82
kidney 84
lung 145
Disorders 14, 84, 148
neurodegenerative 84
neurological 148
DNA 8, 32, 70, 86, 110, 111, 121, 122, 129, 130, 135
double-strand 122
encoded 86
tagged single-stranded 8
Drug(s) 3, 7, 32, 33, 58, 81, 82, 83, 84, 87, 120
anti-diabetic 120
anti-malignant 84
antitumor 7
bio-sensorial 87
delivery system (DDS) 3, 33, 82, 83
retention 84
Dye-sensitized solar cell (DSSCs) 10, 94, 99, 101, 102, 103, 182, 183, 195

E

Electrocatalysis 6, 14, 20, 33, 53, 54, 129
Electrocatalysts 27, 185
Electrochemical 6, 45, 46, 48, 55, 127, 132, 133, 134, 135
graphite process 48
luminescence (ECL) 6, 45, 46, 132, 133, 134, 135
sensors 55, 127
Electrochemiluminescence 46, 132, 134
remote 46
solid-state 134
Electrochemistry method 32
Electrode 25, 32, 45, 129, 130, 132, 179, 180, 181, 183, 195
anodic 25
Electrodialysis 145
Electrolytes 10, 102, 181
flammable 10
Electronic 21, 32, 78, 180, 181, 185
conductivity 180, 181, 185
industry 78
properties 21, 32
Electron(s) 94, 95, 96, 99, 100, 122, 123, 125, 126, 127, 156, 162, 178, 179, 183, 194
conjugated 194
photo-generated 162
transport process 156
Electron transfer 45, 46, 101, 111, 131, 132
photo-induced 46, 111
process 45, 46
Electron transmission 115, 116, 117
photo-induced 115
Electrostatic forces 151, 162
Energy 9, 29, 123, 184, 186, 192, 193
density 186
devices 192, 193
loss 123
sources 184
thermal 29
transformation 9
Energy transfer 9, 111, 117, 123, 126
chemiluminescence resonance 123, 126
fluorescence resonance 9, 111, 117
Energy transmission 9, 58, 111, 116, 117, 122, 185
interaction 9
luminescence resonance 122
pathways 116
Erythrocytes 145

F

Fabrication techniques 54
Faradic efficiency 99
Flexible electronic devices 196
Fluorescein isothiocyanate 158
Fluorescence, wavelength-dependent 54
Fluorescence 6, 8, 9, 41, 47, 53, 57, 66, 81, 83, 85, 111, 113, 115, 116, 117, 119, 120, 122, 130, 147, 155, 157
activity 81
emission 6, 9, 119, 157
emission spectroscopy 53
imaging 85

intensity 6, 47, 57, 147, 155
mechanism 41
microscopy 83
monochromator 41
quenching mechanism 113
resonance energy transfer (FRET) 9, 111, 117
sensitivity 115
sensors 8, 66, 113, 147
signal 116, 117, 120, 122
spectroscopy 57
techniques 130
Fluorescence response 113, 118, 121
sensor's 118
Fluorescent properties 81
FRET process 119
FRET quenching 119, 121, 122
mechanism 119, 121, 122
process 121
Freundlich isotherm framework 152

G

Gene therapy (GT) 82, 86, 87
Glass work-functions 183
Glassy carbon electrode (GCE) 127, 128, 129, 131, 132, 135
Graphene 3, 55, 111, 112, 113, 115, 117, 118, 122, 124, 126, 127, 128, 129, 130, 134, 135, 178, 179, 181
oxide 129, 135, 178
QDs 3, 111, 113, 115
quantum dot (GQDs) 55, 112, 117, 118, 122, 124, 126, 127, 128, 129, 130, 134, 178, 179, 181
Green quantum dots 157

H

Hazardous 26, 179
chemicals 26
properties 179
Hazards, environmental 66
Heavy metal ions sensing 112, 127
Hexagonal mesoporous silicon (HMS) 67
Homocysteine 121
Hydrocarbon 28
Hydrogen 23, 31, 69, 117, 123, 124, 130, 131, 134, 196
fuels 196
peroxide 23, 31, 69, 117, 123, 124, 130, 131, 134
Hydrophilicity 146
Hydrothermal 5, 23, 30, 54, 59, 62, 70, 80, 95, 150, 164, 182, 193
breakdown process 62
carbonization 70
procedure 59
process 5, 62, 150, 182
pyrolysis 70
synthesis 30, 95

I

IFE mechanism 115
Imaging techniques 82, 84
Immunoassays 8
Immunosensor 8, 130, 133
electrochemical 130
ultrasensitive 133
Implantation combined methodology (ICM) 181
Interface, hydrophilic 194
Internal filter effect (IFE) 115, 120
Ion detection 135, 164, 157
heavy metal 135, 164
Ions 3, 115, 155
aluminum 115
cadmium 3
chromium 155
Irradiation, blue emission 69

L

Layered double-hydroxides (LDHs) 7
LEDs, blue 10
Light 67, 69, 72, 99, 103, 123, 146
absorption 67, 72, 99, 103, 146

discharge 123
ultraviolet 69
Light-emitting 2, 6, 10, 12, 20, 33, 180, 195, 196, 199
devices 10
diodes (LEDs) 2, 6, 10, 12, 20, 33, 180, 195, 196, 199
Lincomycin 135
Linear scan voltammetry (LSV) 128
Linkages 149, 150
chemical 149
Lipofectamine 86
Long persistence phosphors (LPP) 182
Luminescence 45, 118, 154, 155, 156, 179
blue 156
intensity 118
quenched 154
semiconductor materials 179
sensitivity 118

M

Macromolecules, large 83
Magnetic 7, 82, 85, 149
elements 149
resonance imaging (MRI) 7, 82, 85
Mechanism 9, 118, 121, 122, 151, 158, 160
charge transfer 9
endothermic adsorption 151
fluorescent 121
light radiation 122
non-radiative resonance 122
of heavy metal/metalloids detection 158
photocatalytic degradation 160
photoinduced electron transfer 118
Metal 25, 57, 145, 161, 162, 185
catalysts 25
detection 57
organic framework 161, 162
oxide corrosion 185
removal 145
Metallic ion(s) 13, 121, 151, 152, 153, 154, 155, 198
absorption 152
detection 155
inorganic 153
quenchers 121
Microwave 4, 20, 21, 24, 28, 29, 33, 39, 161, 164
irradiation 39, 161, 164
pyrolysis method 20, 21, 28, 29, 33
pyrolytic procedure 4
radiation 24, 28, 29
Multicolor imaging technology 40

N

NMR strategy 11
Nuclear magnetic resonance (NMR) 10, 11
Nucleation process 30

O

Oxidation 27, 129
activity 129
process 27
Oxidizing agent 23, 27, 28
Oxygen 3, 7, 43, 45, 11, 28, 45, 80, 98, 151, 162, 178, 183, 185, 194, 196
adsorbed 194
dissolved 43, 162
evolution reaction (OER) 7, 178, 185
reduction reactions (ORR) 7, 45

P

Perovskite solar cells (PSCs) 94, 182, 183
Photocatalytic 94, 96, 98, 99, 146, 148, 159, 162, 164, 184
activity 94, 96, 98, 99, 146, 148, 159, 162, 164
systems 96
water splitting 184
Photodegradation ability 198
Photodesorption 194
Photodetectors 192, 193, 194, 199
broadband 194
devices 192

Photoelectrocatalysis 21
Photogenerated electrons 95, 101
Photoinduced electron transfer (PET) 46, 111, 118, 120
Photoluminescence 5, 33, 38, 40, 46, 53, 54, 55, 56, 64, 68, 119, 120
Pollutants 129, 135, 144, 145, 146, 147, 148, 150, 159, 160
  adsorbed 147
  emerging 144
  environmental 135
  inorganic 148
Pollution, heavy metal 112, 145
Polymer(s) 21, 30, 53, 82, 118, 129, 178, 179, 185, 186
  dots (PDs) 53, 178, 179
  imprinted 118, 129
Powder 69, 101, 182, 195
  X-ray diffraction (PXRD) 69
  conversion efficiency (PCE) 101, 182, 195
Properties 38, 45, 61, 110, 153, 162, 180, 181, 184
  electrical 38, 45, 181
  electrocatalytic 45
  electrochemical 180, 184
  electrochemiluminescent 110
  hydrophilic 61
  luminescent 153
  photocatalytic 162
Prothrombin time assays 8
Pyrolysis 127, 193

Q

QD-based sensors 119
Quantum confinement 2, 6
  effect (QCE) 6
  principle 2
Quantum dots (QDs) 1, 2, 3, 12, 25, 26, 27, 28, 29, 30, 31, 94, 122, 159, 179, 182, 183
  sensitized Solar Cells (QDSSCs) 182, 183

R

Radiation 122, 123
  electromagnetic 122
Radiative electron-hole destruction 126
Raman spectroscopy 72
Reaction 7, 130, 133
  biorecognition 133
  electrocatalytic 7, 130
Reactive oxygen species (ROS) 8, 56, 147, 148, 159, 160

S

Screen-printed carbon electrode (SPCE) 55, 132
Sensing techniques 111
Sensitizers 94, 95, 99, 101, 182
  photoanode 101
Sensor(s) 1, 3, 9, 55, 110, 111, 112, 113, 120, 121, 122, 126, 127, 129, 132, 133, 199
  architectures 110
  bimodal 126
  constructed 55
  electronic 199
  fluorescent 121
Solar cells 94, 101
  fabricated 101
  silicon 94
  thin-coating 94
  applications 94
Solar energy generation 94

T

Techniques 3, 4, 8, 10, 11, 14, 23, 24, 25, 53, 54, 56, 95, 98, 127, 129, 131, 151
  arc discharge 25
  calcination 151
  chronoamperometry 129
  cutting-edge electrode modification 127
  electrochemical 98, 131
Therapies, photodynamic 8

U

UV irradiation 161, 194

W

Water splitting 12, 94, 98
  electrochemical 12
  photoelectrochemical 94, 98
Water treatment 1
White light-emitting diodes (WLEDs) 196

X

X-ray diffraction (XRD) 10, 11, 54, 69

Made in the USA
Columbia, SC
07 April 2026

64c5e400-f641-4c10-91d3-a210e63f6967R01